Applied Soil Physical Properties, Drainage, and Irrigation Strategies

Lambert B. McCarty • Lewis Ray Hubbard, Jr.
Virgil Quisenberry

Applied Soil Physical Properties, Drainage, and Irrigation Strategies

Springer

Lambert B. McCarty
Clemson University
Clemson, SC, USA

Lewis Ray Hubbard, Jr.
Clemson University
Clemson, SC, USA

Virgil Quisenberry
Clemson University
Clemson, SC, USA

ISBN 978-3-319-79593-5 ISBN 978-3-319-24226-2 (eBook)
DOI 10.1007/978-3-319-24226-2

Springer Cham Heidelberg New York Dordrecht London

Softcover re-print of the Hardcover 1st edition 2016

Printed on acid-free paper

Springer International Publishing AG Switzerland is part of Springer Science+Business Media (www.springer.com)

Preface

Many excellent texts are available on the theoretical science and mathematics of soil physics. These cover in detail the principles and theories behind the science and how many of the equations were derived. However, most of these texts are heavily mathematics based, requiring calculus and differential equation training in order to understand how these were derived and are functional.

The goal of *Applied Soil Physics* is to demystify the complicated math necessary to derive many of the formulas used in soil physics and to concentrate on the applications of these. We avoid complicated mathematics in our approach, focusing on how to use these in actual field and laboratory situations with numerous examples of how practitioners can successfully use the information covered in this book.

Four chapters are included: (1) Soil Physical Properties; (2) Soil Drainage; (3) Rootzone Selection and Modifications; and (4) Water Management and Conservation. Chapter 1 covers the basics of soil physical properties which will be applied in subsequent chapters. Chapter 2 covers the principles and practices of necessary calculations when determining appropriate and sufficient drainage for a particular situation and site. Chapter 3 covers the science of determining an appropriate rootzone profile for playability and sufficient drainage while Chap. 4 covers irrigation practices to maximize water management and conservation.

Our wish is to provide a useful text to help students, architects, field designers, construction supervisors, governing boards, greens committee chairs and members, as well as other interested parties on how to scientifically design, test, and construct a successful facility that meets the playability needs of the participants yet provide the necessary moisture management for field supervisors. We welcome your comments and suggestions and wish you the best in applying the science of soil physics.

Clemson, SC, USA Lambert B. McCarty
Clemson, SC, USA Lewis Ray Hubbard, Jr.
Clemson, SC, USA Virgil Quisenberry

Introduction

Soil physical properties influence almost everything related to plant and soil science. From drainage and aerification, to rootzone media selection, and environmentally designed irrigation strategies, proper understanding and knowledge of soil physical properties is essential to these and other principles and practices of soil science.

Most soil physics textbooks are heavily mathematically based to aid in explanation of complex theories and laws related to various aspects of the science. However, students and practitioners often lack this in-depth mathematical background and, in reality, do not need such training to understand and apply most soil physical practices. The goal of the first chapter is to introduce and explain those soil physical properties that influence most end users. Chapters 2, 3, and 4 then apply the principles covered in Chap. 1 in regard to soil drainage (Chap. 2), rootzone profile selection (Chap. 3), and sound irrigation strategies and practices (Chap. 4).

With over 90 combined years in soil science teaching, research, and demonstration, the authors cover and explain each topic in sufficient details so these can be readily applied but not so immersed in mathematics that the readers become disinterested. We welcome any comments related to the text and hope you find it as useful in practice as we found in researching and writing it.

Acknowledgements

Sincere appreciation is extended to Mr. Philip Brown, MS, who thoroughly reviewed the manuscript and made many helpful suggestions.

Abbreviations

Δ	Change in
η	Absolute viscosity of the medium
θ	Soil water content
θ_{fc}	Volumetric water content at field capacity
θ_g	Gravimetric water content
θ_v	Volumetric water content
θ_{wp}	Volumetric water content at wilting point
μ	Micro
π	Pi
ρ_b	Bulk density
ρ_s	Particle density
Ψ	Water potential
Ψ_g	Gravitational potential
Ψ_o	Osmotic potential
$\Psi_{\mathbf{p}}$	Pressure (or matric) potential
Ψ_t	Total water potential
%	Percent
~	Approximate
A	Cross-section of column
ac	Acre
ac-ft	Acre foot
ac-in	Acre inch
adj SAR	Adjusted sodium absorption ratio
AEP	Air entry point
ASTM	American Society for Testing and Materials
atm	Atmosphere
ATRI	Australian Turfgrass Research Institute
C	Centigrade or Celsius
Ca^{+2}	Calcium
$CaCO_3$	Calcium carbonate

$CaSO_4$	Calcium sulfate (aka gypsum)
cb	Centibar
cc	Cubic centimeters
CEC	Cation exchange capacity
Cl^-	Chloride
cm	Centimeter
CO_3^{-2}	Carbonates
cos	Cosine
D	Water depth
d	Density of the medium
D	Diameter of the particle or its pore
d^l	Density of the particle
D_e	Equivalent depth of water or void ratio
dH	Hydraulic head length (total)
dist	Distance
dL	Core length
dS	DeciSiemens
DU	Distribution uniformity
D_x	Gradation index based on diameter of particles
EC	Electrical conductivity
EC_{dw}	Electrical conductivity of a saturated paste extract by the plant grown
ESP	Exchangeable sodium percentage
ET	Evapotranspiration
etc.	Etcetera
ET_p	Potential or pan evapotranspiration
F	Fahrenheit
f_a	Aeration porosity
FDR	Frequency domain reflectometry
ft	Foot
f_t	Total porosity
f_w	Water-filled porosity
g	Acceleration due to gravity
g	Gram
gal	Gallon
h	Height
H_2O	Water
ha	Hectare
HCO_3^-	Bicarbonates
HDPE	High density polyethylene
Hg	Mercury
hr	Hour
i.e.	For example

in	Inch
ISTRC	International Sports Turf Research Center
J	Joules
K	Temperature constant
K^+	Potassium
K_c	Crop coefficient
kg	Kilogram
kPa	KiloPascals
K_{sat}	Saturated hydraulic conductivity
L	Liter
lb	Pound
m	Meter
M_a	Mass of air
mb	Millibars
mg	Milligrams
Mg^{+2}	Magnesium
MH_z	Megahertz
min	Minute
ml	Milliliter
mm	Millimeter
mmhos	Millimhos
mN	Millinewtons
MPa	Megapascals
M_s	Mass of solids
M_t	Total mass
M_w	Mass of water
N	Nitrogen
Na^+	Sodium
NO_3^-	Nitrate
P	Phosphorus
ppm	Parts per million
psi	Pounds per square inch
PVC	Polyvinyl chloride
Q	Quantity of water
r	Radius
RH	Relative humidity
RSC	Residual sodium carbonate
S	Slope
s	Second
s	Degree of saturation
SAR	Sodium adsorption ratio
SMRC	Soil moisture retention curve
SO_4^{-2}	Sulfate
TDR	Time domain reflectometry

TDS	Total dissolved salts
TSS	Total suspended solids
UC	University of California
USDA	United States Department of Agriculture
USGA	United States Golf Association
V	Velocity of the falling particle
v	Volume
V_a	Volume of air
V_b	Dry specific volume
V_e	Void ratio
V_s	Volume of solids
V_t	Total volume
V_v	Volume of voids
V_w	Volume of water
WDPT	Water drop penetration time
wt	Weight

Contents

Author Biographies

Lambert B. McCarty is a Professor of Agriculture and Environmental Sciences specializing in turfgrass science and management at Clemson University in Clemson, South Carolina. A native of Batesburg, SC, he received a B.S. degree from Clemson University in Agronomy and Soils, M.S. from North Carolina State University in Crop and Soil Science, and Ph.D. from Clemson University in Plant Physiology and Plant Pathology. Dr. McCarty spent nine years as a turfgrass specialist at the University of Florida in Gainesville. While at the University of Florida, he oversaw the design and construction of the state-of-the-art research and education turfgrass facility, "The Envirotron." He also was author or coauthor of the books *Best Management Practices for Florida Golf Courses*, *Weeds of Southern Turfgrasses*, and *Florida Lawn Handbook*.

Since 1996, he has been at Clemson University and is involved in research, extension, and teaching activities. He has published numerous articles dealing with all phases of turfgrass management and is in constant demand for his presentations and worldwide agronomic knowledge and experiences. He is currently coauthor of the books *Color Atlas of Turfgrass Weeds*, *Best Golf Course Management Practices, Southern Lawns, Home Lawn Management in SC, Managing Bermudagrass Turf, Sod Production in the Southern United States, Fundamentals of Turfgrass and Agricultural Chemistry, Designing and Maintaining Bermudagrass Sports Fields in the United States*, and *Weed Control in Turf and Ornamentals*, is a coauthor for the GCSAA seminars, *Weed Control, Advanced Weed Management, Managing Bentgrass in Heat-Stress Environments, and Managing Ultradwarf Bermudagrass Golf Greens*, and is active in a number of professional societies. He also teaches a course *Advanced Turfgrass Management* and has been a Certified Professional Agronomist for over 25 years. He is a Fellow, Crop Science Society of America, and Fred V. Grau Turfgrass Scientist Award winner. bmccrty@clemson.edu

Lewis Ray Hubbard, Jr. has been Clemson University's Experiment Station Engineer since 1993. He earned his Ph.D. in Plant and Environmental Sciences from Clemson, with emphases in Soil Physics and Turfgrass, and a minor in Statistics. He also has B.S. degrees in Civil Engineering and Agricultural Engineering from Clemson. Dr. Hubbard has been a Registered Professional Engineer since 1989. He has made numerous scientific and public service presentations, research demonstrations, and classroom lectures. He grew up and lives in Anderson, SC, with his wife Pat and two children, Trey and Anna. lhbbrd@clemson.edu.

Virgil Quisenberry is a Professor *emeritus* specializing in Soil Physics at Clemson University. A native of Kentucky, he received a B.S., M.S., and Ph.D. from the University of Kentucky in Soils and Agriculture. He spent 37 years as a professor at Clemson University dealing with teaching and research in the area of Soils, specifically Soil Physics. He continues to provide soils training to Certified Agronomists (CPAg) and Certified Crop Advisors (CCA) and is active on several graduate student committees. vqsnbrr@clemson.edu.

Chapter 1
Soil Physical and Moisture Properties

Soil is a mixture of mineral and organic matter. A soil commonly consists of primary minerals and secondary minerals. Elementary textbooks often describe an idealized soil as consisting approximately 50 % by volume solids, 25 % by volume water, and approximately 25 % by air (Fig. 1.1). These idealized soils also contain organic matter, up to 5 %. Soils can be quite variable in porosity plus water and air content, yet, still be deemed productive. The type, size, and relative proportions of the mineral components and the amount and nature of the organic fraction affect the soil's physical and chemical properties. These properties in turn determine the soil's capacity to hold water, its nutrient availability, its susceptibility to compaction, its ability to drain, and several other characteristics. When evaluating soils for potential use in traffic situations such as golf greens, sand capping fairways, or sports fields, the following physical tests are used to indicate their potential successful use:

1. Particle size analysis,
2. Physical analysis including bulk density and soil porosity,
3. Saturated hydraulic conductivity,
4. Soil moisture retention or characterization curves

1.1 Soil Physical Properties

Many soil physical properties, such as the capacity to retain water, are influenced by the size distribution of its particles. For the purposes of quantifying and describing soil texture, soil mineral particles are subdivided into three fractions based on the average diameter of the particle: sand, silt, and clay. The relative proportion of these fractions in a soil determine its texture. Texture is not influenced by a soils organic matter content.

L.B. McCarty et al., *Applied Soil Physical Properties, Drainage, and Irrigation Strategies*, DOI 10.1007/978-3-319-24226-2_1

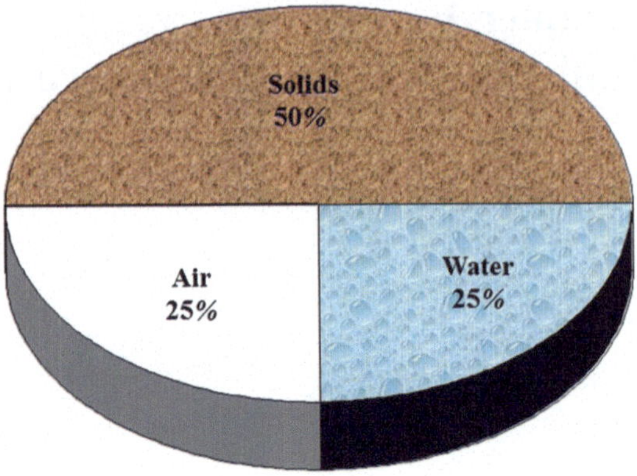

Fig. 1.1 "Ideal" soils are *composed* of approximately 50 % solids (minerals), 25 % each of water and air (pores), and possibly a small (≤5 %) fraction of organic matter

Table 1.1 Particle-size classifications as determined by the United States Department of Agriculture (USDA)

Textural name	Textural subclass	Particle-size range (*mm*)	U.S. standard (sieve number)	Sieve opening (*mm*)	Particles (g^{-1})	Typical settling velocity
Gravel	Gravel	>4.76	4	4.76	<2	$20\,cm\,s^{-1}$
	Fine gravel	2.00 to 4.76	10	2.00	11	$3\,cm\,s^{-1}$
Sand	Very coarse sand	1.00 to 2.00	18	1.00	90	$1\,cm\,s^{-1}$
	Coarse sand	0.50 to 1.00	35	0.50	720	$13\,cm\,min^{-1}$
	Medium sand	0.25 to 0.50	60	0.25	5,700	$3\,cm\,min^{-1}$
	Fine sand	0.10 to 0.25	140	0.10	46,000	$31\,cm\,h^{-1}$
	Very fine sand	0.05 to 0.10	270	0.05	722,000	$6\,cm\,h^{-1}$
Silt	–	0.002 to 0.05	–	–	5,776,000	$1.3\,mm\,h^{-1}$
Clay	–	<0.002	–	–	90,260,853,000	$<1.3\,mm\,h^{-1}$

Soil Particle Analysis

According to the U.S. Department of Agriculture (USDA) classification system, particles greater than 2 *mm* in diameter are classified as cobbles, stones, or gravel dependent upon their size. Sand particles have diameters between 0.05 and 2.0 *mm*, silt 0.002–0.05 *mm*, and clay <0.002 *mm* (Table 1.1). Subdivisions within the sand fraction are: very coarse; coarse; medium; fine; and, very fine. The United States Golf Association (USGA) utilizes the USDA system with a slight modification in the classification of very fine sand (USGA, 0.05–0.15 *mm* versus USDA, 0.05–0.10 *mm*) and fine sand (USGA, 0.15–0.25 *mm* versus USDA, 0.10–0.25 *mm*). Sands are subdivided due to the differing affects the various particle sizes have on soil properties. For example, infiltration is commonly more rapid in coarse sand than in very fine sand due to the larger pores in the larger coarse sand profiles.

The sand fraction of soil particle analysis consists mostly of primary minerals such as quartz, feldspars, micas, and other weather-resistant minerals. Silts are mostly weathered sands and primary minerals, plus additional minerals susceptible to weathering. When dry, silt particles feel smooth like powder or flour. Clay particles tend to be flat or plate-like rather than spherical like many sand particles. Clays are composed of layers of various crystal lattice groups. For example, a 1:1 clay consists of 1 tetrahedral layer combined with one octahedral layer. A 2:1 clay consists of 1 tetrahedral layer between two octahedral layers. While 1:1 clays do not shrink and swell when wet and dry, some 2:1 clays can expand. For example, in temperate soils, clay minerals can include kaolinite (1:1 nonexpanding), montmorillonite (2:1 expanding), vermiculite (2:1 limited expansion), illite (2:1 nonexpanding), and chlorite (2:2 nonexpanding). Expanding (2:1) clays create cracks when they dry and form very hard soil clods due to the enormous area of contact between the plate-shaped clay particles. In tropical regions, because of heavy precipitation and high temperature, long-term intense weathering and leaching have left few clay minerals in many soil profiles. What remains are brightly colored reddish sesquioxides (iron and aluminum oxides). Surface areas of clay particles are many times greater than those of sand or silt; thus, they are capable of adsorbing much more water, and become soft and sticky when wet.

Determining Particle Size

Soil particle-size analysis is based on **sieving** and **sedimentation methods**. A sample of soil is dispersed and particles larger than silt (i.e., gravel and sand) are separated into their various size groups, as outlined in Table 1.1, by the use of sieves (Fig. 1.2). Effective particle size is the distance between wires in a square grid fabric of woven wire making up the bottom of a sieve through which particles with smaller effective diameters pass. The weight of each group is found to determine a percentage of total sample weight. Because silt and clay are so small, sieving cannot accurately separate these two from each other and sedimentation is often used to perform this separation.

Particle-size analysis provides a general description of physical and textural soil properties and is the basis for assigning the soil's textural class name (i.e., sand, sandy loam, clay, etc.). Once the percentages of sand, silt, and clay in a soil have been identified, the soil's textural class can be determined by using the USDA textural triangle (Fig. 1.3). Twelve soil textural classes make up the USDA textural triangle. Some familiar soil textural names are loamy sand, sandy loam, loam, and silt loam.

To determine soil textural class from sieve and sedimentation analysis, first find the percent of sand along the base of the triangle and follow the corresponding diagonal line up and to the left. Then find the percent of clay on the left leg of the triangle and draw a horizontal line toward the right leg of the triangle. The

Fig. 1.2 Soil particles larger than silt are separated into their various size groups by the use of sieves

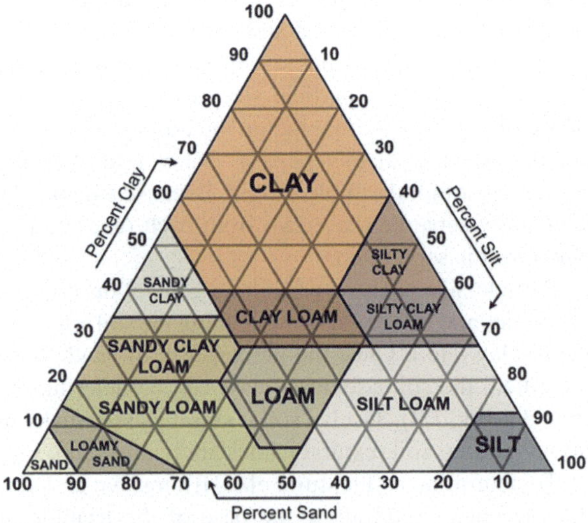

Fig. 1.3 The United States Department of Agriculture's *textural triangle* is used to determine the textural class based on sand, silt, and clay content

intersection of the sand and clay percentage lines indicates the textural class of the soil. To confirm the texture, the percent silt can be used; it should intersect the triangle at the same point as the sand and the clay.

Example If the particle-size distribution of a soil is 40 % sand, 40 % silt, and 20 % clay, what is its textural classification based on the textural triangle?

Lines drawn on the soil textural triangle (Fig. 1.3) for a soil with 40 % sand and 20 % clay, intersect at the center of the *loam*.

Example What are the percent sand, silt and clay of a soil containing the following separates and what is its textural classification?

- 48 *g* of particles <0.002 *mm* in diameter,
- 84 *g* of particles >0.05 but <2.0 *mm* in diameter,
- 148 *g* of particles between 0.002 and 0.05 *mm* in diameter

$$84g + 148g + 48g = 280g\,\text{total}$$

Sand (0.05 to 2.0 *mm*):

$$\frac{84\,g}{280\,g} \times 100 = 30\%$$

Silt (0.002 to 0.05 *mm*):

$$\frac{148\,g}{280\,g} \times 100 = 53\%$$

Clay (<0.002 *mm*):

$$\frac{48\,g}{280\,g} \times 100 = 17\%$$

From the soil textural triangle (Fig. 1.3), a soil with 30 % sand, 53 % silt, and 17 % clay is a *silt loam*.

As mentioned, due to the physical limitations of mechanical sieves for separating smaller particles, silt and clay percentages are determined based on the rate of settling of these in a suspension. This method involves the proportionality of clay and silt settling rates to their size (particle diameter). The larger the particle (i.e., sand or gravel), the quicker it will settle in a suspension (i.e., water) solution. Conversely, the smaller the particles (i.e., silt and clay), the slower settling occurs. This is described mathematically by **Stokes equation**. Sand generally takes 2 or 3 min to settle, while silt takes several hours and clay requires up to several days. Fine clay particles are so small that they can collide with water molecules and this keeps them suspended perpetually; this is called **Brownian motion** and is the reason why a puddle with clay in it rarely clears.

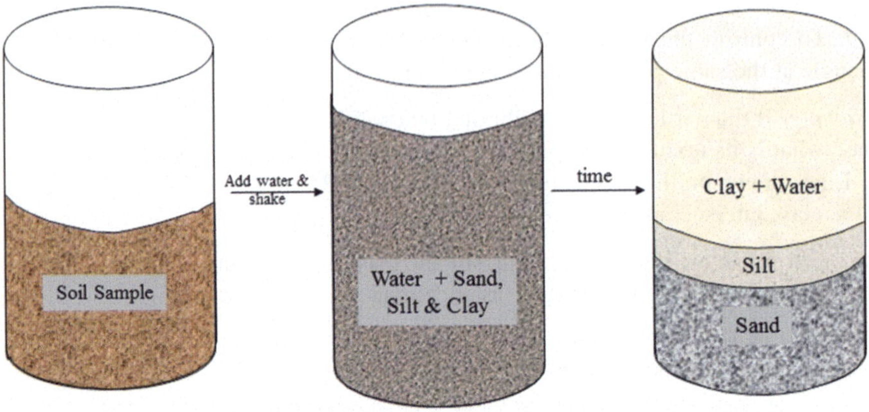

Fig. 1.4 Sand, silt, and clay components of a soil can be separated utilizing Stokes law which calculates the velocity of a falling particle through a suspension based on the particle's radius and velocity, as well as the viscosity of the fluid

According to Stokes equation the velocity of a spherical particle settling under the influence of gravity in a fluid of a given density and viscosity is proportional to the square of the particle's radius. This governs the method of sedimentation analysis. A particle falling in a fluid will encounter a frictional resistance proportional to the product of its radius and velocity, as well as the viscosity of the fluid (Fig. 1.4). Since Stokes equation is based on the rate a spherical object falls, it is assumed the soil particles are smooth spheres, which can be incorrect. Other assumptions when using Stokes equation are (a) terminal velocity is instantaneous; (b) resistance when settling is due to fluid viscosity and is not influenced by the cylinder wall; (c) no interaction occurs between particles; (d) the soil particle density is 2.65 $g\ cm^{-3}$ and, (e) no variation occurs in the temperature of the fluid from top to bottom.

$$\text{Stokes Equation}: \quad V = \frac{g(d^1 - d)D^2}{18\eta}$$

where: V = velocity of the falling particle ($cm\ s^{-1}$)
d^1 = density of the particle ($g\ cm^{-3}$), (2.65 $g\ cm^{-3}$ for most mineral soils)
d = density of the medium ($g\ cm^{-3}$), (0.997 $g\ cm^{-3}$ for water at 25°C)
g = acceleration due to gravity (980 $cm\ s^{-2}$)
D = diameter of the particle (cm)
η = absolute viscosity of the medium ($dyne\ s\ cm^{-2}$)

Subtracting fluid medium density from particle density accounts for the buoyancy that reduces the effective weight of particles in suspension.

In the equation, g, d^1, and d are constants. If the temperature is constant, the viscosity of water is also constant (0.01 at 20°C, for example). By substituting these values into one equation, typical fall rates for various-sized particles can be calculated as:

$$\text{velocity or V}\ (cm\ s^{-1}) = \text{KD}^2$$

where K equals the constant, taking into account density, viscosity, gravity, and temperature. At 20 C (68 F), K is approximately 8,711. At 25 C (77 F), K is approximately 10,000. D = particle diameter.

Example Determine the time required for all particles (all sand) larger than 0.05 *mm* (0.005 *cm*) to fall 10 *cm* in a suspension at 25°C.

$$\begin{aligned}\text{velocity or V}\ (cm\ s^{-1}) &= \text{KD}^2\\ &= 10{,}000 \times (0.005)^2\\ &= 0.25\,cm\,s^{-1}\end{aligned}$$

With this velocity, the time required for particles to fall 10 *cm* can be determined using the relationship between velocity, time, and distance:

$$\begin{aligned}\text{velocity or V}\ (cm\,s^{-1}) &= \frac{\text{distance}\ (cm)}{\text{time}\ (s)}\\ \text{time or T}\ (s) &= \frac{\text{distance}\ (cm)}{\text{velocity}\ (cm\ s^{-1})}\end{aligned}$$

Therefore:

$$\text{time} = \frac{\text{distance}}{\text{velocity}} = \frac{10\,cm}{0.25\,cm\,s^{-1}} = 40\,s$$

From this example, at the end of 40 s, the suspension above 10 *cm* depth in a container is free of all particles 0.05 *mm* in diameter or larger. In other words, it is free of all sand. This same process can be repeated for other textural components such as silt. If the amount of soil originally suspended is known, the proportions of sand, silt, and clay can be determined by measuring the amount of material remaining in suspension after a specific time has elapsed.

From Stokes equation, the settling rate in water of a particle with a diameter of 0.05 *mm* (lower limit of sand) at 25°C, is 0.25 $cm\ s^{-1}$, and with a diameter of 0.002 *mm* (upper limit of clay), the rate is only 0.004 $cm\ s^{-1}$. Sand, therefore, has been calculated to settle in a 7.25 *in* (18.4 *cm*) high cylinder beaker in approximately 74 *s*. If a thoroughly distributed sample is placed in a 1,000 *ml* cylinder beaker 7.25 *in* (18 *cm*) in height, it should retain the silt and clay fractions in suspension longer than 74 *s* (approximately 2 *h* for silt and "days" for clay), enabling the soil scientist to separate these from the larger-diameter sand particles. When using this method it is important to disperse the clay particles, if the particles are allowed to stick together they will behave like a larger particle and fall at a more rapid rate and results will be inaccurate. Commonly, a dispersing agent containing sodium such as sodium hexametaphosphate is used to ensure clay particle dispersal.

Example

(a) How fast would the smallest sand particle fall in water at 25 C?
Smallest sand particle is 0.05 *mm* or 0.005 *cm*

$$\text{velocity} = \text{KD}^2 = (10,000) \times (0.005)^2 = 0.25\ cm\ s^{-1}$$

(b) Determine the time required for the smallest sand to fall 10 *cm*.

$$\text{time} = \frac{\text{distance}}{\text{velocity}} = \frac{10\,cm}{0.25\,cm\,s^{-1}} = 40\,s$$

(c) Determine the time needed for all sand (diameter >0.50 *mm*), silt (diameter >0.002 *mm*), and coarse clay (diameter >0.001 *mm*) particles to settle to a depth of 20 *cm* in an aqueous solution at 25C.

sand:

$$\text{velocity} = \text{KD}^2 = (10,000) \times (0.005)^2 = 0.25\ cm\ s^{-1}$$

$$\text{time} = \frac{\text{distance}}{\text{velocity}} = \frac{20\,cm}{0.25\,cm\,s^{-1}} = 80\,s = 1.33\,min$$

silt:

$$\text{velocity} = \text{KD}^2 = (10,000) \times (0.0002)^2 = 0.0004\,cm\ s^{-1}$$

$$\text{time} = \frac{\text{distance}}{\text{velocity}} = \frac{20\,cm}{0.0004\,cm\,s^{-1}} = 50{,}000\,s = 833\,min = 13.9\,h$$

coarse clay:

$$\text{velocity} = \text{KD}^2 = (10,000) \times (0.0001)^2 = 0.0001 cm\ s^{-1}$$

$$\text{time} = \frac{\text{distance}}{\text{velocity}} = \frac{20\,cm}{0.0001\,cm\,s^{-1}} = 200{,}000\,s = 3,333\,min = 55.6\,h = 2.3\,\text{days}$$

Using the knowledge gained regarding settling rates of the various soil separates, the percent sand, silt and clay can be quantified using a **hydrometer**. A hydrometer is a device used to measure suspension density, thus reflecting the amount of particles that remain in suspension after a certain settling time (Fig. 1.5). A hydrometer with a Bouyoucos scale in grams per liter ($g\ L^{-1}$) is used to determine the amount of soil in suspension. The greater the density of a suspension (from the presence of soil), the greater the buoyant force on the hydrometer, therefore the higher the hydrometer will sit in the suspension, and the higher the Bouyoucos scale reading. As particles settle out of the suspension, density decreases and a lower reading is obtained. Since temperature influences the settling rate, a temperature correction must be made if the suspension temperature differs from the temperature at which the hydrometer is calibrated.

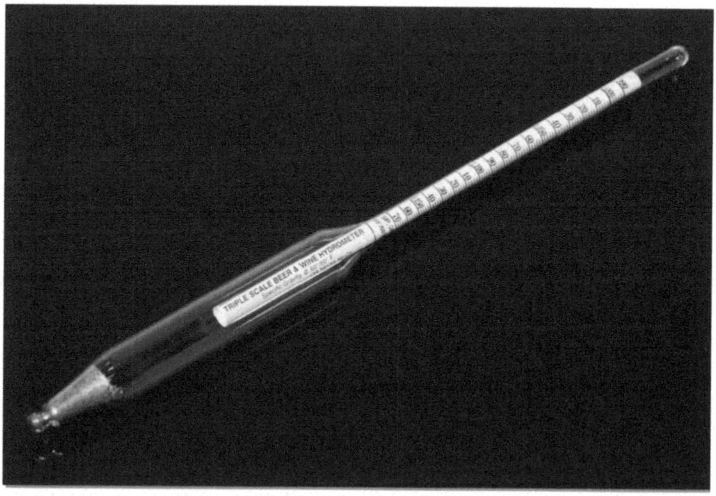

Fig. 1.5 Hydrometer used to separate soil fractions based on its buoyancy in water with soil particles

Example From the following data, calculate the texture of a 50 *g* soil sample.

Hydrometer Reading (temperature corrected)	$g\ L^{-1}$
40 *s*	31
~7 *h*	16

1. $$\% \text{ silt and clay} = \frac{40\,s \text{ reading}}{\text{dry weight soil } (g)} \times 100 = \frac{31}{50} \times 100 = 62\%$$

   (after 40 *s*, all of the sand has settled out leaving just silt and clay in suspension)

2. $$\% \text{ clay} = \frac{7\,h \text{ reading}}{\text{dry weight soil } (g)} \times 100 = \frac{16}{50} \times 100 = 32\%$$

   (after ~7 *h*, it is assumed the sand and silt has settled out, leaving just clay in suspension)

3. $\%\ \text{silt} = \%(\text{silt} + \text{clay}) - \%\text{clay}$
$$= 62\% - 32\%$$
$$= 30\%$$
4. $\%\ \text{sand} = 100\% - \%(\text{silt} + \text{clay})$
$$= 100\% - 62\%$$
$$= 38\%$$

This soil (38 % sand + 30 % silt + 32 % clay) is classified by the textural triangle as a *clay loam.*

Soil Particle and Bulk Density

Two important measurements of soils are **particle density** (ρ_s) and **bulk density** (ρ_b). Particle density is the average density of soil particles and is defined as mass (or weight) of dry soil per unit volume of soil solids, not including pore volumes occupied by air or water (Table 1.2). Mineral components (sands, silts, clays) have higher particle densities than organic matter. Particle density varies little, with most mineral soils within a narrow range of 2.60 to 2.75 $g\ cm^{-3}$ (or $g\ cc^{-1}$). A particle density of 2.65 $g\ cm^{-3}$ is often used as an assumed particle density by soil scientists for various calculations. For comparison, water, concrete, steel, and lead have densities of 1.0, 2.4, 7.7, and 11.3 $g\ cm^{-3}$, respectively, while organic matter has a lower particle density of 1.1 to 1.4 $g\ cm^{-3}$.

Bulk density can be used as a measure of soil compaction (or density) if some information about the soil texture is known, and is defined as the mass (weight) of dry soil per given unit volume, including both solids and pores occupied by air and water (Tables 1.2 and 1.3). Bulk density, unlike particle density, is an indicator of pore space volume in addition to soil solids (Fig. 1.6).

When considering soil compaction, bulk density can be misleading, for example a clay soil that feels compact may have a bulk density of 1.4 $g\ cm^{-3}$ but a relatively loose sandy soil may have a bulk density of 1.7 $g\ cm^{-3}$. Clay soil has many small pores giving it a high porosity, whereas sandy soil has a few large pores with an overall lower porosity. Bulk densities of soils generally range from 1.0 to 1.9 $g\ cm^{-3}$ (Table 1.4). Clay, clay loam, and silt loam soils normally range from 1.0 to 1.6 $g\ cm^{-3}$, while sands and sandy loams normally range between 1.4 and 1.8 $g\ cm^{-3}$. Organic soils have extremely low bulk densities (0.2 to 0.8 $g\ cm^{-3}$) due to low particle densities and large amounts of pore space.

Example Determine the weight of an acre (*ac*) of soil 6 *in* (15 *cm*) deep with an average bulk density (ρ_b) of 1.5 $g\ cm^{-3}$.

Table 1.2 Common measurements of soil mass-volume relationships. Refer to Fig. 1.7 to view this relationship

Measurement	Symbol	Unit	Mass-volume relationship	Equation
Particle density	ρ_s	$g\ cm^{-3}$	$\frac{M_s}{V_s}$	$\frac{\text{mass (or weight) of dry soil }(g)}{\text{volume of soil solids }(cm^3)}$; $2.65\ g\ cm^{-3}$ is often used for ρ_s
Bulk density	ρ_b	$g\ cm^{-3}$	$\frac{M_s}{V_t} = \frac{M_s}{V_s + V_a + V_w}$	$\frac{\text{mass (or weight) of dry soil }(g)}{\text{volume of dry soil (solids \& pores) }(cm^3)}$
Dry specific volume	V_b	$g\ cm^{-3}$	$\frac{V_t}{M_s}$	$\frac{1}{\rho_b}$
Solid space	–	%	$\frac{V_s}{V_t} \times 100$	$\frac{\rho_b}{\rho_s} \times 100$
Porosity (total)	f_t	$cm^3\ cm^{-3}$ or %	$\frac{V_v}{V_t} = 1 - \frac{V_s}{V_t}$	$\left(1 - \frac{\rho_b}{\rho_s}\right) \times 100$ or $100 - \%$ solid space $= \frac{V_a + V_w}{V_s + V_a + V_w}$
Volumetric water content (or soil water content by volume)	θ_v	$cm^3\ cm^{-3}$ or %	$\frac{V_w}{V_t} = \frac{V_w}{V_s + V_v} = \frac{V_w}{V_s + V_a + V_w}$	$\theta_g \times \rho_b$ or $\frac{\text{wet soil weight }(g) - \text{dry soil weight }(g)}{\text{volume of soil }(cm^3)} \times 100$
Gravimetric water content (or soil water content by weight)	θ_g	$g\ g^{-1}$ or %	$\frac{M_w}{M_s}$	$\frac{\text{wet soil weight} - \text{dry soil weight }(g)}{\text{dry soil weight }(g)} = \frac{\text{weight of water }(g)}{\text{dry soil weight }(g)} = \frac{\text{mass water }(g)}{\text{mass soil }(g)}$
Aeration porosity (air-filled or non-capillary porosity)	f_a	$cm^3\ cm^{-3}$ or %	$\frac{V_a}{V_t} = \frac{V_a}{V_s + V_a + V_w}$	$f_t - \theta_v$ or $\left(1 - \frac{\rho_b}{\rho_s}\right) - \theta_v$
Water-filled porosity (capillary porosity)	f_w	$cm^3\ cm^{-3}$ or %	$\frac{V_w}{V_v} = \frac{V_w}{V_a + V_w}$	$\frac{\text{percent water by volume}}{\text{total porosity}} = \frac{\theta v}{f_t} = \frac{\text{soil water content} \times \rho b}{\text{total porosity}}$
Equivalent depth of water	D_e	cm	$\theta_v \times h = \frac{V_w}{V_t} \times h$	$\theta_v \times$ soil depth
Degree of saturation	s	%	$\frac{V_w}{V_v} \times 100$	$\frac{V_w}{V_a + V_w} \times 100$

(continued)

Table 1.2 (continued)

Measurement	Symbol	Unit	Mass-volume relationship	Equation
Void ratio	V_e	%	$\frac{V_v}{V_s}$	$\frac{V_a + V_w}{V_t - V_v} = \frac{V_v/V_t}{1 - V_v/V_t}$
Volume solids	V_s	cm^3	$\frac{M_s}{\rho_s}$	$\frac{\text{mass (or weight) of dry soil } (g)}{\text{particle density, } \rho_s \text{ (or 2.65 } g\ cm^3)}$
Volume voids	V_v	cm^3	$V_w + V_a$	water volume (saturation) – water volume (dry) = $V_t - V_s$
Volume air	V_a	cm^3	$V_t - (V_w + V_s)$	$V_v - V_w$ or $V_s - V_w$
Volume water	V_w	cm^3	$\frac{M_w}{\rho_w}$ or $\frac{V_w}{\rho_w}$	$\frac{\text{wet soil weight} - \text{dry soil weight } (g)}{\text{density of water, } \rho_w \text{ (or 1 } g\ cm^3)}$ or $\frac{V_v - V_a}{\text{density of water, } \rho_w \text{ (or 1 } g\ cm^3)}$

Table 1.3 Common measurements of soil parameters and their equations

Measurement	Equation
Volume of a soil core (cm^3)	$h(\pi)r^2$ h = core height (*cm*); r = core radius (or ½ core diameter, *cm*); π = pi (~3.142)
Stokes Law: soil settling in water (V, $cm\ s^{-1}$) (Fig. 1.4)	kD^2 D = particle diameter (*cm*), k = temperature dependent constant, often 10,000
Darcy's Law: water flow in soil (Q, $cm^3 time^{-1}$) (Fig. 1.20)	$K = \left(\frac{Q}{At} \times \frac{L}{dH}\right)$ or $Q = \frac{KA(dH)}{L}$ $K = K_{sat}$ = saturated hydraulic conductivity (*cm* time^{-1}); A = cross-section area of column (cm^2); T = time (*sec*) for water to pass through core; L = length (or height) of soil column (*cm*); H = water head above soil core (*cm*); dH = hydraulic head (*cm*) above and within soil core, $H + L$
water potential (Ψ_t)	$\Psi_t = \Psi_g + \Psi_p + \Psi_o$ Ψ_g = gravitational potential; Ψ_p = pressure (or matric) potential; Ψ_o = osmotic potential
Field capacity (%)	Moisture at −33 *kPa* (loams and clays); −10 *kPa* (sand)
Permanent wilting point (%)	Moisture at −1,500 *kPa* (−15,000 *cm*, −15 *bar*); −300 *kPa* (sand)
Aeration Porosity (aka non-capillary porosity, %) at 40 *cm*	$\frac{\text{saturated soil weight } (g) - 40\ cm \text{ tension soil weight } (g)}{\text{volume of soil } (cm^3)} \times 100$
Water-filled Porosity (aka capillary porosity, %) at 40 *cm*	$\frac{\text{soil weight at 40 } cm \text{ tension } (g) - \text{oven dry soil weight } (g)}{\text{volume of soil } (cm^3)} \times 100$
Water retention or θ_g (%) at 40 *cm*	$\frac{\text{soil weight at 40 } cm \text{ tension } (g) - \text{oven dry soil weight } (g)}{\text{oven dry soil weight } (g)} \times 100$

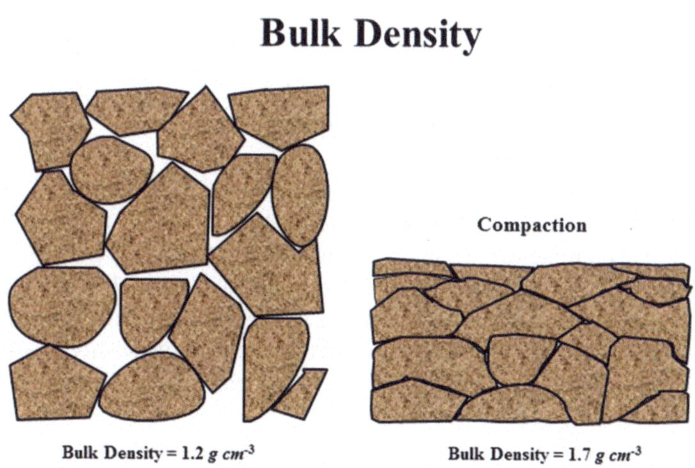

Fig. 1.6 Compacted soils (*right*) have higher bulk densities (number of particles in a given volume); thus, they have slower water infiltration and percolation compared to noncompacted soils (*left*)

Table 1.4 Typical bulk density and total porosity values for various soil textural classes and amendments

Soil textural class	Bulk density ($g\ cm^{-3}$)	Porosity (%)
Sands or compact clay	1.4 to 1.8	Low (32 to 47)
• Coarse sand	• 1.55[a]	42
• Medium sand	• 1.55[a]	42
• Fine sand	• 1.55[a]	42
Loam	1.2 to 1.6	Medium (39 to 55)
Loose silt loams or clay	1.0 to 1.4	High (47 to 62)
Organic soils	0.2 to 1.0	Very high (62 to 92)
Amendments:		
• Vitrified clay	• 0.84	57
• Zeolite	• 0.48 to 0.87	61
• Diatomaceous earth	• 0.39 to 0.59	72
• Calcined clay	• 0.56 to 0.64	73
• Sphagnum peat	• 0.15	74

[a]Bulk density is not affected by particle diameter if the particle density is the same and the packing is the same. For example, a barrel of golf balls and a barrel of soft balls will have the same porosity, if they are packed the same

$$6\,in \times \frac{1\,ft}{12\,in} \times \frac{43{,}560\,ft^2}{ac} \times \frac{1.5\,g}{cm^3} \times \frac{1\,lb}{454\,g} \times \frac{28{,}320\,cm^3}{ft^3} = \frac{2{,}037{,}917\,lb}{ac\,(6\,in\ \text{deep})}$$

Therefore, a typical acre of this particular soil 6 *in* deep weighs approximately 2,000,000 *lb* (908,000 *kg*), assuming an average bulk density of 1.5 $g\ cm^{-3}$.

Bulk density values also are useful to convert soil weight to volume.

Example What is the volume (m^3) of 500 *kg* of soil with a bulk density of 1.30 $g\ cm^{-3}$?

$$500\,kg \times \frac{1000\,g}{kg} \times \frac{cm^3}{1.30\,g} \times \frac{m^3}{(1{,}000{,}000\,cm)^3} = 0.3846\,m^3$$

A common source of increased soil compressibility (or bulk density) occurs when heavy machinery is used to alter the subsoil when it is wet, causing long-term damage. Similar compaction occurs when unrestricted play (or traffic) is allowed when soils are wet. The increased wetness of the soil acts as a lubricant between the soil particles, allowing them to slide closer together, reducing the pore space. The resulting compaction increases the soil's bulk density and its ability to allow water infiltration and drainage at the soil surface is reduced.

Example

1. What is the bulk density of a soil sample that weighs 120 g and occupies a volume of 75 cm^3?

$$\text{bulk density}\,(\rho_b) = \frac{\text{dry weight}\ (g)}{\text{volume of soil}\ (cm^3)} = \frac{120\,g}{75\,cm^3} = 1.60\,g\,cm^3$$

2. A 7.5 cm diameter and 7.5 cm high soil cylinder weighs 75 g empty and 505 g when full of dry soil. What is the bulk density of this soil?

step1: determine the mass of dry soil,

$$\begin{aligned}\text{dry weight of soil} &= (\text{weight of soil} + \text{container}) - (\text{container weight})\\ &= (505\,g) - 75\,g\\ &= 430\,g\end{aligned}$$

step2: determine the volume of soil. This equals the volume of the container, thus,

$$\text{volume} = \mathrm{h}\pi\mathrm{r}^2$$

where, h = height of the container (7.5 cm in this case)

$\pi = pi$(or~3.142)

r = radius (or one-half diameter of a circle) (3.75 cm in this case)

$$= 7.5\,cm \times \pi \times (3.75\,cm)^2$$

$$= 331\,cm^3$$

step3: determine the bulk density,

$$\text{bulk density}(\rho_b) = \frac{\text{dry weight}\ (g)}{\text{volume of soil}\ (cm^3)} = \frac{430\,g}{331\,cm^3} = 1.30\,g\,cm^3$$

Example From the following data, calculate the bulk density of the soil.

(a) weight of soil core + container = 700 g

(b) weight of container = 150 g

(c) container size of 7.08 cm diameter, height of 7.62 cm

step 1: volume of container $= \pi r^2 h$

$$= 3.142 \times (3.54\,cm)^2 \times 7.62\,cm$$
$$= 300\,cm^3$$

step 2: weight of soil : $= 700g - 150g$

$$= 550\,g$$

step 3: bulk density (ρ_b) : $= \dfrac{\text{dry weight of soil } (g)}{\text{volume of soil } (cm^3)}$

$$= \frac{550\,g}{300\,cm^3}$$
$$= 1.8\,g\,cm^{-3}$$

Compacted soils have less pore space and thus have higher bulk densities (Fig. 1.6). This results in slower infiltration and percolation. Noncompacted sandy soils can have infiltration and percolation rates as high as 4 *ft* (1.3 *m*) per h, while compacted clay loams have significantly lower rates, often <1 *in* (2.5 *cm*) per *h*. Modern golf green and sports field designs integrate compaction-resistant soil mixes of predominantly medium- and fine-sized sands with organic matter and other materials. The sands in these mixes tend to not excessively slide together, thus, withstand soil compaction, and the organic matter assists in water retention and nutrient exchange. These mixes withstand compaction while retaining enough moisture to maintain plant survival. Ideally the bulk density range for golf greens is between 1.35 and 1.55 $g\,cm^{-3}$, with an optimum of approximately 1.40 $g\,cm^{-3}$.

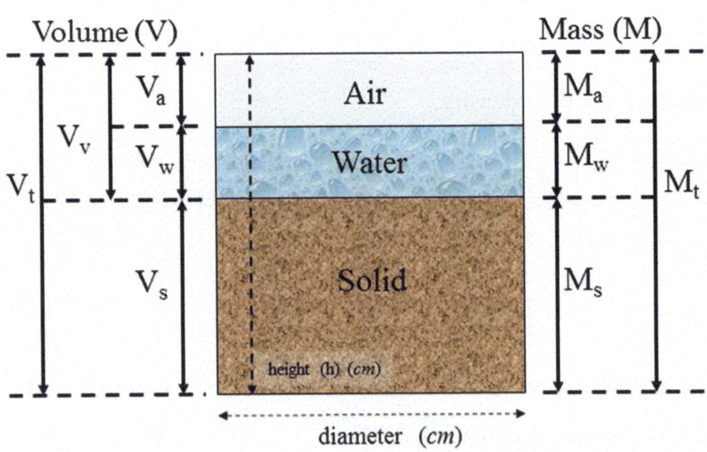

Fig. 1.7 Relationship of volume and mass of a soil sample separated into capillary (water), non-capillary (air) pores, and solid mass. V = volume, M = mass, a = air, w = water, s = solid, v = voids, t = total

To reduce bulk density in a compacted soil, according to the bulk density equation, either the soil mass has to be reduced or soil volume increased. In traditional agriculture, increasing soil volume is a common means of relieving soil compaction (Fig. 1.8). This is achieved by a number of soil mechanical means such as plowing, harrowing, cultivating, roto-tilling, sub-soiling, etc. These implements generally increase the volume which the existing soil mass occupies. The cultivation responses are typically effective until the soil mass settles back to its original volume from additional traffic or rainfall.

In other commodities, such as turf, increasing soil volume is considered counterproductive to the purpose of the playing surface. Turf managers, therefore, reduce compaction (bulk density) by reducing soil mass. This is accomplished by punching holes and removing soil cores from the site (termed **coring** or **aerification**) (Fig. 1.9). Other devices punch or penetrate the soil but do not remove a core. Though many positive reasons exist for doing this "solid tine" aerifying, it doesn't reduce soil compaction as it has little effect on soil mass or volume since a core isn't removed. When solid-tine aerifying, soil mass (compaction) is actually repositioned along the sides of the holes as well as at the bottoms.

Fig. 1.8 Decreasing soil compaction requires reducing soil mass or increasing soil volume. In traditional agriculture, relieving compaction is often performed by increasing soil volume using various mechanical devices such as plows, harrows, cultivators, roto-tillers, sub-soilers, rippers, etc.

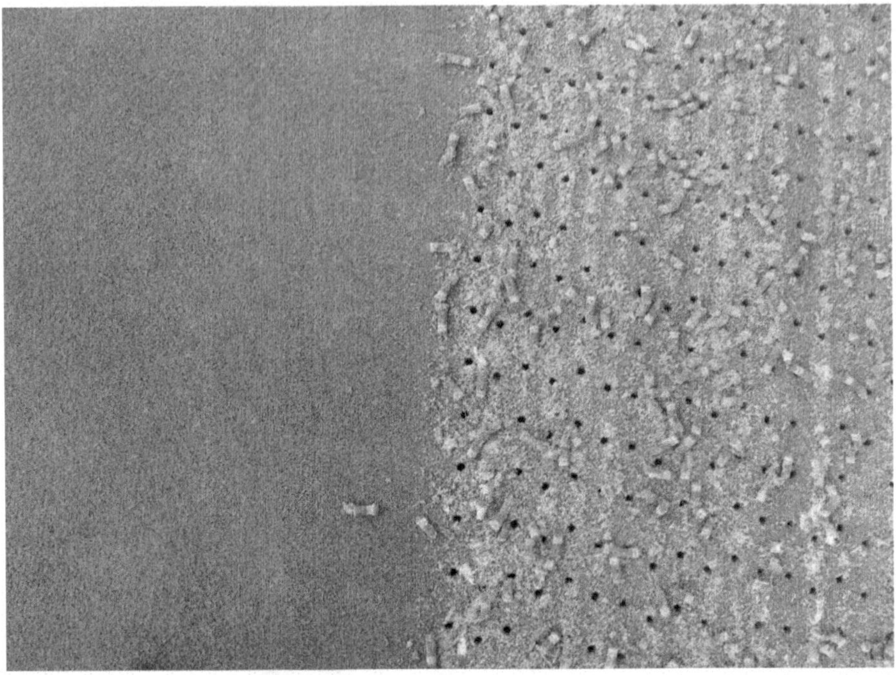

Fig. 1.9 When relieving soil compaction in commodities such as turf, increasing soil volume is often impractical as it would disrupt the playing surface. Soil compaction, therefore, is often reduced by decreasing soil mass through various practices such as soil aerification (or coring) and removal of the cores

Soil Porosity

Soil porosity, **pore space**, or **void space** is the percentage of total soil volume not occupied by solid particles, or the percentage of total soil volume occupied by air and water. The size of individual pores depends on the size of soil particles and the arrangement of how they are packed together. In dry soils, pores are mostly filled with air. In moist soils, pore spaces contain both air and water. If solid particles lie close together, porosity is low. If they are arranged in porous aggregates, as often found in medium-textured soils high in organic matter, pore space per unit volume will be high. Organic matter increases soil porosity by promoting soil particle aggregation and improving soil structure. Porosity can be determined from bulk density by the equation (Table 1.2):

$$\text{total porosity}, \mathrm{f_t}(\%) = \left(1 - \frac{\text{bulk density}, \rho_\mathrm{b}}{\text{particle density}, \rho_\mathrm{s}}\right) \times 100$$

or

$$\text{total porosity, } f_t(\%) = 100 - \left(\frac{\text{bulk density}, \rho_b}{\text{particle density}, \rho_s} \times 100\right)$$

Generally, particle density is usually assumed to be 2.65 $g\ cm^{-3}$ and bulk density is determined on an undisturbed soil core. Sandy soils generally have a total pore space between 32 and 47 %, while finer-textured soils vary in total pore space from 42 to 62 %.

Example A soil core 5 *cm* in diameter and 6 *cm* in height weighs 220 *g* when collected and 190 *g* when oven-dried (M_s). Following grinding, a soil-water volume was 171 *ml* after being poured into 100 *ml* of water. Calculated soil bulk density (ρ_b), particle density (ρ_S), and percent pore space (f_t) (1 *ml* water = 1 cm^3 water).

$$\begin{aligned}
\text{core volume } (V_t) &= \pi r^2 h \\
&= (3.14)(2.5\ cm)^2(6\,cm) \\
&= 117.8\,cm^3 \\
\text{volume water displaced by soil } (V_s) &= 171\,ml - 100\,ml \\
&= 71\,ml \text{ or } 71\,cm^3 \\
\text{bulk density } (\rho_b) &= \frac{M_s}{V_t} \\
&= \frac{190\,g}{117.8\,cm^3} \\
&= 1.61\,g\,cm^{-3} \\
\text{particle density } (\rho_s) &= \frac{M_s}{V_s} \\
&= \frac{190\,g}{71\,cm^3} \\
&= 2.67\,g\,cm^{-3} \\
\text{total porosity} (f_t) &= \left(1 - \frac{\text{bulk density}, \rho_b}{\text{particle density}, \rho_s}\right) \times 100 \\
&= \left(1 - \frac{1.61\,g\,cm^{-1}}{2.67\,g\,cm^{-1}}\right) \times 100 \\
&= 40\%
\end{aligned}$$

Example From the following information, calculate total porosity (f_t) and particle density (ρ_s) (1 *g* water = 1 cm^3 water): core volume = 98.2 cm^{-3}; soil saturated weight = 185 *g*; soil oven dry weight = 150 *g*.

$$\begin{aligned}
\text{void volume } (\mathrm{V_v}) &= \text{water volume (saturation)} - \text{water volume (dry)} \\
&= 185\,g - 150\,g \\
&= 35\,g \text{ or } 35\,ml^3 \text{ water} \\
\text{total porosity } (\mathrm{f_t}) &= \frac{\text{total pore volume } (\mathrm{V_v})}{\text{total volume } (\mathrm{V_t})} \\
&= \frac{35\ cm^3}{98.2\ cm^3} \\
&= 0.36\ \text{(or 36\%)} \\
\text{bulk density } (\rho_b) &= \frac{\mathrm{M_s}}{\mathrm{V_t}} \\
&= \frac{150\,g}{98.2\,cm^3} \\
&= 1.53\ g\,cm^{-3} \\
\text{particle density } (\rho_s) &= \frac{\mathrm{M_s}}{\mathrm{V_s}} \\
&= \frac{150\,g}{63.2\,cm^3} \\
&= 2.37\,g\,cm^{-3}
\end{aligned}$$

Example What is the total porosity (f_t) of a sand soil with a bulk density of 1.50 $g\ cm^{-3}$ assuming a particle density of 2.65 $g\ cm^{-3}$?

$$\begin{aligned}
\text{total porosity, } \mathrm{f_t}(\%) &= \left(1 - \frac{\text{bulk density}, \rho_b}{\text{particle density}, \rho_s}\right) \times 100 \\
&= \left(1 - \frac{1.50\,g\,cm^{-1}}{2.65\,g\,cm^{-1}}\right) \times 100 \\
&= 43\%
\end{aligned}$$

Example A container with dry soil has a 3.5 *cm* radius and is 7 *cm* high. A graduated cylinder contains 100 *ml* water when full. Water was slowly added to the soil container until saturation. Following soil saturation, 33 *ml* remained in the graduated cylinder. Calculate soil porosity.

$$
\begin{aligned}
\text{soil volume } (V_t) &= \pi r^2 h \\
&= (3.14)(3.5\,cm)^2(7\,cm) \\
&= 269.3\,cm^3 \\
\text{void volume } (V_v) &= 100\,g - 33\,g \\
&= 67\,g \text{ or } 67\,cm^3 \\
\text{total porosity } (f_t) &= \frac{V_v}{V_t} \times 100 \\
&= \frac{67\,cm^3}{269.3\,cm^3} \times 100 \\
&= 24.9\%
\end{aligned}
$$

Total soil porosity includes small and large pores. Small pores, called **capillary** or **micropores**, hold water against the pull of gravity and are responsible for the soil's water-holding capacity. Larger **noncapillary** or **macropores** drain rapidly and are typically air-filled. The relative proportion of macro- and microporosity is primarily determined by soil texture and structure. Sandy soils typically have a high proportion of macropores and relatively few micropores. With sands, regardless of the packing arrangement, large voids will be present among the large sand particles. Clay and silt particles, however, pack together because these particles are very small and can be platelike (flat). Since clay particles are so small, the voids among the particles are small.

The suggested porosity range for golf greens is 35 to 55 % total pore space with an optimum range of 40 to 55 % by volume (refer to Chap. 3). Capillary porosity is usually between 15 and 25 % and noncapillary porosity between 15 and 35 %. Ideally, capillary and noncapillary pore space should be equal at 25 % of the total soil volume. Minimum air-filled porosity at which soils will support good turfgrass growth is between 10 and 15 %. Lower porosity values indicate excessive soil compaction. These porosity ranges are for a rootzone mix that has been compacted, allowed to percolate water for 24 *h*, and then exposed to a 30 *cm* tension (or suction). Water-retention capacity at 30 *cm* tension for oven-dry soils typically ranges from 12 to 25 % by weight, with 18 % (1.8 cm^3 water held per 10 *cm* soil) being optimum.

Calculating Soil Porosity

As discussed, in calculating total pore space or porosity, two density measurements of soils, **particle density** (ρ_s) and **bulk density** (ρ_b), must be known. By knowing these two variables, the total solid space makeup of a soil can be determined. From here, total solid space is subtracted from 100 to indicate **total pore space** (Table 1.2).

Example What is the porosity of a soil core 10 *cm* high, 6 *cm* in diameter, and that weighs 500 *g* when dry? Assume a particle density of 2.65 $g\ cm^{-3}$:

$$
\begin{aligned}
\text{volume of soil core } (\mathrm{V_t}) &= \mathrm{h}\pi\mathrm{r}^2 \\
&= (10\,cm)(3.142)(6\,cm \div 2)^2 \\
&= 283\,cm^3 \\
\text{bulk density of soil } (\rho_\mathrm{b}) &= \frac{\text{mass}}{\text{volume}} \\
&= \frac{500\,g}{283\,cm^3} \\
&= 1.77\,g\,cm^{-3} \\
\text{total porosity, } \mathrm{f_t}(\%) &= \left(1 - \frac{\text{bulk density}, \rho_\mathrm{b}}{\text{particle density}, \rho_\mathrm{s}}\right) \times 100 \\
&= \left(1 - \frac{1.77\,g\,cm^{-1}}{2.65\,g\,cm^{-1}}\right) \times 100 \\
&= 33\%
\end{aligned}
$$

The next step is determining what percentage of pore space is actually filled with water and what portion is filled with air. To determine this, additional variables must be calculated. The first is the **water content of soil** by weight; in other words, the weight of water in a soil in relation to the total weight of the soil (termed **gravimetric water content,** $\boldsymbol{\theta_g}$). To perform this, a sample of the moist soil is weighed, dried at 105 °C for 5 hours or until its weight remains constant, and then reweighed. This provides the weight of the water which is then divided by the weight of the soil. Mathematically it is determined as:

$$
\theta_\mathrm{g} = \frac{\text{mass}_{\textit{wet soil}} - \text{mass}_{\textit{dry soil}}}{\text{mass}_{\textit{dry soil}}}
$$

(soil water weight = mass of wet soil − mass of dry soil)

Example A soil sample weighed 20 *g* fresh from the field. After drying in an oven for 24 *h* at 105 °C, it weighed 15 *g*. Determine its gravimetric water content (θ_g).

$$
\begin{aligned}
\text{gravimetric water content, } \theta_\mathrm{g} &= \frac{\text{mass}_{\textit{wet soil}} - \text{mass}_{\textit{dry soil}}}{\text{mass}_{\textit{dry soil}}} \\
&= \frac{20\,g - 15\,g}{15\,g} \\
&= \frac{5\,g\,\text{water}}{15\,g\,\text{soil}} \\
&= 33\,g\ \text{water}\ g^{-1}\text{soil (or 33\%)}
\end{aligned}
$$

Example What is the gravimetric water content (θ_g) of a soil weighing 80 g when moist and 65 g when dry?

$$\text{gravimetric water content } (\theta_g) = \frac{\text{wet soil wt} - \text{dry soil wt } (100)}{\text{dry soil wt}}$$

$$= \frac{80\,g - 65\,g(100)}{65\,g} = 23 g\,g^{-1}$$

Next, the water content of a soil, by volume (often called the **volumetric water content, θ_v**), is determined. This value is simply found by multiplying the water content by weight (θ_g) by the bulk density (ρ_b) of the soil (Table 1.2). Alternatively, the volumetric water content can be calculated by dividing the volume of water in the soil by the total volume of the soil. Since 1 g of water = 1 cm^3 of water, the volume of water is the same as the weight of water. This value is usually expressed in inches of water per inch of soil (or cm^3 water cm^{-3} soil). For example, a volumetric water content of 10 % refers to 1 *in* (2.5 *cm*) of water in 10 *in* (25 *cm*) of soil or 1.2 *in* (3 *cm*) of water in 1 *ft* (30 *cm*) of soil. The units of depth may be used to describe volumetric measurements since cross-sectional area is the same for both water volume and total soil volume.

$$\text{volumetric water content, } (\theta_v)\,(cm^3\,cm^{-3}) = \frac{\text{volume of water in soil } (cm^3)}{\text{volume of soil } (cm^3)}$$

$$= \text{water by weight, } \theta_g \times \text{bulk density, } \rho_b$$

Example Find the volumetric water content of a soil with a bulk density (ρ_b) of 1.5 $g\ cm^{-3}$ and a water content (θ_g) of 0.3 $g\ g^{-1}$ dry soil on a gravimetric basis.

$$\text{volumetric water content } (\theta_v) = \text{gravimetric water content } (\theta_g) \times \text{bulk density } (\rho_b)$$

$$= 0.3\,g\,g^{-1} \times \frac{1.5\,g\,cm^{-3}}{1\,g\,cm^{-3}}$$

$$= 0.45\,cm^3 \text{ water } cm^{-3} \text{ dry soil (or 45\%)}$$

The portion of the total soil porosity filled by air (**aeration porosity**) is then determined by the following:

$$\text{air filled (aeration or non-capillary) porosity}$$
$$= \text{total soil porosity} - \text{volumetric water content}$$

or

$$\text{air filled (aeration or non-capillary) porosity} = 1 - \left(\frac{\text{bulk density, } \rho_b}{\text{particle density, } \rho_s}\right) - \theta_v$$

Water-filled (or **capillary**) **porosity** is then determined by simply subtracting air-filled porosity from the total soil porosity.

Example If a soil has a bulk density (ρ_b) of 1.50 $g\ cm^{-3}$ and gravimetric water content (θ_g) of 0.26 $g\ g^{-1}$, what is the percent water by volume (or volumetric water content, θ_v) and what is the percent water-filled porosity (f_w)?

$$\text{volumetric water content } (\theta_v) = \text{water wt}, \theta_g \times \frac{\text{bulk density}, \rho_b}{\text{water density}}$$

$$= 0.26\, g\, g^{-1} \times \frac{1.50\, g\, cm^{-3}}{1\, g\, cm^{-3}}$$

$$= 0.39\, cm^3 \text{water } cm^{-3} \text{dry soil}$$

$$\text{total porosity}, f_t(\%) = \left(1 - \frac{\text{bulk density}, \rho_b}{\text{particle density}, \rho_s}\right) \times 100$$

$$= \left(1 - \frac{1.50\, g\, cm^{-3}}{2.65\, g\, cm^{-3}}\right) \times 100$$

$$= 43\%$$

$$\text{water-filled porosity } (f_w) = \frac{\%\text{water by volume}, \theta_v}{\text{total porosity}, f_t} \times 100$$

$$= \frac{39\%}{43\%} \times 100$$

$$= 91\%$$

The classical laboratory method of determining soil porosity involves measuring the water retention capacity of a saturated sample held at a tension of 30 or 40 *cm* at 15 atmospheres (*atm*). Water removed by this tension is considered to be that which occupies noncapillary (or air-filled) pore space, and retained water is considered to occupy capillary pore space (Table 1.3).

Example A soil core 5 *cm* in diameter and 30 *cm* in length indicated a soil wet weight of 918.42 *g* and oven dry weight of 876.50 *g*. Calculate gravimetric water content (θ_g), bulk density (ρ_b), volumetric water content (θ_v), equivalent depth of water of the core; percent total porosity, air filled porosity (f_a), percent saturation (s), and equivalent depth of water required to saturate the column.

$$\text{gravimetric water content } (\theta_g) = \frac{\text{mass}_{wet\,soil} - \text{mass}_{dry\,soil}}{\text{mass}_{dry\,soil}}$$

$$= \frac{918.42 - 876.50\,g}{876.50\,g}$$

$$= 0.048\,g\,\text{water}\,g^{-1}\,\text{soil}$$

$$\text{bulk denisty}(\rho_b) = \frac{\text{mass dry soil}}{\text{volume soil}}$$

$$= \frac{876.50\,g}{\pi(2.5\,cm)^2\,(30\,cm)}$$

$$= 1.49\,g\,cm^{-3}$$

$$\text{volumetric water content } (\theta_v) = \frac{\theta_g \times \rho_b}{\text{water density } (g\,cm^{-3})}$$

$$= \frac{(0.048\,g\,g^{-1})\,(1.49\,g\,cm^{-3})}{1\,g\,cm}$$

$$= 0.072\,cm^3\text{water}\ cm^{-3}\text{soil}$$

$$\text{equivalent depth of water } (D_e), = (\theta_v)\,(\text{soil depth})$$

$$\text{(amount water occupying core)} = (0.072\ cm^3 cm^{-3})\,(30\ cm)$$

$$= 2.16\,cm$$

$$\text{total porosity } (f_t) = 1 - \frac{\text{bulk density}, \rho_b}{\text{particle density}, \rho_s}$$

$$= 1 - \frac{1.49\,g\,cm^{-3}}{2.65\,g\,cm^{-3}}$$

$$= 0.438\,(\text{or}\,43.8\%)$$

$$\text{aeration porosity } (f_a) = \left(1 - \frac{\text{bulk density}, \rho_b}{\text{particle density}, \rho_s}\right) - \theta_v$$

$$= \left(1 - \frac{1.49\,g\,cm^{-3}}{2.65\,g\,cm^{-3}}\right) - 0.072$$

$$= 0.366\,(\text{or}\,36.6\%)$$

$$\text{water-filled porosity } (f_w) = \frac{\text{volumetric water content}, \theta_v}{\text{total porostiy}, f_t}$$

$$= \frac{0.072\,cm^3\,cm^{-3}}{0.438\,cm^3\,cm^{-3}}$$

$$= 0.164\,(\text{or}\,16.4\%)$$

$$\text{equivalent depth of water to saturate column} = (f_t - \theta_v)(\text{soil depth})$$

$$= (0.438\ cm^3 cm^{-3} - 0.072\,cm^3\,cm^{-3})(30\ cm)$$

$$= 10.98\,cm$$

Example If a soil weighs 30 g at field capacity, 27.2 g at wilting point, 25 g at air dry, and 24.2 g when oven dry, what is the water by weight of each?

$$\text{percent water at field capacity} = \frac{\text{weight1} - \text{weight2}}{\text{over-dry weight}} \times 100$$
$$= \frac{30\,g - 24.2\,g}{24.2\,g} \times 100$$
$$= 24\,g\,\text{water}\,g^{-1}\text{soil (or 24\%)}$$
$$\text{percent water at wilting point} = \frac{27.2\,g - 24.2\,g}{24.2\,g} \times 100$$
$$= 12\,g\,\text{water}\,g^{-1}\text{soil (or 12\%)}$$
$$\text{percent available water} = \frac{30\,g - 27.2\,g}{24.2\,g} \times 100$$
$$= 12\,g\ \text{water}\,g^{-1}\text{soil(or 12\%)}$$

Example

1. A 25 *cm* thick layer of soil with a volumetric water content (θ_v) of 0.20 $cm^3\ cm^{-3}$ has the following equivalent depth of water.

$$\text{equivalent depth of water, } D_e = \text{depth of soil } (cm) \times \text{water by volume } (cm^3 cm^{-3})$$
$$= 25\,cm \times 0.20\,cm^3\text{water}\,cm^{-3}\text{soil}$$
$$= 5\,cm\,\text{of water}$$

2. How much irrigation is needed to wet a dry soil 15 *cm* in depth to 36 $cm^3\ cm^{-3}$ water by volume?

$$\frac{0.36\,cm\ \text{water}}{cm\,\text{soil}} \times 15\,cm\ \text{soil depth} = 5.4\,cm\,\text{water}$$

3. How deep will a 3.18 *cm* rainfall penetrate this soil currently at a θ_v of 0.20 cm^3 water cm^{-3} soil and with a field capacity at 36 cm^3 water cm^{-3} soil?

$$\text{water penetration} = \frac{\text{depth water applied}}{\text{difference in } \theta_v}$$
$$= \frac{3.18\,cm}{(0.36 - 0.20)}$$
$$= 19.8\,cm$$

Example A cube of soil measures 10 x 10 x 10 *cm* (Fig. 1.10). Its total mass is 1,525 *g*, of which 200 *g* is water. Assume density of water (ρ_w) = 1 $g\ cm^{-3}$ and particle density (ρ_s) = 2.65 $g\ cm^{-3}$. Find: Gravimetric water content (θ_m), volumetric water content (θ_v), soil bulk density (ρ_b), total soil porosity (f), air-filled porosity (f_a), degree of saturation (s), equivalent depth of water (D_e), and gravimetric water content if saturated (θ_{ms}).

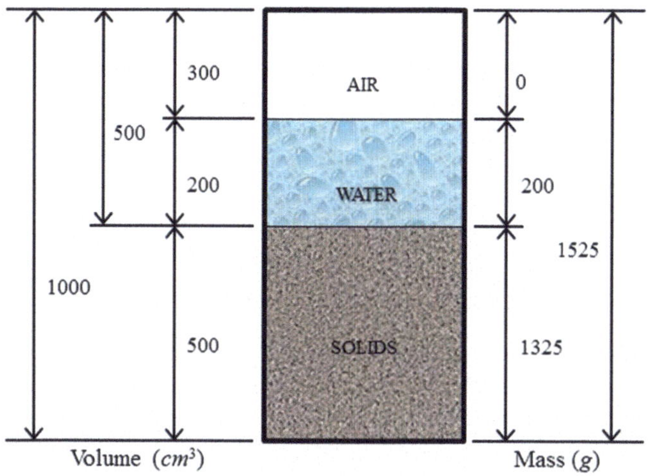

Fig. 1.10 Example chapter problem relating the volume and mass of a soil sample

total mass, M_t
$$= 1525g(\text{given})$$

total volume, V_t
$$= (10cm \times 10cm \times 10cm)$$
$$= 1000cm^3$$

mass water, M_w
$$= 200g(\text{given})$$

mass solids, M_s
$$= M_t - (M_w + M_a)$$
$$= 1525 - (200 + 0)$$
$$= 1325g$$

volume water, V_w
$$= M_w \div \rho_w$$
$$= \frac{200\,g}{1\,g\,cm^{-3}}$$
$$= 200cm^3$$

volume solids, V_s
$$= M_s \div \rho_s$$
$$= \frac{1325\,g}{2.65\,g\,cm^{-3}}$$
$$= 500cm^3$$

volume air, V_a
$$= V_t - (V_w + V_s)$$
$$= 1000 - (200 + 500)$$
$$= 300cm^3$$

volume voids, V_v
$$= V_w + V_a$$
$$= (200 + 300)$$
$$= 500cm^3$$

gravimetric water content, θ_m
$$= M_w \div M_s$$
$$= \frac{200\,g}{1325\,g}$$
$$= 0.151$$

volumetric water content, θ_v
$$= V_w \div V_t$$
$$= \frac{200\,cm^3}{1000\,cm^3}$$
$$= 0.20\,(\text{or } 20\%)$$

soil bulk density, ρ_b
$$= M_s \div V_t$$
$$= \frac{1325\,g}{1000\,cm^3}$$
$$= 1.325g\ cm^{-3}$$

total soil porosity, f_t
$$= V_v \div V_t$$
$$= \frac{500\,cm^3}{1000\,cm^3}$$
$$= 0.50\,(\text{or } 50\%)$$

air-filled porosity, f_a
$$= V_a \div V_t$$
$$= \frac{300\,cm^3}{1000\,cm^3}$$
$$= 0.30(\text{or } 30\%)$$

degree of saturation, s
$$= V_w \div V_v$$
$$= \frac{200\,cm^3\,\text{water}}{500\,cm^3\,\text{soil}}$$
$$= 0.40\,(\text{or } 40\%)$$

saturated gravimetric water content, θ_{ms}
$$= M_{w(sat)} \div M_s$$
$$= \frac{500\,g}{1325\,g}$$
$$= 0.377\,(\text{or } 38\%)$$

saturated volumetric water content, $\theta_{v\ (sat)}$
$$= V_{w(sat)} \div Vt$$
$$= \frac{500\,cm^3}{1000\,cm^3}$$
$$= 0.50\,(\text{or } 50\%)$$

What depth of water would need to be added to saturate this soil?

$$
\begin{aligned}
D_e &= \theta_v \times \Delta z & D_{e(sat)} &= \theta_{v(sat)} \times \Delta z \\
&= 0.20 \times 10cm & &= 0.50 \times 10cm \\
&= 2.0cm & &= 5.0cm
\end{aligned}
$$

$$\text{therefore}: \quad 5.0\ cm - 2.0\ cm = 3.0\ cm$$

Water Potential and Mathematical Units

When performing mathematical calculations concerning water flow, potential or pressure terms are used. Tensions, stress, and suction are some of the terms used to express potential. The more common mathematical units associated with these terms include: *bars*, centimeters of water (*cm* H_2O), centimeters of mercury (*cm* Hg), inches of water (*in* H_2O), atmospheres (*atm*), centibars (*cb*), millibars (*mb*), Joules per kilogram ($J\ kg^{-1}$), kilopascals (*kPa*), megapascals (*MPa*), pounds per square inch (*psi*), ergs per gram ($ergs\ g^{-1}$), and dynes per square centimeter ($dynes\ cm^{-2}$). *Bars* and kilopascals (*kPa*) are commonly used units. Relationships between units include:

1 bar = 1020 *cm* H_2O (or ~1000 *cm* H_2O)	1 *kPa* = 1 *cb*
= 75.01 *cm* Hg	= 0.001 *MPa* = 1000*Pa*
= 401.4 *in* H_2O @ 4°C	= 10 *cm* H_2O
= 0.9869 *atm* (or ~1 *atm*)	= 0.75 *cm* Hg @ 0°C
= 100 *cb* = 1000 *mb*	= 10 *mbar* = 0.01 *bar*
= 100 $joules\ kg^{-1}$	= 1 $J\ kg^{-1}$
= 14.50 *psi*	= 0.0099 *atm* (or ~0.01 *atm*)
= $10^6 ergs\ g^{-1}$ = $10^6 dynes\ cm^{-2}$	= 0.145 *psi*
× 14.5 = *psi*	= 10,000 $dynes\ cm^{-2}$
× 1019.7 = $g\ cm^{-2}$	× 1 = $J\ kg^{-1}$
× 29.53 = *in* Hg @ 0°C	× 1 = 0.01 *bar*
× 75 = *cm* Hg @ 0°C	× 0.01 = *bar*
× 0.10 = *MPa*	× 0.145 = *psi*
× 100 = *kPa*	× 4.01 = *in* H_2O @ 4°C
× 100,000 = *Pa*	× 10.2 = *cm* H_2O @ 4°C

Numerous web sites exist dealing with unit conversions. One is www.unitconversion.org.

1.2 Soil Moisture Properties

Describing Soil Moisture

Several descriptions of soil moisture exist, including; saturation, field capacity, wilting point, and permanent wilting point. Soil is **saturated** when all the pores are filled with water. However, a field soil is rarely saturated where all pores are completely filled with water. Often a soil appears saturated since water infiltration ceases and runoff occurs. In actuality, air is trapped in the soil and cannot escape, thus does not allow any water to enter. Some have recommended the word **satiated** be used when a soil will take no more water.

As a soil is thoroughly wetted, water begins to drain downward (if soil physical conditions permit). This water is often referred to as **gravitational water**.

After drainage has removed water from the macropores and the two forces of gravity and capillary tension become equalized, the soil is at **field capacity** (which is equivalent to **water-holding capacity**), with water normally occupying 20 to 35 % of the total volume (Fig. 1.11). Any water in excess of field capacity will drain due to gravitational pull. Having excessive soil water for an extended period is undesirable, therefore the primary function of drainage is the removal of this excess gravitational water from the soil. Field capacity is essentially the upper limit of storable water in the soil once free drainage has occurred after rainfall or heavy irrigation.

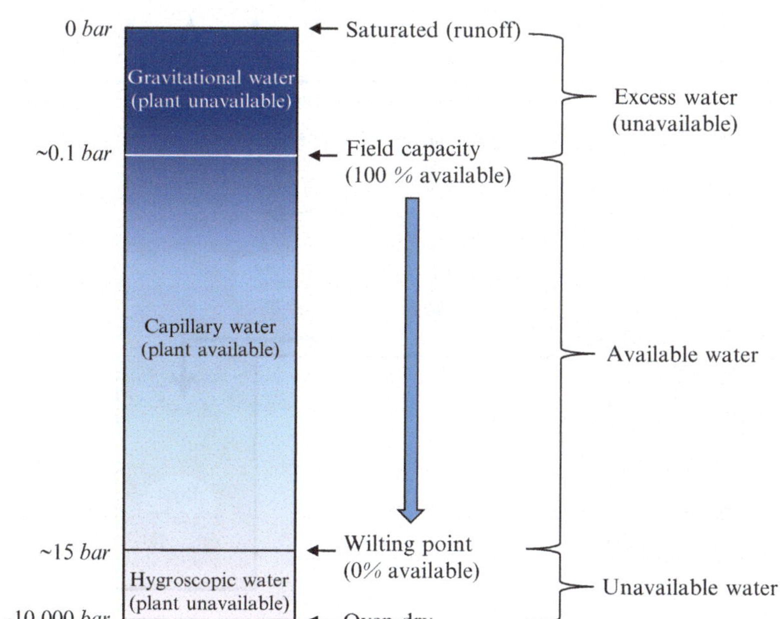

Fig. 1.11 The mathematical relationship between soil water tension (expressed as *bars*) and available and unavailable soil moisture

In constructed golf greens and sports fields, field capacity is determined by applying a force of 30 or 40 *cm* of tension to the soil to simulate the gravitational force on a rootzone 12 to 16 *in* (30 to 40 *cm*) deep. In native soils, field capacity is usually determined by applying tension from 0.05 (for sand soils) to 0.15 (for loam soils) *bar* [10 to 33 kilopascal (*kPa*)] [note: field soils do not drain to a tension of 1/3 bar within the time period associated with field capacity]. The higher the clay and organic matter content of the soil, the greater the water holding capacity and therefore the water content at field capacity. Sand-textured soils may have as little as 0.07 $g\ g^{-1}$ water at field capacity, a loamy soil may have 0.25 $g\ g^{-1}$ water at field capacity, whereas clay soils may have as much as 0.4 $g\ g^{-1}$ water at field capacity.

Soil does not stay at field capacity very long, i.e., soil water is always dynamic. Evaporation of water from the soil surface and soil water absorption by plant roots decrease the soil water content. Forces of soil adsorption and capillarity pull at water molecules and hold them in smaller micropores (Fig. 1.12). As particle and pore size decrease, these combined forces strengthen. At some point, roots can no longer take up water from the smallest pores as it is held too tightly, and plants will start to wilt. This is the wilting point of the soil. A soil is considered to be at the **permanent wilting point** when its water content yields a severely wilted plant which is unable to recover, even after irrigation. Water held at a tension of −15 *bar* (−1,500 *kPa*, −15,000 *cm* water) is often considered the lower limit of available

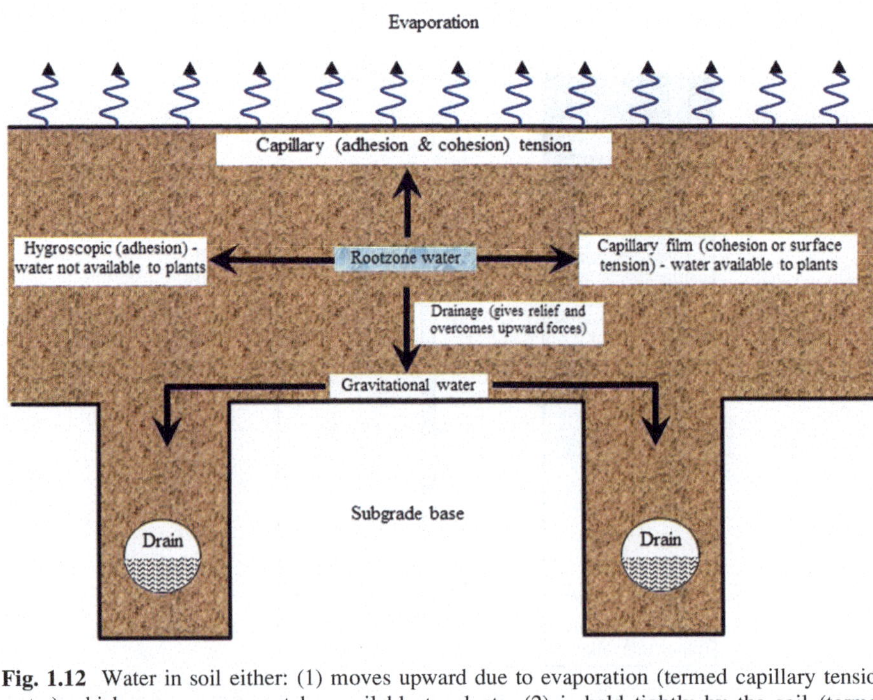

Fig. 1.12 Water in soil either: (1) moves upward due to evaporation (termed capillary tension water) which may or may not be available to plants; (2) is held tightly by the soil (termed hygroscopic water) and is unavailable to plants; (3) forms capillary film which is available to plants; or, (4) drains as gravitational water

water for heavier soils while $-300\ kPa$ ($-3\ bar$, $-3{,}000\ cm$) is often used for clean sand soil. In sandy soils the amount of water held in this way is small in proportion to the total, but in clayey soils it can be a large percentage of the total soil water content. These values are estimates, and actual values depend on the soil texture, structure, and the type of plant growing in the soil.

Plant-Available Water

The amount of water held in the soil between field capacity and wilting point is termed **plant-available water** (Fig. 1.11). Soils with a high percentage of silt have the greatest plant-available water content, as much as $0.25\ cm^3\ cm^{-3}$ or $3\ in\ ft^{-1}$ of soil ($25\ cm\ m^{-1}$). Sand-textured soils have less available water than silt-textured soils due to the presence of few micropores. The plant-available water content of clay-textured soils is also less than silty soils because a larger percentage of the water in a clay soil is held too tightly to be used by plants. A water-retention capacity between 12 and 25 % by volume is desirable, with an ideal capacity of 18 %. This translates to the equivalent depth of water being 0.18 *in* held in^{-1} of soil ($18\ cm\ m^{-1}$) based on the following formula used to calculate the equivalent depth of water in a soil:

$$\text{equivalent depth of water}(D_e) = \text{volumetric water content}(\theta_v) \times \text{soil depth}$$

From this equation, if the available water content of the soil (or volume metric water content) and the depth of turfgrass rooting are known, then the amount of water the plant has access to can be estimated.

Example What is the depth of water for a 35 *cm* thick layer of soil with a volumetric water content (θ_v) of 0.25?

$$\begin{aligned}\text{equivalent depth of water, } D_e &= \text{depth of soil } (cm) \times \text{water by volume, } \theta_v (cm^3 \text{ water } cm^{-3}\text{soil})\\ &= 35\ \ cm \text{ soil} \times 0.25 cm^3 \text{ water } cm^{-3} \text{ soil}\\ &= 8.75\ \ cm \text{ of water}\end{aligned}$$

Example If a bentgrass green has an average rooting depth of 2 *in* (5 *cm*), and the soil it is grown on has a field capacity of 21.4 % and a permanent wilting point of 14.6 %, how much water is available to the grass from the soil?

$$\begin{aligned}\text{total water available} &= \frac{\text{rooting depth} \times cm^3 cm^{-3} \text{available water by volume, } \theta_v}{100}\\ &= \frac{\text{rooting depth} \times [\text{field capacity} - \text{permanent wilting point}]}{100}\\ &= \frac{2\ in \times [21.4 - 14.6]}{100}\\ &= 0.136\ in\ (\text{or } 3.45\ mm)\end{aligned}$$

Therefore, 0.136 *in* (3.45 *mm*) of water is available to the grass from the soil. If water use (or ET) by the turf is 0.10 *in* (2.5 *mm*) per day, then daily irrigation is needed so moisture stress of the plant does not occur.

Soil scientists measure the amount of water left in the soil at various tensions (or pressures) to create a **moisture release** (or **retention**) **curve** (covered later in this chapter). Different soils have different-shaped moisture release curves with different heights of perched water tables. The shape and characteristics of this curve reflect the particle-size distribution of a soil and degree of soil compaction. Moisture release curves indicate water movement and amounts at various tensions, allowing soil scientists to predict soil water behavior at various depths and drying points which will be covered in greater detail later in the chapter.

1.3 Soil-Water Relationships

Water Potential (Ψ)

In order for water to move in soil, work must be done on the water by the previously discussed forces to account for adhesive and cohesive forces. As in all natural systems, movement of a material such as water is dependent on energy gradients. To predict the movement of water in soil, the energy potential of water is considered. Soil water potential is an expression of the energy state of water in soil and needs to be known or estimated to describe water flux or how much work a plant must expend to extract water from the soil. Water always moves from a point of high total potential (or energy) to a point of lower total potential. The fundamental forces acting on soil water are gravitational, matric, and osmotic. Total water potential (Ψ_t) is the sum of gravitational potential (Ψ_g), pressure (or turgor, or matric) potential (Ψ_p), and osmotic (or solute) potential (Ψ_o). Other forces act on water, but are almost always considered insignificant.

$$\Psi_t = \Psi_g + \Psi_p + \Psi_o$$

Gravitational potential (designated as Ψ_g) of soil water at a point is determined by the elevation of the point relative to an arbitrary reference level. Just as work is needed to raise a body against the earth's gravitational force, work is needed for water to move in soil, depending on its position in the gravitational force field. Water above the reference elevation has positive (+) gravitational potential. Water below the reference elevation has negative (–) gravitational potential. If, for example, this reference elevation is set at the top of the water table, the gravitational potential there is zero and the gravitation potential is positive above that elevation. If the soil surface is set as the reference elevation, the gravitational potential below the surface is negative with respect to that reference elevation. Gravitational potential is independent of the chemical and pressure conditions of water. It depends solely on relative elevation.

Pressure potential (designated as Ψ_p) of soil water is determined by the comparison to water at atmospheric pressure (i.e., at a free-water surface). Soil water at a hydrostatic pressure greater than atmospheric has a positive pressure potential. For example, water below a free-water surface (such as a groundwater table) has a positive pressure potential. Water at the free-water surface has zero pressure potential. At hydrostatic pressure less than atmospheric (in other words, under *suction* or *tension*), such as that risen in soil pores, a negative pressure potential occurs. Negative pressure potential is often termed **capillary** or **matric potential** (sometimes designated as Ψ_m). Water above the free-water surface is held by capillary and adsorptive forces. This water has a negative pressure (or matric) potential.

Osmotic (or **solute**) **potential** (designated as Ψ_o) is created by the presence of solutes (or salts) in the water solution. The more solutes present, the more their molecules (or ions) are attracted to water and the lower the vapor pressure of soil water. Due to solute presence, a greater amount of tension or suction (or work) is required by plant roots to extract water from soil. Soil salts and fertilizer (such as nitrogen or potassium) are common sources of these solute salts.

Water potential is generally expressed as *bar* or megapascal (*MPa*) where 1 *MPa* equals 10 *bar*. To obtain pounds per square inch ($lb\ in^{-2}$ or *psi*), *bar* can be multiplied by 14.7 while *MPa* values are multiplied by 147. For example, a cell with a turgor pressure of 4 *bar* equals 59 *psi*, a significant pressure.

Classic soil water potentials have been defined as 0 *bar* when the soil is saturated and up to −15 *bar* when soils are so dry that plants are considered to be permanently wilted. For ***plant*** water potential, −5 to 10 *bar* often represents fully turgid plants and ranges to −20 *bar* for severely wilted ones. In the ***atmosphere*** 0 *bar* represents water-saturated air, or 100 %, while −1,000 *bar* is a very low relative humidity (i.e., arid). Due to this gradient, water will move from a site of high-water potential (i.e., 0 *bar*) in soil to one of lower potential (i.e., negative value) in air (Fig. 1.13). However, when soils are so dry their water potential values are less than that of the root, plants cannot extract sufficient water from the soil and they begin to wilt.

The difference in soil water content between field capacity (0.1 *bar* for example) and permanent wilting point (−15 *bar* for example) is the amount of **available water** (Fig. 1.11). For example, consider a soil with a field capacity of 0.35 cm^3 water cm^{-3} soil and a permanent wilting point of 0.15 cm^3 water cm^{-3} soil. The difference, 0.20 cm^3 water cm^{-3} soil, is the amount of water, expressed as a percent of total volume, which is potentially available to the plant. This can be used to calculate the amount of available water in a given rootzone. However, one cannot normally allow the soil to dry out close to permanent wilting point. Doing so and making even the slightest error can result in plant death. Also the turf begins to display symptoms of drought stress much earlier than when the permanent wilting point is reached which then usually necessitates immediate action in the form of irrigation.

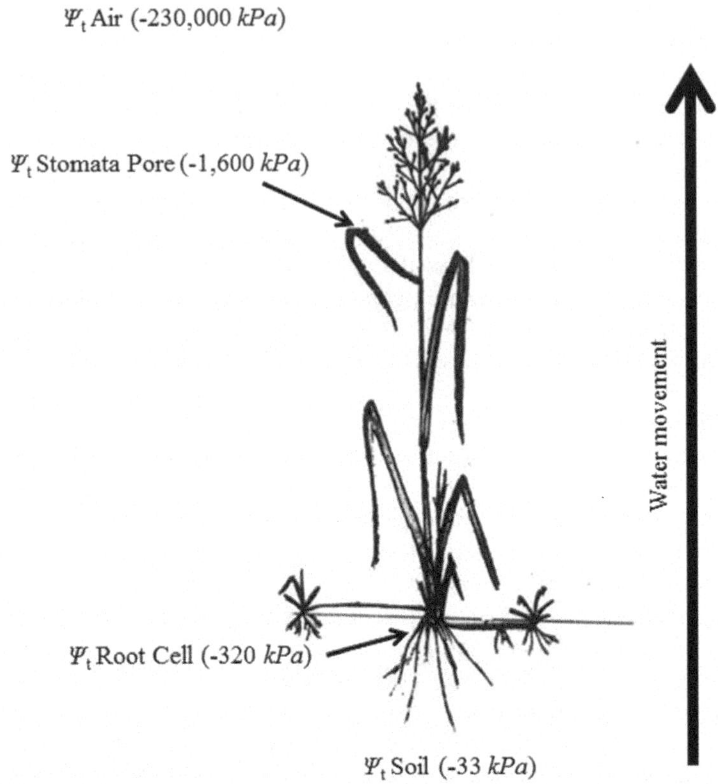

Fig. 1.13 Water movement from soil with less moisture stress (higher water potential at $-33\,kPa$), through roots, stems, and eventually released through leaf stomata into the drier atmosphere (lowest water potential at $-230{,}000\,kPa$).

Example A soil solution has a Ψ_t of $-0.3\,MPa$ (or $-3\,bar$) and the root cell has a Ψ_t of $-0.6\,MPa$ (or $-6\,bar$). Since the Ψ_t of the root is less than that of the soil (-6 vs $-3\,bar$), water can move from the soil into the root. If the Ψ_t in leaves is $-0.8\,MPa$ ($-8\,bar$), water will move from the roots through the crown and leaf sheaths into the leaf blades. If the relative humidity of the atmosphere is 50 %, this is equivalent to about $\Psi_t = -21.6\,MPa$ ($-216\,bar$). This means the force drawing water from the grass leaves is: $\Psi_t = -0.8\,MPa - (-21.6\,MPa) = 20.4\,MPa$ or $204\,bar$ or $2{,}999\,psi$, a truly awesome transpiration force.

Water movement follows a gradient from highest to lowest total water potential. In plants, this is somewhat analogous to a kerosene lantern where the wick constantly transmits fuel from its container-filled bottom to its top where the burning flame consumes it. Similar routes occur in plants where moisture in the soil is higher than the plant and atmosphere so a gradient develops along where water moves from the soil, into plant roots, through the stems and eventually released from leaf stomata into the drier atmosphere (Fig. 1.13). With water potential, the more negative a value is, less moisture it contains.

Infiltration and Percolation Rates

Infiltration rate refers to the quantity of water that can pass through the soil surface in a given time. **Percolation rate** refers to the quantity of water moving downward through the soil profile in a given period. The infiltration rate, in comparison to the application rate, determines whether applied water enters the soil, runs off (if sloped), or puddles. The rate at which water enters soil is, in part, dependent on soil texture and structure and the impact of these factors on soil porosity. Soils that have a high proportion of noncapillary (or macro) pores have high infiltration rates. Coarse sandy soils have high infiltration rates because the large sand particles result in an abundance of noncapillary pores. Fine-textured soils (having more silt and/or clay) may also have high infiltration rates if good structure results in the presence of large pores between structural aggregates. More often, however, fine-textured soils have low infiltration rates because most of the pores are small and accept water slowly. Fingers of preferential pathway flow may develop at the wetting front due to natural soil cracks, animal burrowing and/or from air pressure increases as the wetting front moves downward. The infiltration rate of a soil is also highest when it is dry. As a soil becomes wetter, infiltration rates decrease until a steady state is reached.

Infiltration and percolation rates of soils are critical as they determine playability after rainfall. Although relatively high initial percolation rates can be achieved on sand-based soils, due to natural soil settling, percolation rates usually decrease over time. Surface soil compaction from player traffic and maintenance machinery also decreases infiltration rates. Macropores may become blocked by silt and clay which can be inadvertently added in soil amendments and irrigation water, or by wind-blown soil and dust. Sand-sized particles from sandstone are not stable and may break down into finer sized grains with traffic. Excessive accumulation of soil organic matter in sand-based greens also decreases macroporosity and the percolation rate (Fig. 1.14). However, because a rapid percolation rate is generally correlated with a low water-holding capacity, a balance between percolation and water-holding capacity is needed.

In native soil areas, infiltration rates not only determine playability after rainfall, but also dictate the rate at which irrigation can be supplied. The precipitation rate of irrigation sprinklers must be less than the soil's infiltration rate so irrigation water does not run off. With sand based soils, which allow more rapid infiltration, this is mostly not a big issue.

Water Movement in Soil

Soil water movement is dependent on (1) water being pulled downward by **gravity** (or **hydraulic gradient**), (2) water adhering to itself due to hydrogen bonding (called **surface tension** or **cohesion**), and (3) water sticking to other surfaces, such as soil particles (referred to as **adhesion**, **adsorption**, or **hygroscopic moisture**) (Fig. 1.12). Simply stated: soil water moves from a higher total potential to a lower total potential.

Fig. 1.14 Water remaining for an extended period in a golf cup often indicates undesirable fine-textured soils were used in the initial construction. This often leads to reduced internal drainage, low soil oxygen levels, further soil compaction, and weakened plants

Gravity (or Hydraulic Gradient). Gravity is the constant downward-pulling force on water. Water at the soil surface has 50 *cm* more gravitational potential than water 50 *cm* below the soil surface. Thus water at the soil surface would move downward, if gravity were the only consideration, which it is not. Gravitational potential is only one of the components of total potential.

Surface Tension (or Cohesion). Since they are polar molecules and possess hydrogen-bonding characteristics, water molecules are attracted to each other in all directions and are attracted much more to each other than to adjacent air molecules. Surface tension is created when water meets air, causing water molecules to shrink, pulling the surface of a water drop together, thus, becoming round and beading up. The smaller the drop of water, the stronger the surface tension becomes, and the more difficult it becomes to break this tension. The degree of the water molecules bonding together determines the surface tension and, for example, is often strong enough to support insects walking across it. The surface of water can be bent slightly by gently touching it; however, if the force applied is too great, the surface will break. Wet soils have less cohesive tension than dry soils.

Water Adhesion (or Adsorption) to Soil Particles. As soil particle and pore sizes decrease, an increase in attraction (or adhesion) occurs between soil particles and soil water. In most instances, water adheres to soil particles very strongly, much more so than to other water molecules. This adhesive force "pulls" or holds water against the force of gravity. The adhesive forces can, in fact, pull water in all directions away from a water source in soil, independent of gravity (Fig. 1.12). If a dry column of soil is placed in contact with free water, moisture will rise into the soil. However, capillary water movement is generally limited in coarse textured (sandy) soils (rarely more than 4 *in*, 100 *mm*) from its source. Furthermore, the smaller the soil pore spaces, the slower water will move laterally (sideways) and the more tightly the water is held. Soils with a larger pore size (i.e., coarse-textured sand) hold less moisture at a given potential than soils held with a smaller pore size (i.e., silt or clay).

Capillarity

Water is held in soils in two forms, **adhesion** and **cohesion** (aka, **surface tension**). Adhesion causes a surface film around soil particles while surface tension is where water is held in soil pores. When gravity pulls water downward in soil, adhesion and cohesion forces act against it, attempting to hold water molecules near the soil particles or in pore spaces.

Capillary tension is the combined force of surface tension and adhesion that retains water in small soil pores against the pull of gravity. This retained water held in the soil against gravity is collectively referred to as **capillary water**. It includes the film of water left around soil particles and water in capillary soil pores after gravitational water has drained. Once water molecules wet a particle, they seek another dry surface on which to cling; as this water moves, it pulls additional water along with it. This pulling action produces a negative pressure or vacuum. Capillary action is represented in Fig. 1.15 where water has been absorbed from the base of

Fig. 1.15 Water height in a sand:peat pile due to capillary action from moisture below. This moisture height is determined by the soil's capillary forces (adhesion and cohesion) against the force of gravity

the sand:peat pile and has risen to the point where the force of gravity equals that of soil surface adhesion and cohesion.

A similar reaction can be seen with a sponge and water. When dry, contact with water (capillary tension) causes a sponge to wet-up both upward and sideways. Likewise, if the sponge is highly saturated, some water will initially drip out of it, but a certain amount will be held. This is capillarity and adsorption forces causing the sponge to hold water and is in a way, very similar to soils. The extent to which capillary action works depends on the size of spaces or channels formed when soil particles pack together. The smaller the channels (i.e., more compacted or finer textured the soil), the greater the capillary action.

Similar capillary forces are seen with drinking straws. When a straw is placed in a liquid, the liquid in the straw is slightly higher than the surface of the drink outside the straw. This is due to the water molecules inside the straw adhering to the sides of the straw, and their cohesive properties allowing them to draw up other water molecules with them (refer to Chap. 2 for additional information). If different straws of varying diameters are observed, the smaller the diameter of the straw, the higher the water will rise in it. In fine diameter straws, gravity has less impact on the surface tension due to a lower surface area of water being exposed, thus the water can rise higher. These same principles can be seen in soils, small pores will draw water higher up than soils with large pores.

Cohesion or surface tension, is weaker than adhesion and is the first force to break under the pull of gravity. Therefore, surface tension is the limiting factor in the amount of water a pore can hold. Water in large pores is mostly held by surface tension, as it is too far away from pore edges to be influenced by adhesive forces. Gravity, therefore, can more easily pull the water downward in larger pores, causing them to drain first. The remaining water is attracted more strongly to the edges of soil particles by the stronger adhesive forces. Water held by adhesion requires greater gravitational forces to remove it. In a saturated soil, all the pore spaces are filled with water; additional water has nowhere to go and gravity pulls water downward, causing soil to drain. Larger pores eventually empty out and drainage stops. As soil water content decreases due to the pull of gravity, the drainage rate decreases at an exponential rate, as the remaining water is held tighter in narrower pores. Water in small (or capillary) pores is retained by capillary forces, allowing it to be used by plants.

As soil moisture levels change, so do the soil's strength and stability. At saturation, soil particles are not held together by capillary tension (a combination of water surface tension and soil adhesive forces), and the soil becomes very unstable until it becomes progressively soft enough to form mud, easily compacting or washing away. As soil moisture is progressively removed, water in smaller pores is held much tighter by capillary tension, soil particles become more tightly bonded together, and the soil becomes stronger. As the soil dries out, it can become unstable again, as all particles become free from each other, forming dust which can easily be blown away. An example involves beach sand. Immediately following an incoming wave, the sand is saturated and has little resistance to applied pressure such as walking, resulting in and deep footprinting (Fig. 1.16). However, after

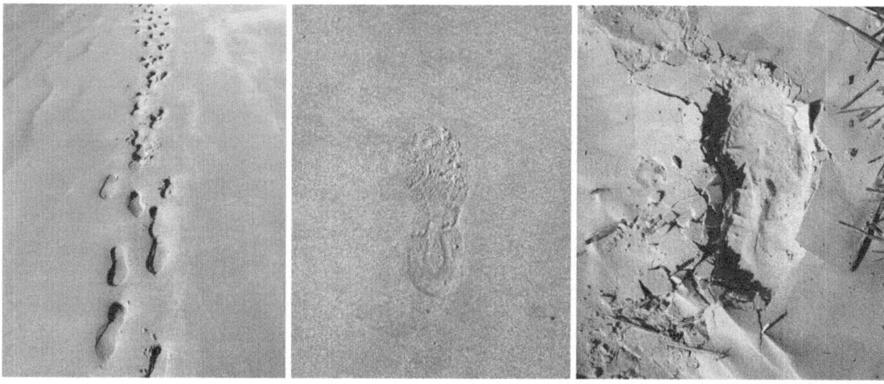

Fig. 1.16 As soil moisture levels change, so do a soil's strength and stability. For example, beach sand, when wet, easily compacts from foot traffic (*left*). As moisture is removed, soil particles become more tightly bonded together and resistance to compression increases (*middle*). As soil moisture is depleted, the soil once again becomes unstable, losing its ability to resist compaction (*right*)

several minutes, the excessive moisture drains below the surface, leaving the sand close to field capacity. Walking across this sand yields little to no foot tracks as soil particles become more tightly bonded together. As the sand then progressively dries out, water particles are no longer present to help bind sand particles, resistance (stability) to applied pressure is lost, and foot printing occurs again.

Perched Water Table

A **perched water table** is a zone of saturated soil just above the interface of a finer-textured soil over a coarser-textured soil. This is often a desirable condition, created by design for golf greens and sports fields, where finer sand is placed over a coarser aggregate layer such as gravel (Fig. 1.17). Water will not drain from the upper finer-textured soil until a sufficient depth of water (air entry value) develops above the coarser soil. Then, the weight of any additional water cannot be contained by the capillary retention forces, and water starts flowing downward.

This saturated zone develops as the adhesive force of the finer-diameter soil particles on water is greater than the drainage force due to gravity. The finer textured and/or more compacted a soil becomes, or the greater the difference in particle-size distribution between the soils, the harder it is for water to cross over from one surface to another, and the higher the perched water table. Conversely, a coarser soil will have a shallower perched water table as will rootzones with round gravel particles vs. angular-shaped gravel. If the upper finer soil depth is not greater than the depth of the perched water table, the whole soil profile remains saturated and will not drain.

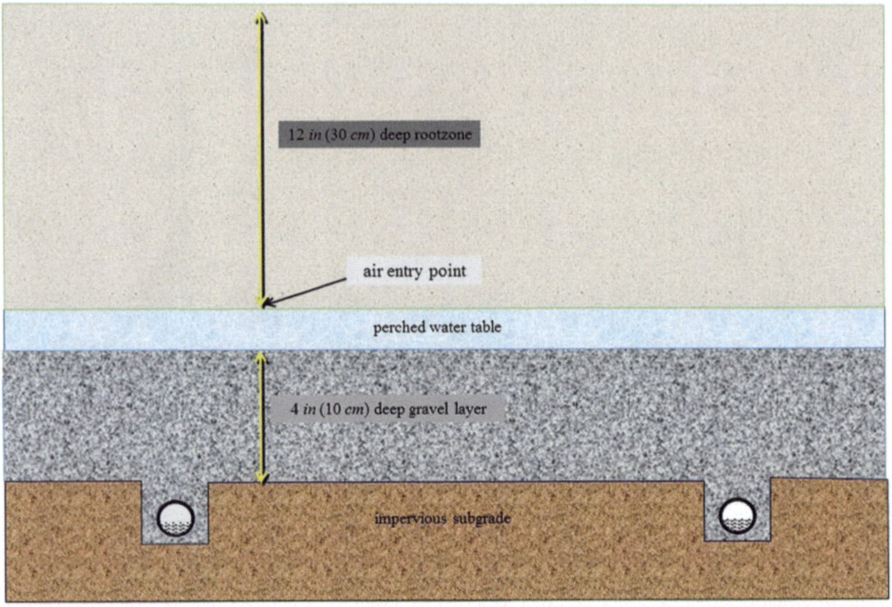

Fig. 1.17 Desirable flat perched water table formed when an appropriately sized rootzone sand is placed over a gravel layer, such as in USGA specification golf greens. This has been shown the best means of conserving sufficient water when needed yet provide adequate drainage following heavy rainfall

By creating an adequate perched water table, fast-draining, low-compacted sand can be used as a rootzone. When the correct depth of sands are placed over the correctly sized gravel layer, the perched water table provides a reservoir of water for the grass to use, but the complete profile still drains sufficiently. Drainage flow will continue from the sand into the gravel due to the combined forces of gravity and the adhesive forces at the contact points between the sand and gravel particles. This "flow" continues until the surface tension in the pores of sand and the adhesive forces of sand particles in contact with the gravel equal those forces (gravity and adhesion) pulling the water down.

At this point, equilibrium is reached and a saturated perched water table exists above the sand/gravel interface. Optimally, this zone of saturated soil should extend no more than about 6 *in* (15*cm*) up from the interface. Unless additional water is added to the system above, no further water moves downward out of the perched water table.

Saturated and Unsaturated Hydraulic Conductivity

The rate of water movement through a soil is referred to as the flux or flux density. The flux of water (saturated and unsaturated) across a plane in soil can be determined from the capacity of the soil to allow water movement and the gradient across that plane, i.e., flux = hydraulic conductivity x the gradient. Although commonly stated in the industry, "hydraulic conductivity is the rate of water movement through a soil," is technically incorrect.

Darcy's Equation:

$$\frac{Q}{AT} = K_{sat} \times \frac{dH}{L}$$

$$\textbf{flux} = \textbf{conductivity} \times \textbf{gradient}$$

where: K_{sat} (or K) = saturated hydraulic conductivity ($cm\ s^{-1}$)
Q = quantity of water (cm^3) passing through the soil core
A = cross-sectional area (cm^2) of the soil core
T = time (*sec*) required for the water to pass through the core
L = length (*cm*) of the soil core
dH = head (*cm*) of water imposed on the core (length of soil core + height of water above soil core)

This incorrect statement can be illustrated by having a new golf green with a saturated hydraulic conductivity of 20 $in\ h^{-1}$ (51 $cm\ h^{-1}$). No doubt water will move rapidly through this green, but will it be at 20 $in\ h^{-1}$ (51 $cm\ h^{-1}$)? From the above equation, the flux could only be 20 $in\ h^{-1}$ (51 $cm\ h^{-1}$) if K = 20 in h^{-1} (51 $cm\ h^{-1}$) and the hydraulic gradient = 1 (change of 1 *in* pressure per 1 *in* decrease in elevation). This means the green would have to remain completely saturated (top to bottom) with no water standing on the soil surface.

Saturated hydraulic conductivity (designated as K_{sat}) is defined as the proportional relationship of water flow through a saturated soil in response to a given difference in head (pressure). It combines infiltration and percolation. Large pores allow for high saturated hydraulic conductivities. Sandy soils, therefore, generally have much higher saturated hydraulic conductivities than clay-type soils. However, clay soils with strong structure may have substantial macroporosity, resulting in elevated saturated hydraulic conductivities.

Hydraulic conductivity for rootzone mixes can be determined in the laboratory. Combinations of sand, soil, and/or organic material (usually peat) can be mixed in various ratios for testing purposes. The USGA guidelines for rootzone mixes recommend a saturated hydraulic conductivity (K_{sat}) between 6 and 24 $in\ h^{-1}$ (15 and 61 $cm\ h^{-1}$). The K_{sat} of rootzone mixes will decrease over time due to compaction and organic matter accumulation (Fig. 1.18). Laboratory measurements of K_{sat} can also be determined on soil cores taken from established soils using specialized equipment and extreme care.

On an established turfgrass site, an infiltrometer is often used to determine K_{sat} (Fig. 1.19). Water is added to the single- or double-ring infiltrometer and, after a period of time, the depth of water absorbed is measured. Hydraulic conductivity of established turf sites can also be determined by extracting intact soil cores, taking them to the laboratory, subjecting them to a hydraulic head, and using **Darcy's**

Fig. 1.18 Layers such as thatch (*left*) often disrupt normal surface and subsurface drainage which not only slows or delays play but also weakens turf due to soil compaction, reduced soil oxygen, and increased disease incidence

Fig. 1.19 A double-ring infiltrometer being used to measure hydraulic conductivity (internal drainage) of an established turf site

Law. Darcy's equation is used for calculating hydraulic conductivity under saturated soil conditions (Fig. 1.20).

$$K_{sat} = \frac{Q}{AT} \times \frac{L}{dH}$$

Darcy's equation describes that water flow through soil is directly proportional to soil conductivity (K), height of the water column above the soil surface (dH), area

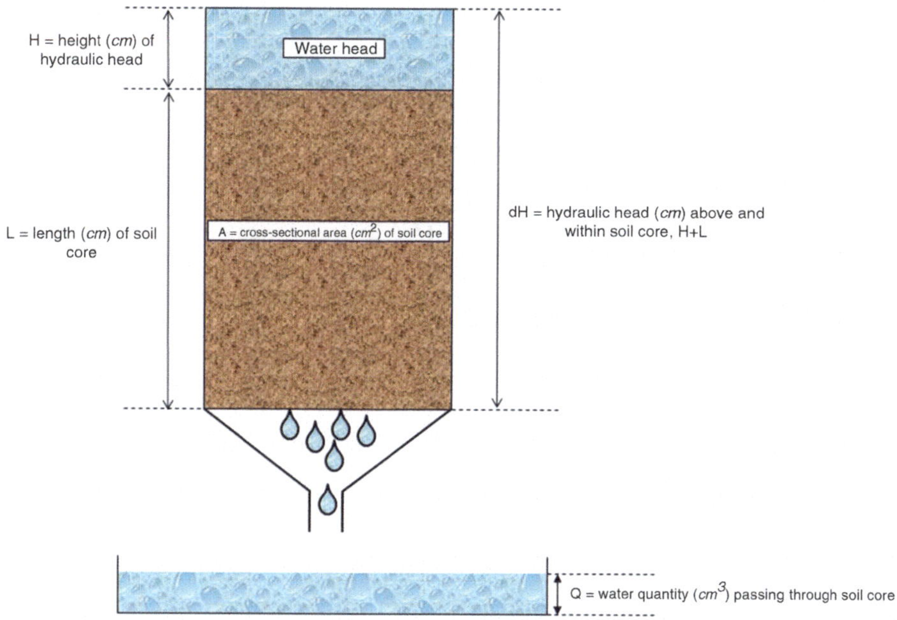

Fig. 1.20 Water flow through soil (Q) is directly proportional to soil permeability (K); height of hydraulic head above soil core (H); height of hydraulic head above and within the soil core (dH); area of soil column (A) and time (T); and inversely proportional to the height of the soil column (L) as determined by Darcy's Law

of the soil column (A) and time (T), and is inversely proportional to the height of the soil column (dL) (Fig. 1.20). The equation is often rearranged as:

$$\frac{Q}{AT} = K_{sat} \times \frac{dH}{L}$$

or

$$Q = \frac{KA \times (L + H)}{L}$$

Example Sand in a 7.6 *cm* column is saturated. The column's cross-sectional area is 20 cm^2 with a 5 *cm* height of water kept above the column. If the K_{sat} of the sand is 25.2 $cm\ h^{-1}$, what is the flow rate of water through this column? ($1\ cm^3 = 0.000264172\ gal = 0.0610237\ in^3$)

dH is the height of water plus the height of the soil core, L + H (7.6 *cm* + 5 *cm*) (Fig. 1.20):

$$
\begin{aligned}
Q &= \frac{KA \times dH}{L} \\
&= \frac{(25.2\,cm\,h^{-1})\,(20\,cm^2)\,(12.6\,cm)}{7.6\,cm} \\
&= 836\,cm^3\,h^{-1} \\
&= \frac{836\,cm^3}{h^{-1}} \times \frac{0.000264172\,gal}{cm^3} \\
&= 0.221\,gal\,h^{-1} \\
&= \frac{836\,cm^3}{h^{-1}} \times \frac{0.0610237\,in^3}{cm^3} \\
&= 51\,in^3\,h^{-1}
\end{aligned}
$$

Example Sand in a 10.2 *cm* column with a cross-sectional area of 31.7 cm^3 and a height of 5 *cm* water imposed on its surface has a flow rate of 1,235 $cm^3\,h^{-1}$. Determines its K_{sat}.

$$
\begin{aligned}
Q &= \frac{KA \times dH}{L} \\
K_{sat} &= \frac{Q}{AT} \times \frac{L}{dH} \\
&= \frac{1,235\,cm^3 h^{-1}}{31.7\,cm^2} \times \frac{10.2\,cm}{15.2\,cm} \\
&= 26.1\,cm\,h^{-1} \\
&= 10.3\,in\,h^{-1}
\end{aligned}
$$

In turf, to predict future field conditions, the samples in question are brought to saturation by wetting them from the bottom up by placing the cores in a water bath. This ensures no air entrapments occur in the samples. Following saturation, the cores are allowed to drain. The samples are then placed on a tension table (or plate) and exposed to 30 to 40 *cm* tension. Once drainage stops at this 30 to 40 *cm* tension, the sample is then compacted to 3.03 $J\,cm^{-2}$ (14.3 $lb\,ft^{-1}$) using a 3 *lb* (1.36 *kg*) hammer dropped 15 times from a height of 30.5 *cm* (12 *in*). This amount of compaction closely correlates with that typically found on golf greens after several years of play. Once compaction is completed, the sample is ready for the saturated hydraulic conductivity test.

Unsaturated hydraulic conductivity is the proportional relationship of water flow through an unsaturated soil in response to a given difference in head. Water rising in the profile from a perched water table or being drawn out of a sand-based green by a native soil collar are examples of unsaturated flow. Overall, water movement in unsaturated soils is much slower than in saturated soils, when subjected to similar differences in head.

1.4 Soil Moisture Retention Curves

A soil moisture retention curve, abbreviated SMRC (also known as a soil moisture release curve, soil water characterization curve, soil water release curve, and others), is a graph of a non-linear function relating soil moisture content and the matric suction (or tension) required to retain that moisture level in the soil. Curves are constructed from a laboratory procedure where a soil sample is subjected to a range of matric suction, typically 0 to 40 *cm* (0 to 16 *in*) for turf purposes, to determine how much moisture is retained at each matric suction point. This suction is referred to as tension (measured in units of negative pressure) or as the water-column equivalent of that negative pressure, also known as matric head or matric potential (measured in units of water depth). Some common equivalent units for tension are: $1\ kPa = 1{,}000\,Pa = 0.001\,MPa = 0.01\,bar = 10\,mbar = 1\,J\ kg^{-1} = 0.0099\,atm = 0.145\,psi = 10\,cm$ water $=$ ~4 *in* water. Also, $1\,bar = 1{,}020\,cm$ water $= 401$ *in* water $= 100\,J\ kg^{-1} = {\sim}1\,atm = 100{,}000\,Pa = 0.1\,MPa = 100\,kPa$.

Large soil pores do not retain water against the tension (or suction) being applied and empty with less tension, while increasing tension is needed as pore size become smaller. Sand soils tend to release water with less tension, while clay soils, with smaller pores and more surface areas with adhesive and osmotic bindings with water, require more tension to release water. SMRC can be used to predict soil water storage, the supply of water to plants (that between field capacity and permanent wilting point), and soil aggregate stability. These curves, however, do not provide adequate information on the flow of water through the soil which is normally determined via saturated hydraulic conductivity (K_{sat}).

Constructing Soil Moisture Retention Curves

Several methods exist for constructing SMRCs. The first was developed in 1907 by Edgar Buckingham in which he took 48 *in* (122 *cm*) columns of soil varying in texture from sand to clay and periodically added water from a side tube to maintain a constant 2 *in* (5 *cm*) depth of water in the bottom of the columns. The column tops were sealed to prevent evaporation. The addition of water was continued for two months or longer to allow the soils, initially at low water contents, to imbibe water upward through capillary suction to approach a state of equilibrium. Moisture content was measured at several heights in the columns. By establishing equilibrium with a constant saturated zone in the bottom of the column, matric tension could be reported by definition as height above the free water surface.

A second method of constructing SMRC referred to as the **hanging water column** or **tension table** (Fig. 1.21). The tension table method simulates the increasing suction energy (or tension) that occurs as a soil dries or drains. At zero tension, the soil is saturated. As tension is increased, the largest pores release moisture and air enters the pore space. The tension required to initiate moisture

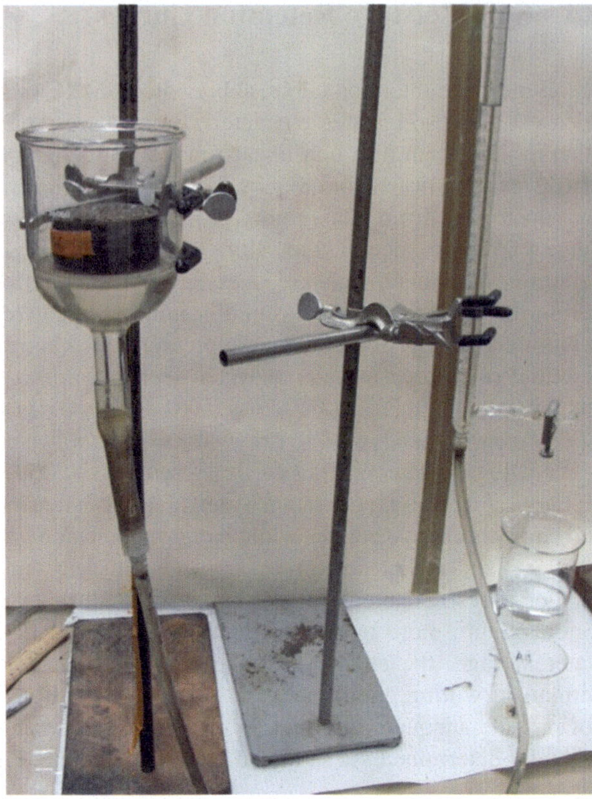

Fig. 1.21 The use of a hanging water column to create soil moisture retention curves for turfgrass soils

removal from the saturated largest pores is defined as the **air-entry point**, sometimes referred to as **critical tension** (Fig. 1.22). Air-entry values tend to be smaller in coarse-textured (or more uniform pore) soil and tend to be bigger (or higher) in finer-textured (wider array of pore sizes) soils. The maximum suction value obtained by the tension table is 100 *kPa* (=1 *bar* = 1,000 *cm*). As water contents decrease, increasing tension is required to remove water which is more aggressively bound in the smaller pores.

The following general procedure is used to construct SMRCs using the hanging water column.

- Mix and pack specified ratios of sand, soil and/or organic matter to be used as rootzone mix, or use samples from an existing rootzone.
- Place sample in water to saturate from the bottom up,
- Place sample on a tension table and subject it to tensions at regular increments, such as 5 *cm* (2 *in*) increments from 0 to 30 or 40 *cm* (12 or 16 *in*) for the depth of a rootzone for a USGA green (Fig. 1.21).
- Measure water content at each tension and report as volumetric moisture content (abbreviated θ_v).

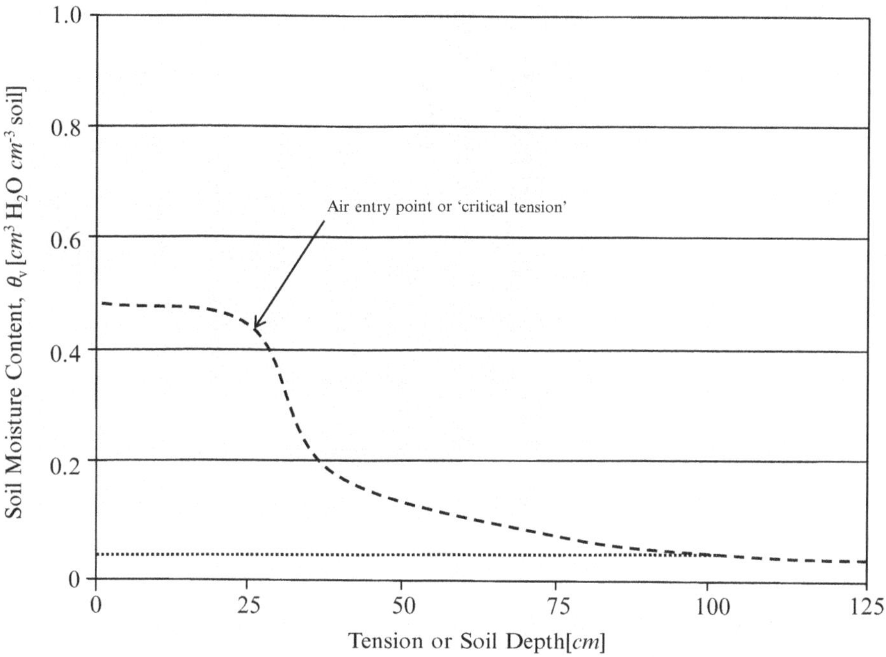

Fig. 1.22 Soil moisture retention curve for a sand soil. In this example, a zone of unavailable water exists in the top 5 *in* (13 *cm*) while a zone of plant available water exists from 5 *in* (13 *cm*) to about 10 *in* (25 *cm*) at which the zone of saturation (or perched water table) starts

– Construct a SMRC graph showing soil volumetric moisture content, θ_v (cm^3 moisture cm^{-3} soil) vs. tension (*cm*).

This method is most often used for evaluating sands and various amendments for potential use in turfgrass situations. Similar results may be obtained by applying pressure via a pressure plate apparatus to the soil rather than suction (tension) using a tension table.

A third method of generating moisture data to construct SMRCs employs a column created by stacking a series of rings. The rings are taped or otherwise covered to make the column watertight (Fig. 1.23). A permeable membrane is placed in the bottom of the column, which is then packed with soil. The soil is saturated by immersion and allowed to drain to a state of equilibrium (generally 24 *h* or more). Rings are then separated with the soil intact. Moist soil from each ring is weighed, dried in an oven, and reweighed to calculate volumetric water content. The tension on the soil is equal to the distance from the bottom of the column which is at atmospheric pressure across the permeable membrane up to the center of each ring. As with the previously described methods, a series of moisture release points are plotted and then connected to form the SMRC. Other methods exist to construct SMRCs but are used mostly for engineering purposes.

Fig. 1.23 Soil columns used to construct soil moisture retention curves at field capacity

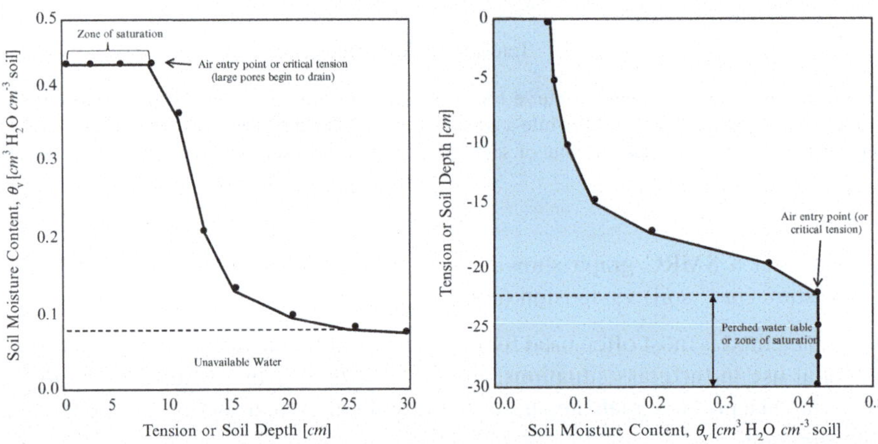

Fig. 1.24 A typical SMRC for a sand being considered for turfgrass use. The *left graph* plots increasing soil water tension (x-axis) against soil water content (y-axis). The *right side* indicates the same SMRC but now soil water content is on the x-axis and soil depth on the y-axis. This indicates that the top 10 *cm* or so of soil will hold little water but θ_v increases with depth through 22 *cm* where a perched water table exists. This area of the graph indicates that 4.6 *cm* effective depth of water-filled porosity

Interpreting Soil Moisture Retention Curves

The shape of the SMRC depends largely on soil particle size distribution and on the degree of soil compaction. Figure 1.24 (**left**) shows a typical SMRC for a sandy soil. To construct this curve, the soil was initially saturated (indicated as 0.42 cm^3 water

cm^{-3} soil) and then subjected to various tension up to 30 *cm*. When small suction was applied, no water drained from the soil because 2.5, 5.0, and even 7.5 *cm* of suction were inadequate to pull water from any pore (the pores were too small) and the soil stayed saturated. As tension was increased to 10 *cm*, the larger pores began to drain. The exact **air entry point** (tension required to drain the largest pores) is actually somewhere between 7.5 and 10.0 *cm* tension in this example. However, since 7.5 *cm* is the last tension recorded where the soil remained saturated (curve was flat), this could be reported as the air entry point from this SMRC.

At 15 *cm* tension, most of the soil water drained since the pores could not hold water against tensions of this magnitude. Much less water was lost between tensions of 20 and 30 *cm*. Generally, one would expect smaller decreases in water content from equal increments of increasing tension and the SMRC would become flatter. The point where only the smaller pores contain water due to capillary (adhesion and surface tension) forces and water movement due to gravity has stopped is **field capacity**. The water content at the point where the curve is judged to become "flat" can sometimes be identified as field capacity. This is often only an assumption, since field capacity is not a true physical property specific to a soil, but rather a descriptive property of a soil profile in a particular environment and under specific wetting and drainage conditions (see earlier). Once water is held (adsorbed) so tightly in soil and its mobility is so low plant roots can no longer extract enough to survive, the soil is described as its **wilting point** (typically the level measured at ~1,500 *kPa* tension) for heavier soils, ~300 *kPa* for sands.

Figure 1.24 (**right**) is the same SMRC as shown in Fig. 1.24 (**left**). The axes have been reoriented by rotating the entire graph 90 degrees counter-clockwise and then "flipping" the graph around the y-axis. Instead of showing water content as a function of tension, Fig. 1.24 (**right**) shows water content as a function of soil depth. If the reference level is set at the soil surface of a 30 *cm* column, then the suction, the pull on the water decreases 1 *cm* for each cm below the soil surface. Thus, 30 *cm* of gravitational pull is equivalent to 30 *cm* of tension from a hanging water column.

It is important to remember SMRCs represent soil water content at a state of equilibrium, where water is not being added and drainage is negligible. In this case, soil tension (matric potential) will equal the pull of gravity (gravitational potential), thus water is held in the profile.

Figure 1.24 (**right**) is the way one might view soil water content versus soil depth in a 30 *cm* (12 *in*) deep golf green or sports field rootzone profile. The shaded area represents soil moisture in the profile at equilibrium. A perched water table exists in the bottom 7.5 *cm* (3 *in*) of the profile from the interface with the gravel layer up to the air entry point. Soil water content decreases rapidly above the perched water table until it is less than 0.10 cm^3 water cm^{-3} soil in the surface 10 *cm* (4 *in*).

Another example of a SMRC constructed from a sandy soil is Fig. 1.25. Note the x-axis (tension) is shown on a logarithmic scale. This is not uncommon especially in fine textured soils such as clays that retain soil water at very high tensions, as a change in tension several orders of magnitude may result in only a small change in soil water content, particularly in the lower range of water content. Extending the tension axis allows the SMRC to be used to provide information on the expected

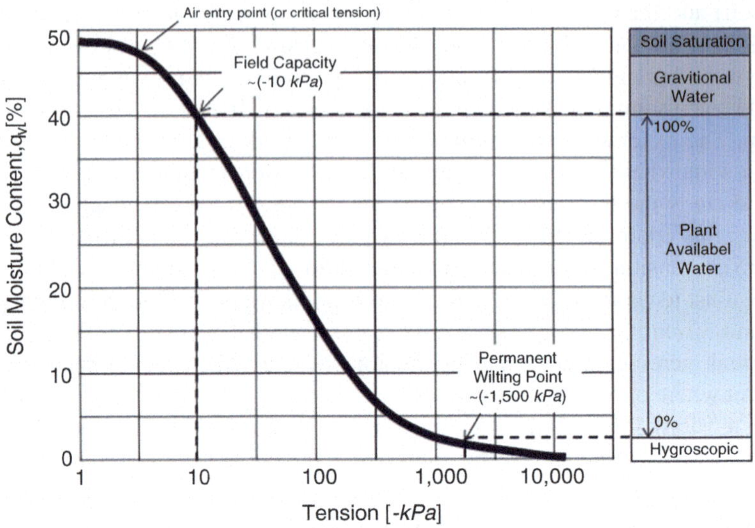

Fig. 1.25 A soil moisture retention curve indicating the zone of soil saturation, gravitational water, the start of available water at field capacity until the wilting point occurs which is the start of unavailable water

availability of water from the soil as transpiration through plants and evaporation dry the soil to the point plants can no longer extract moisture.

Figure 1.25 shows several regions of interest on a SMRC which can be differentiated by assigning values (points on the curve) for air entry point, field capacity, and permanent wilting points. These regions or zones include:

- Zone of soil saturation (perched water table): between zero tension and air entry point,
- Zone of gravitational water: between the air entry point and field capacity, water drains by gravity, before it can be taken up by plants,
- Zone of plant available water: between field capacity and permanent wilting point, where water is retained by capillary forces greater than the pull of gravity but matric suction can be overcome by plant root uptake,
- Zone of hygroscopic water: all water held at tensions greater than permanent wilting point, where water is not available to plants.

In the SMRC in Fig. 1.26, the shaded area above (to the right) of the curve and below the horizontal line at total porosity (which is equal to volumetric water content at saturation) is the aeration (or non-capillary) porosity expected in this soil. Each grid block represents 5 *cm* of tension or equivalent depth of water (horizontal) by 5 % volumetric water content (vertical). These units can be multiplied to indicate each grid block represents 0.25 *cm* effective depth of porosity. Since approximately 14 blocks in this example are above the SMRC and below the horizontal line at total porosity, once could expect (14 x 0.25 = 3.5) 3.5 *cm* effective depth of aeration porosity in the 30 *cm* depth of soil.

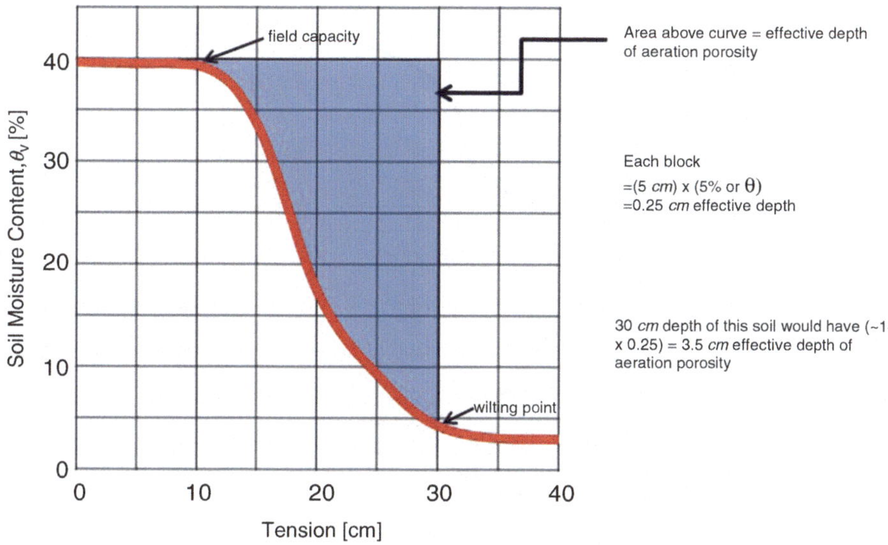

Fig. 1.26 A soil moisture retention curves showing soil water content at various tensions or depths (x-axis). Soil saturation is occurring for the first 10 *cm* tension which is where field capacity starts (also called the air entry point, or top of the capillary fringe). Soil moisture is available until about 32 *cm* which indicates wilting point

The shaded area below the same SMRC in Fig. 1.27 represents water-filled (or capillary) porosity. As with aeration porosity, each block represents 5 *cm* depth by 5 % volumetric water content for a 0.25 *cm* effective depth. Approximately 32 blocks below the curve indicates 8.0 *cm* (32×0.25) equivalent depth of water-filled porosity in the 30 *cm* of soil. Several mathematical models exists which predict water stored in a profile based largely on vertical distribution of soil water tension (or volumetric water content) data but these still require measurements for input variables and still are only a prediction.

As mentioned previously, SMRCs are not a reliable source of field capacity values. However, some soil scientists have endeavored to define the soil water content at field capacity in terms of established numerical values for matric tension. A matric tension of $-33\,kPa$ has been widely used as an acceptable value for tension at **field capacity**. Tension in the $-10\,kPa$ range has been suggested as more appropriate for coarser (sandy) soils. Soil moisture content at $-1{,}500\,kPa$ ($-300\,kPa$ for sands) has been used as a standard for estimating **permanent wilting point**. **Plant available water** is defined as the amount of water held by a soil between field capacity and permanent wilting point. Using these values allow SMRCs to estimate the water available to plants in a soil profile (Fig. 1.25). This method has proven to be useful in estimating plant available water only in coarser soils, where most of this water is held at tensions near field capacity.

It is important to distinguish SMRCs from other graphs of water content as a function of soil depth, usually measured from the soil surface downward (Fig. 1.28). In this graph, the x-axis is the volumetric water content (θ_v, reported

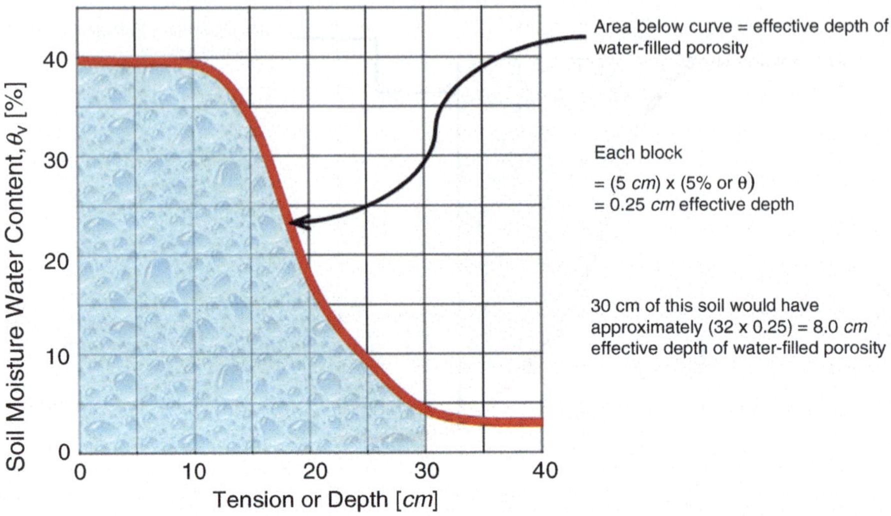

Fig. 1.27 The same moisture retention curve is shown in Fig. 1.24 but indicating an area above the curve as the effective depth of aeration porosity (*left*). This can be calculated by multiplying each 5 *cm* block by 5 % volumetric water content in each block to achieve 0.25 *cm* effective aeration porosity with each block. By multiplying the total number of blocks (14 in this example) by 0.25 *cm*, a total of 3.5 *cm* effective depth of aeration porosity is found

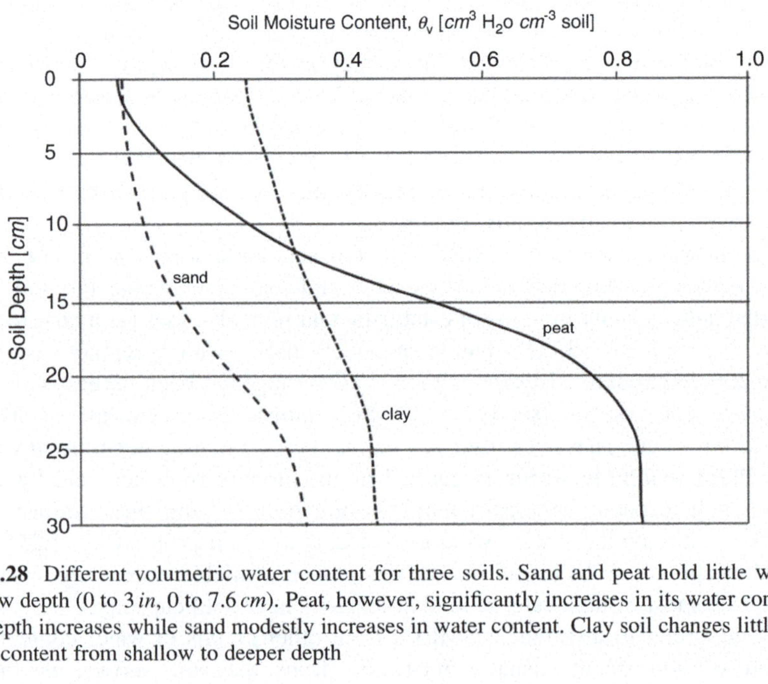

Fig. 1.28 Different volumetric water content for three soils. Sand and peat hold little water at shallow depth (0 to 3 *in*, 0 to 7.6 *cm*). Peat, however, significantly increases in its water content at soil depth increases while sand modestly increases in water content. Clay soil changes little in its water content from shallow to deeper depth

as cm^3 water cm^{-3} soil) while the y-axis is soil depth (*cm*) measured down from the soil surface. The sand and peat soils hold little water (almost 0 %) at the top of the column but increase in soil moisture as soil depth increases. Peat in this example holds considerably more moisture (over 60 % or 0.6 cm^3 moisture cm^{-3} soil) as soil depth increases. The clay soil has more adhesive and osmotic bindings thus is more moderate as its water content changes less from the top of the soil column (about 19 % water content) compared to about 30 % at the 12 *in* (30 *cm*) depth. These graphs are useful in reporting soil water content, but do not necessarily represent soil profiles at equilibrium. These soil depths do not equal soil tension unless the profile is at equilibrium and the curves are correctly oriented on the soil depth axis (as described above).

Information from Soil Moisture Retention Curves

Much useful information can be ascertained from a soil water retention curve. First, soil moisture distribution throughout a soil's profile is quantifiable. As discussed, the top 4 to 6 *in* (5 to 15 *cm*) in the sand example in Fig. 1.24 has a moisture content of less than 0.10 cm^3 water cm^{-3} soil following drainage to field capacity. However, it is not known how much plant available water is present because the soil water content at wilting point is unknown and we don't know how deep the roots will penetrate. The curve does indicate the amount of water available in the bottom 4 *in* (10 *cm*) will be about 0.42 cm^3 water cm^{-3} soil. Roots reaching below about 6 *in* (15 *cm*) will tap into a larger zone of water thus delaying the need to irrigate. The curve also indicates the depth of the perched water table above the gravel layer, which in this example is about 7.5 *cm* (3 *in*).

Secondly, SMRCs can be used to develop various irrigation practices. For example, when "flushing" greens in an attempt to remove salts, bicarbonates, etc., the tendency is often to quickly flood the green. In Fig. 1.29, a grid has been imposed on Fig. 1.24 (**left**) in order to count the blocks to the left of the SMRC to determine the amount of water held throughout the soil profile. This value is then multiplied by effective depth of water in each grid. From Fig. 1.29, approximately 13.75 blocks contain water. Each block has an effective depth of moisture of 0.5 *cm* which was determined by multiplying each 5 cm block by its soil moisture content (0.1 cm^3 water cm^{-3} soil). Therefore, this soil profile contains approximately 6.9 *cm* (13.75 blocks x 0.5 *cm* effective depth of water per block) or 2.7 *in* effective depth of water-filled porosity. In theory, a minimum of 6.9 *cm* (2.7 *in*), under ideal conditions, would be required to flush out all the "old" water. However, irrigation systems are not uniform in their application, so a distribution uniformity test would show how efficient a system is, indicating how much additional moisture would actually be needed to apply 6.9 *cm* (2.7 *in*) moisture uniformly over the green. The most effective way to leach the profile would be to apply the water slowly so the soil

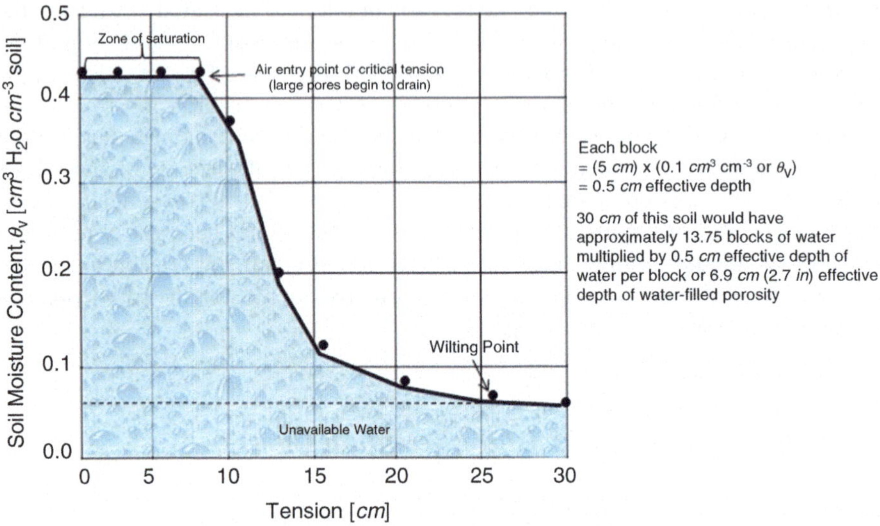

Fig. 1.29 A grid system imposed on Fig. 1.22 (*left*) to indicate how much moisture is in the soil profile. The sum of blocks containing moisture (13.75 in this example) is multiplied by 0.5 % (effective depth) to obtain 6.9 *cm* (2.7 *in*) total water in this rootzone profile

water content does not increase any more than necessary above the curve (the profile stays near field capacity) as shown in Fig. 1.24 (**right**). All the saline water is shown by the shaded area. Any time the soil water content increases above the curve, leaching efficiency will decrease as the added water will only build up the soil's moisture content and eventually puddle, reducing the amount of "bad" water being flushed out. Irrigation application rates should therefore should be as low as possible to produce maximum efficiency (maximum leaching with a minimum amount of water). Since the speed of the irrigation being applied is typically not easily adjusted, to apply the water slowly, a "pulse" system of applying the water would probably be needed. With this, water would be applied until puddling and runoff occurred and then the irrigation stopped. This sequence would be followed until the 6.9 *cm* (2.7 *in*) was applied.

Another common method of "flushing" greens is to plug the drain outlet of a green, saturate the green profile with water, and then release the water by unplugging the drain outlet. Again, if the green is allowed to become saturated, salts, etc., can move upward through the profile to the soil surface. Some salts will drain downward with draining water once the outlet is opened, but some will remain near the surface. If this process were repeated several times, desirable results might be obtained, but the quantity of clean water to accomplish this would be great.

Predicting Soil Compaction

SMRC shape and characteristics are reflections of the particle size distribution of a soil and the degree of compaction that potentially could develop. A soil in question may have a SMRC developed following light compaction while another curve is developed following severe compaction. If these two curves are noticeably different (divergent), the soil will likely compact under use. With different levels of compaction that develop across a turf surface, these areas will behave differently. Areas less compacted may be dry on the surface while a more highly compacted area may be quite moist, making management decisions difficult. Soils exhibiting such characteristics should not be used.

Example Two sands are being considered for a rootzone. Sand 1 meets USGA specifications while Sand 2 does not due to excessive fines (Table 1.5). How could one provide a reasonable prediction of how these two sands would perform in the field? From the particle size analysis and from their two moisture release curves, it can be seen that the sands would behave differently.

Sand 1 is extremely uniform and Fig. 1.30 (**top**) shows closely related curves. It can be expected this sand will not be greatly affected by compaction. The perched water table is about 20 *cm* deep.

The second sand has a much wider sand distribution range (Table 1.5) especially in the fine sand range (0.10 to 0.25 *mm*) (41 % compared to only 4 % for Sand 1). The perched water table is slightly deeper, about 25 *cm* (Fig. 1.30, **bottom**). The noncompacted and compacted SMRCs for Sand 2 are also quite different, indicating water retention at different levels. Generally, the more fines present in a soil sample, the deeper the perched water table.

In addition, from the SMRCs, the slope of the curve between the air entry point at the top of the perched water table and field capacity indicates the uniformity of the soil. The soil is extremely uniform if this area is relatively flat as with Sand 1. With Sand 2, this portion of the curve is fairly steep, indicating this sample has a relatively wide range of particle size.

Using SMRCs and other means to determine appropriate soil depths when constructing sports fields or sand capping an area are discussed in Chap. 2.

Table 1.5 Particle size distribution for two sands being considered for sports turf use

		Sand sample (% by weight)	
Particle size (*mm*)	Description	1	2
>2.0	Gravel	0	0
1.0–2.0	Very coarse sand	0	0.4
0.5–1.0	Coarse sand	3.0	15.9
0.25–0.5	Medium sand	93.0	40.4
0.10–0.25	Fine sand	4.0	41.4
<0.10	Very fine sand + silt + clay	0	1.9

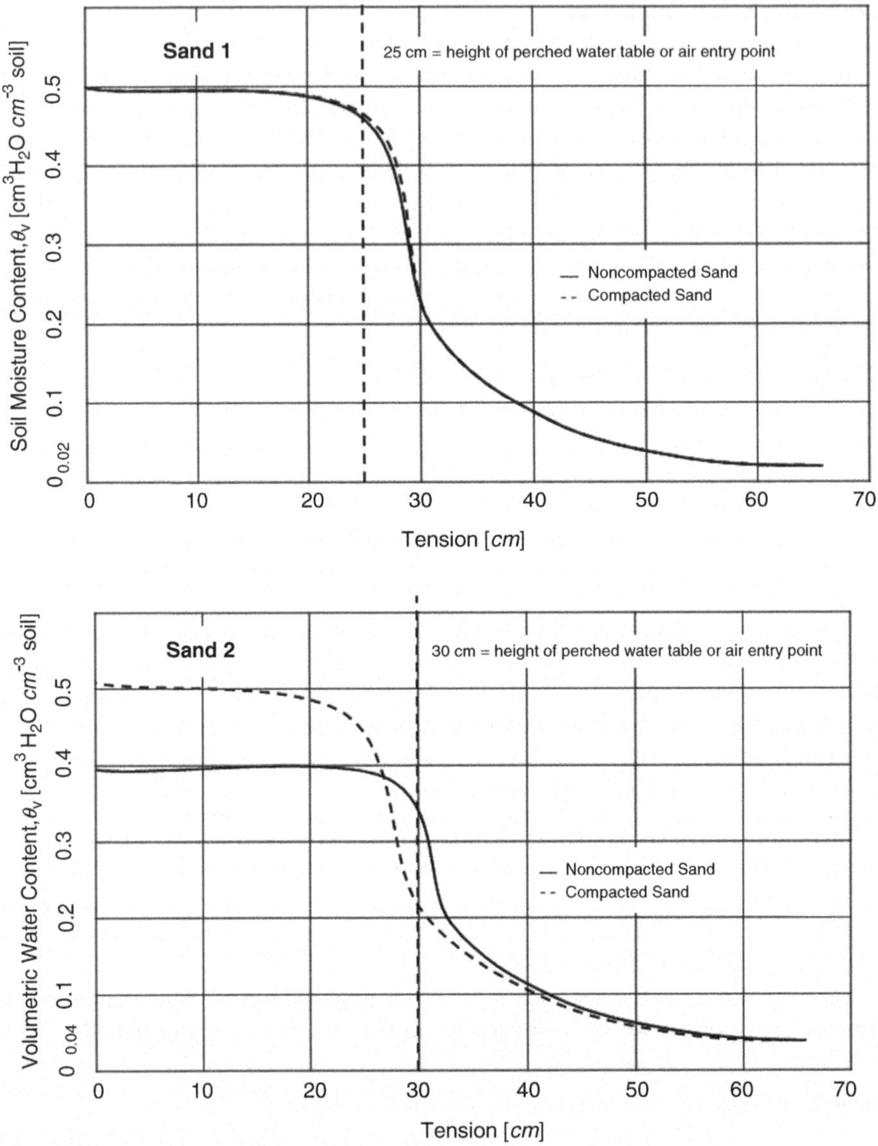

Fig. 1.30 Soil moisture curves being used to predict if a soil will compact. The *top graph* has a uniformed sized sand which meets USGA specification and two curves, one with the sand noncompacted and the other, compacted. Since both curves are very similar (non-divergent), soil compaction should not be a major concern. The *bottom graph* has Sand 2 which does not meet USGA specification as too much fine sand, silt, and clay are present. Under compaction, therefore, this curve is widely different (divergent) than when it is not compacted, indicating it will hold excessive moisture and develop a perched water table about 30 *cm* (almost 12 *in*) up from the bottom of the sand, possibly saturating the rootzone

Evaluating Rootzone Mixes

Constructing soil moisture retention curves allows better characterization or prediction of sands and amendment(s) being considered to construct a desirable rootzone mix. For example, if an amendment being considered releases most of its moisture at relatively low tensions and retains little at a moderate tension, the benefit of adding it to a coarse-textured sand may be limited. Conversely, if an amendment retains significant water at higher tensions, releasing little at low tension, thus making it unavailable to the turf, it also may be unsuitable.

Example Three sands (coarse, medium, and fine) and three soil amendments (zeolite, calcined clay, and sphagnum peat) had SMRCs developed on volumetric water content ($cm^3\ cm^{-3}$) from 0 to 10,000 *cm* tension (Fig. 1.31, **top**). Based on SMRCs, discuss the benefit and limitations of using each for a rootzone.

Soil amendments contained more water (>55 %) at saturation compared to the sands (<45 %) and released this more gradually. Of the three soil amendments, peat moss had the most gradual release (Fig. 1.31, **bottom**). Sphagnum peat moss would be the best amendment to add to the various sands to increase moisture content at various tensions followed by the calcined clay. Zeolite would be the least advantageous material to add to increase water content at various depths.

Example Describe moisture retention of the three sand sizes at various depths in Fig. 1.32 (**top**) and if either of the three amendments being considered (sphagnum peat moss, calcined clay, and/or zeolite) would be beneficial if added on a 20 % by volume. Figure 1.32 (**top**) examines water retention and availability of three sand sizes.

All sands were close to saturation (~45 %) at the bottom of the 30 *cm* (12 *in*) sand column. At shallower soil depths, fine sand retained much of its moisture with ~36 % soil moisture at the surface. Medium and coarse sands held much less moisture throughout the profile with ~17 % soil moisture at the surface for medium sand and <5 % for coarse sand. If used unamended, coarse sand would remain droughty at the soil surface while fine sand would lack adequate aeration except possibly in the upper 5 *cm* (2 *in*) of soil. The medium-sized sand appears as best candidate for being amended based on the balance of air-filled and water-filled pores throughout most of the rootzone.

Figure 1.32 (**bottom**) represents resulting moisture retention at various depths of medium sand amended at 20 % by volume with either sphagnum peat moss, calcined clay, or zeolite. All soil mixtures were nearly saturated at bottom of the 30 *cm* (12 *in*) rootzone depth. However, as soil depth was decreased, zeolite amendment had little effect on increasing soil moisture retention. Calcined clay increased soil moisture retention some once soil depth was below 12 *cm* (~5 *in*) but it also had little effect on increasing soil moisture at shallower depths. Sphagnum peat was the only amendment to increase soil moisture retention at most soil depths when mixed with the medium sand at 20 % by volume. Since a minimum of 15 %

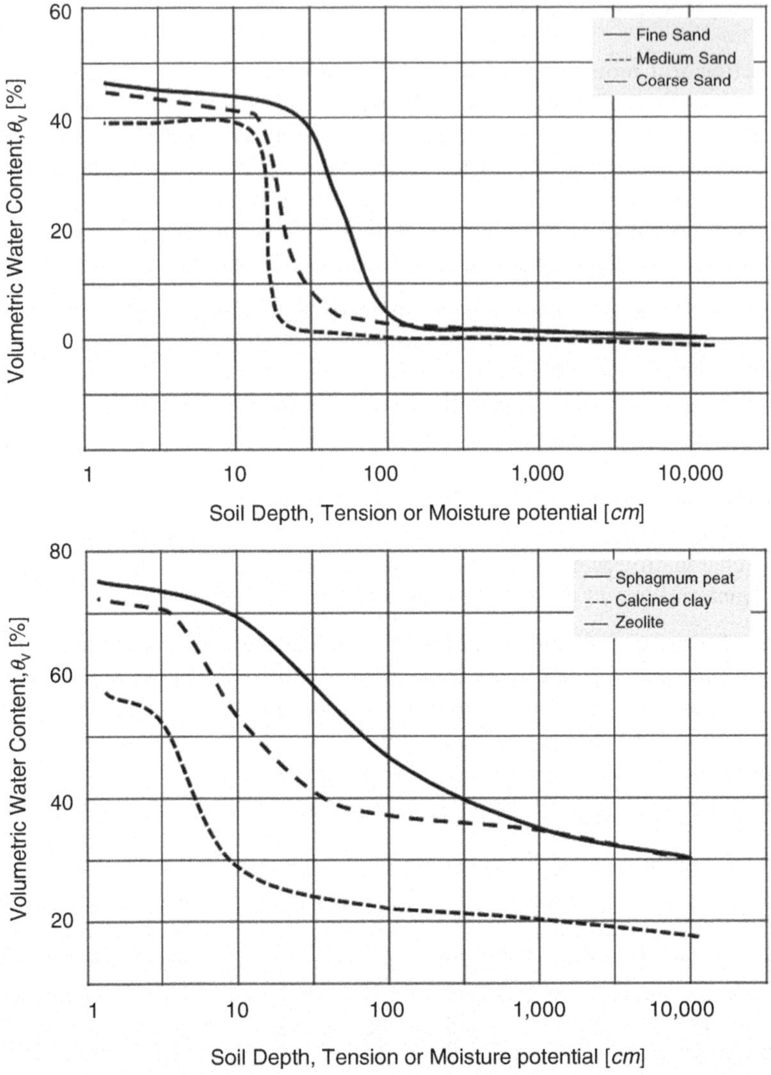

Fig. 1.31 Volumetric water content (%) at various depths (*cm*) for three sand fractions (*top*) compared to three commonly used amendments (*bottom*). The fine sand retained greater soil moisture at various tensions followed by the medium and coarse sands. For the amendments, sphagnum peat held the most moisture at various tensions followed by calcined clay and least by zeolite. When comparing the two graphs, the fine sand would need the least amount of amendment (if any) to increase moisture retention while the coarse sand would benefit most. Sphagnum peat should be considered for the coarse and possibly, medium sand. However, further testing of any rootzone mixture being considered should be made to compare the results with established specifications

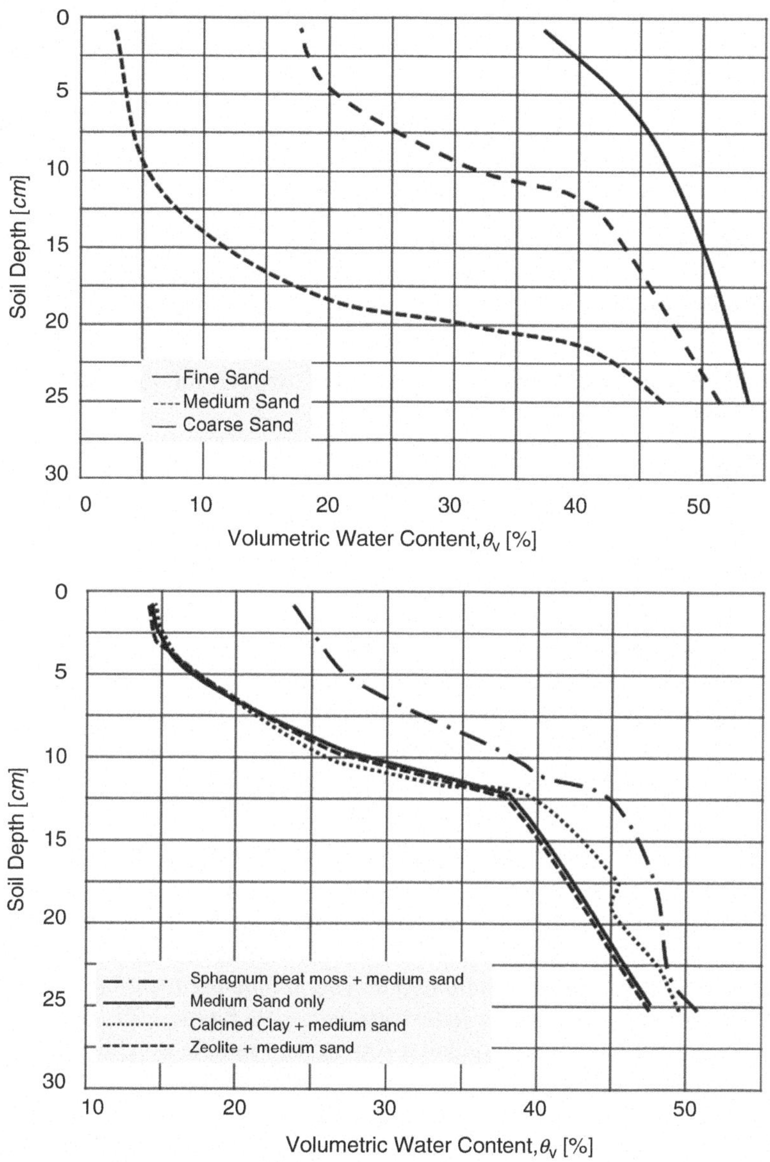

Fig. 1.32 Soil moisture retention at various depths (*cm*) for three sand sizes (top) indicating poor moisture retention by coarse sand and excessive soil moisture for fine sand throughout the 30 *cm* soil profile. In the bottom graph, soil moisture retention as influenced by medium sand amended with 20 % sphagnum peat moss, calcined clay, or zeolite. Medium sand amended with 20 % sphagnum peat moss increased soil moisture about 8 % throughout the profile compared to unamended medium sand, calcined clay increased it about 3 %, especially at deeper soil depths while zeolite had little effect on soil moisture retention on medium sand throughout the soil profile

moisture retention is necessary for seedling establishment, only the medium sized sand amended with 20 % (by volume) peat moss would consistently be above this value.

Another use of SMRCs is to predict the appropriate soil depth needed for sports fields, sand capping golf course fairways, etc. is water content related to tension. This is covered in Chap. 2 on rootzone selection.

1.5 Questions

1. Using the textural triangle, determine the texture for the following soils (*answers*):

   (a) 25 % sand, 30 % silt, 45 % clay (*clay*)
   (b) 40 % sand, 30 % silt, 30 % clay (*clay loam*)
   (c) 60 % sand, 10 % silt, 30 % clay (*sandy clay loam*)
   (d) 70 % sand, 12 % silt, 18 % clay (*sandy loam*)
   (e) 90 % sand, 5 % silt, 5 % clay (*sand*)
   (f) 80 % sand, 15 % silt, 5 % clay (*loamy sand*)
   (g) 10 % sand, 85 % silt, 5 % clay (*silt*)
   (h) 5 % sand, 75 % silt, 20 % clay (*silt loam*)
   (i) 40 % sand, 40 % silt, 20 % clay (*loam*)
   (j) 55 % sand, 5 % silt, 40 % clay (*sandy clay*)
   (k) 10 % sand, 60 % silt, 40 % clay (*silty clay loam*)
   (l) 5 % sand, 45 % silt, 50 % clay (*silty clay*)

2. Define **matric potential**. (*Matric potential is how tightly water is held (or absorbed) in soil).*
3. Define and explain how one measures **field capacity**. (*Soil water following gravitational drainage. Measured after free drainage ceases or in the lab, as the amount of moisture remaining following an imposed tension (or pressure) of −33 kPa (or −1/3 bar) for loam soils and −10 kPa (or −0.1 bar)* for sands).
4. Determine the bulk density (ρ_b) of a 400 cm^3 soil sample weighting 575 g when oven dried.

$$\text{bulk density } (\rho_b) = \frac{\text{mass solids } (\text{M}_s)}{\text{soil volume } (\text{V}_t)}$$

$$= \frac{575\ g}{400\ cm^3}$$

$$= 1.44\ g\ cm^{-3}$$

5. Determine the gravimetric (θ_g) and volumetric (θ_v) water content of a soil with a wet weight of 100 g and dry weight of 75 g and a bulk density (ρ_b) of 1.4 $g\ cm^{-3}$.

$$\text{gravimetric water content } (\theta_g) = \frac{\text{mass water } (M_w)}{\text{mass solids } (M_s)}$$
$$= \frac{100\ g - 75\ g}{75\ g}$$
$$= 0.33\ g\ water\ g^{-1} soil\ (or\ 33\%)$$
$$\text{volumetric water content } (\theta_g) = \theta_g \times \rho_b$$
$$= 0.33\ g \times 1.4\ g\ cm^{-3}$$
$$= 0.46\ cm^3 water\ cm^{-3} soil\ (or\ 46\%)$$

6. Determine the percentage of water at field capacity, at wilting point, and available water for a soil containing 85 *g* at field capacity, 71 *g* at wilting point and 58 *g* after oven drying.

$$\%\text{water at field capacity} = \frac{\text{field capacity wt} - \text{oven dry wt}}{\text{oven dry wt}}$$
$$= \frac{85\ g - 58\ g}{58\ g}$$
$$= 0.47\ g\ water\ g^{-1} soil\ (or\ 47\%)$$
$$\%\text{water at wilting point} = \frac{\text{wilting point wt} - \text{oven dry wt}}{\text{oven dry wt}}$$
$$= \frac{71\ g - 58\ g}{58\ g}$$
$$= 0.22\ g\ water\ g^{-1} soil\ (or\ 22\%)$$
$$\%\text{available water} = \frac{\text{field capacity wt} - \text{wilting point wt}}{\text{oven dry wt}}$$
$$= \frac{85\ g - 71\ g}{58\ g}$$
$$= 0.24\ g\ water\ g^{-1} soil\ (or\ 24\%)$$

7. Using Stokes equation, determine the time required for all particles larger than 0.05 *mm* in diameter to settle 15 *cm* in a suspension with a temperature at 25 °C.

$$\text{velocity } (V) = K \times D^2$$
$$= 10,000 \times (0.005\ cm)^2$$
$$= 0.25\ cm\ s^{-1}$$
$$\text{settling time } (T) = \frac{\text{distance}}{V}$$
$$= \frac{15\ cm}{0.25\ cm\ s^{-1}}$$
$$= 60\ s\ (or\ 1\ min)$$

8. Determine the time needed for all sand, silt and coarse clay particles (0.0001 *mm* diameter) to settle in an aqueous solution at 25 °C to a depth of 25 *cm*.

$$\begin{aligned}
\text{sand : velocity} &= \mathrm{K} \times \mathrm{D}^2 \\
&= 10,000 \times (0.005\ cm)^2 \\
&= 0.25\ cm\ s^{-1} \\
\text{settling time (sand)} &= \frac{\text{distance}}{\mathrm{V}} \\
&= \frac{25\ cm}{0.25\ cm\ s^{-1}} \\
&= 100\ s \\
\text{silt : velocity} &= \mathrm{K} \times \mathrm{D}^2 \\
&= 10,000 \times (0.0002\ cm)^2 \\
&= 0.0004\ cm\ s^{-1} \\
\text{settling time (silt)} &= \frac{\text{distance}}{\mathrm{V}} \\
&= \frac{25\ cm}{0.0004\ cm\ s^{-1}} \\
&= 62,500\ s\ (or\ 17.4\ h) \\
\text{clay : velocity} &= \mathrm{K} \times \mathrm{D}^2 \\
&= 10,000 \times (0.0001\ cm)^2 \\
&= 0.0001\ cm\ s^{-1} \\
\text{settling time (clay)} &= \frac{\text{distance}}{\mathrm{V}} \\
&= \frac{25\ cm}{0.0001\ cm\ s^{-1}} \\
&= 250,000\ s\ (or\ 69.4\ h\ or\ 2.9\ days)
\end{aligned}$$

9. Calculate the texture of a 75 *g* sample from the following hydrometer data: after 40 *s* the hydrometer reading was 55 $g\ L^{-1}$, after 8 *h* it was 15 $g\ L^{-1}$.

$$\begin{aligned}
\%\text{silt and clay} &= \frac{40\ s\ \text{reading}}{\text{soil dry weight}} \times 100 \\
&= \frac{55\ g\ L^{-1}}{75g} \times 100 \\
&= 73.3\%
\end{aligned}$$

$$\%\text{clay} = \frac{8\,h\,\text{reading}}{\text{soil dry weight}} \times 100$$
$$= \frac{15 g\,L^{-1}}{75 g} \times 100$$
$$= 20\%$$
$$\%\text{silt} = \%(\text{silt} + \text{clay}) - \%\text{clay}$$
$$= 75 - 20$$
$$= 55\%$$
$$\%\text{sand} = 100\% - \%(\text{silt} + \text{clay})$$
$$= 100 - 75$$
$$= 25\%$$

$$\text{Soil texture } (25\%\text{sand} + 55\%\text{silt} + 20\%\text{clay}) = \textit{silt loam}$$

10. A block-shaped container 10 x 10 x 10 *cm* has a soil wet weight of 1460 *g* and dry weight of 1200 *g*. Determine volumetric water content (θ_v), wet and dry soil bulk density, soil porosity, and percent soil saturation.

$$\text{soil volume } (V_t) = 10 \times 10 \times 10\ cm$$
$$= 1000\ cm^3$$
$$\text{water volume } (V_w) = \frac{\text{mass of water}}{\text{density of water}}$$
$$= \frac{\text{soil wet weight} - \text{dry weight}}{\text{density of water}}$$
$$= \frac{1460\ g - 1200\ g}{1\ g\ cm^3}$$
$$= 260\ cm^3$$
$$\text{volumetric water content } (\theta_v) = \frac{V_w}{V_t}$$
$$= \frac{260\ cm^3}{1000\ cm^3}$$
$$= 0.26\ g\ water\ g^{-1} soil$$
$$\text{wet bulk density} = \frac{\text{mass wet soil}}{\text{volume of soil } (V_t)}$$
$$= \frac{1460\ g}{1000\ cm^3}$$
$$= 1.46\ g\ cm^3$$

$$\text{dry Bulk density } (\rho_b) = \frac{M_s}{V_t} = \frac{1200\ g}{1000\ cm^3} = 1.20\ g\ cm^3$$

$$\text{mass of solids } (M_s) = \text{wet soil mass} - \text{mass of water} = 1460\ g - 260\ g = 1200\ g$$

$$\text{volume of solids } (V_s) = \frac{\text{mass solids}}{\text{particle density } (\rho_s)} = \frac{1200\ g}{2.65\ g\ cm^3} = 452.8\ cm^3$$

$$\text{volume of voids } (V_v) = \text{total volume} - \text{volume of solids} = 1000\ cm^3 - 452.8\ cm^3 = 547.2\ cm^3$$

$$\text{total porosity } (f_t) = \frac{V_v}{V_t} = \frac{547.2\ cm^3}{1000\ cm^3} = 0.55\ (or\ 55\%)$$

$$\text{degree of saturation (s)or capillary porosity } (f_w) = \frac{\text{volume water } (V_w)}{\text{volume fluids } (V_v)} = \frac{260\ cm^3}{547.2\ cm} = 47.5\%$$

11. Calculate the bulk density (ρ_b) of a 400 cm^3 soil sample weighing 600 g with a 0.1 $g\,g^{-1}$ (10 %) moisture content by weight.

$$\text{moisture content } (\theta_g) = \text{wet wt} - \text{dry wt soil}$$

$$0.10 = \frac{600\ g - \text{soil dry weight}}{\text{soil dry weight}}$$

$$(0.10 \times \text{soil dry wt}) = 600\ g - \text{soil dry weight}$$

$$(1.0 \times \text{soil dry wt}) + (0.10 \times \text{soil dry wt}) = 600\ g$$

$$1.10 \times \text{soil dry wt} = 600\ g$$

$$\text{soil dry wt } (M_s) = \frac{600\ g}{1.10}$$

$$\text{dry weight } (M_s) = 545.5\ g$$

$$\text{bulk density } (\rho_b) = \frac{545\ g}{400\ cm^3} = 1.36\ g\ cm^3$$

12. Calculate the volume of a soil sample that is 12 % (0.12 $g\ g^{-1}$) moisture by weight, weighs 650 g and has a bulk density of 1.3 $g\ cm^3$.

$$\text{moisture content } (\theta_g) = \frac{\text{wet weight} - \text{dry weight soil}}{\text{soil dry weight}}$$

$$0.12\ g\ g^{-1} = \frac{650\ g - \text{soil dry weight}}{\text{soil dry weight}}$$

$$(0.12 \times \text{soil dry weight}) = 650\ g - \text{soil dry weight}$$

$$(1.0 \times \text{soil dry wt}) + (0.12 \times \text{soil dry wt}) = 650\ g$$

$$1.12 \times \text{soil dry weight} = 650\ g$$

$$\text{soil dry weight } (M_s) = \frac{650\ g}{1.12}$$

$$= 580.4\ g$$

$$\text{soil volume } (V_t) = \frac{580.4\ g}{1.3\ g\ cm^{-3}}$$

$$= 446.4\ cm^3$$

13. Calculate total porosity (f_t) for a soil sample with a bulk density (ρ_b) of 1.35 $g\ cm^{-3}$ (assume particle density, ρ_s, = 2.65 $g\ cm^{-3}$).

$$\text{total porosity } (f_t) = 1 - (\rho_b \div \rho_s) \times 100$$

$$= 1 - (1.35\ g\ cm^{-3} \div 2.65\ g\ cm^{-3}) \times 100$$

$$= 49\%$$

14. Calculate total porosity (f_t) for a 250 cm^3 soil sample containing 140 cm^3 water when saturated.

$$\text{total porosity } (f_t) = V_v \div V_t$$

$$= 140\ cm^3 \div 250\ cm^3$$

$$= 56\%$$

15. Determine the bulk density (ρ_b) of a soil sample with a total porosity (f_t) of 45 % (assume particle density, ρ_s, = 2.65 $g\ cm^{-3}$).

$$\text{total porosity } (f_t) = 1 - [(\rho_b \div \rho_s)] \times 100$$

$$\text{therefore, } \rho_b = [1 - (f_t)] \times \rho_s$$

$$= [1 - 0.55] \times 2.65\ g\ cm^{-3}$$

$$= 1.46\ g\ cm^{-3}$$

16. What is the percent water by volume and total pore space filled by water for a soil with a bulk density (ρ_b) of 1.50 $g\ cm^{-3}$ and gravimetric water content (θ_g) of 25 %?

$$\text{total porosity } (f_t) = 1 - [(\rho_b \div \rho_s)] \times 100$$
$$= 1 - [(1.50\ g\ cm^{-3} \div 2.65\ g\ cm^{-3})] \times 100$$
$$= 43\%$$
$$\%\text{water by volume} = \frac{(\theta_g) \times (\rho_b)}{\text{density of water}}$$
$$= \frac{0.25 \times 1.50\ \ g\ cm^{-3}}{1.0\ g\ cm^{-3}}$$
$$= 37.5\%$$
$$\text{water} - \text{filled (capillary) porosity } (f_w) = \frac{\text{volumetric water content}, \theta_v}{\text{total porosity, } f_t}$$
$$= \frac{0.375}{0.43}$$
$$= 87\%$$

17. Determine the particle density (ρ_s) of a soil sample with a bulk density (ρ_b) of 1.55 $g\ cm^{-3}$ and total porosity (f_t) of 40 %.

$$\text{total porosity } (f_t) = 1 - (\rho_b \div \rho_s) \times 100$$
$$\text{therefore, bulk density } (\rho_s) = -\rho_b \div (f_t - 1) \times 100$$
$$= -1.55 \div (0.40 - 1) \times 100$$
$$= 2.58\ g\ cm^{-3}$$

18. A soil sample contains 22 kg of soil with a gravimetric water content (θ_g) of 0.18. Determine the volume of water in the container.

$$\text{gravimetric water content } (\theta_g) = \frac{\text{mass water } (M_w)}{\text{mass solids } (M_s)}$$
$$0.18\ g = \frac{M_w}{22 - M_w}$$
$$M_w = 22 \times 0.18 - (0.18 \times M_w)$$
$$1.18 \times M_w = 3.96\ kg$$
$$M_w = \frac{3.96\ kg}{1.18}$$
$$= 3.356\ kg \text{ or } 3356\ g$$
$$\text{volume of water } (V_w) = \frac{M_w}{\text{density of water}}$$
$$= \frac{3356\ g}{1\ g\ cm^3}$$
$$= 3356\ cm^3$$

19. Calculate the oven dry weight (M_s) of a 350 cm^3 soil sample with a bulk density (ρ_b) of 1.42 $g\ cm^{-3}$.

$$\text{oven dry weight } (M_s) = 350\ cm^3 \times 1.42\ g\ cm^{-3}$$
$$= 497\ g$$

20. What is the bulk density of the following soil in a container 5 cm in diameter and 2.5 cm in height that weighs 95 g with wet soil, dry soil alone weight = 75.0 g.

$$\text{bulk density } (\rho_b) : \text{dry soil} = \frac{\text{mass dry soil}}{\text{volume soil}}$$
$$= \frac{75.0\ g}{\pi(2.5\ cm)^2\ (2.5\ cm)}$$
$$= 1.53\ g\ cm^{-3}$$
$$\text{bulk density } (\rho_b) : \text{wet soil} = \frac{\text{mass wet soil}}{\text{volume soil}}$$
$$= \frac{95\ g}{\pi(2.5\ cm)^2\ (2.5\ cm)}$$
$$= 1.94\ g\ cm^{-3}$$

21. A soil core 5 cm in diameter and 13 cm long has a wet soil weight of 445.3 g and oven dry weight of 399.5 g. Assuming field capacity has been reached, calculate gravimetric water content (θ_g), bulk density, volumetric water content (θ_v), equivalent depth of water, percent total porosity, capillary porosity (or saturation), non-capillary porosity, and equivalent depth of water required to saturate the soil column.

$$\text{gravimetric water content } (\theta_g) = \frac{\text{mass}_{wet\ soil} - \text{mass}_{dry\ soil}}{\text{mass}_{dry\ soil}}$$
$$= \frac{445.3\ g - 399.5\ g}{399.5\ g}$$
$$= 0.115\ g\ water\ g^{-1} soil$$
$$\text{bulk density } (\rho_b) = \frac{\text{mass dry soil}}{\text{volume soil}}$$
$$= \frac{399.5\ g}{\pi(2.5\ cm)^2(13\ cm)}$$
$$= 1.57\ g\ cm^{-3}$$
$$\text{volumetric water content } (\theta_v) = \frac{\theta_g \times \rho_b}{\text{water density } (g\ cm^{-3})}$$
$$= \frac{(0.115\ g\ g^{-1})(1.57\ g\ cm^{-3})}{1\ g\ cm^{-3}}$$
$$= \frac{0.18\ \text{cm}^3\text{water}}{cm^3 soil}$$

$$\begin{aligned}
\text{equivalent depth of water, } (\mathrm{D_e}) &= (\theta_v)(\text{soil depth}) \\
(\text{amount water occupying core}) &= (0.18\ cm^3 cm^{-3})(13\ cm) \\
&= 2.34\ cm \\
\text{total porosity } (\mathrm{f_t}) &= 1 - \frac{\text{bulk density}, \rho_b}{\text{particle density}, \rho_s} \\
&= 1 - \frac{1.57\ g\ cm^{-3}}{2.65\ g\ cm^{-3}} \\
&= 0.41\ (or\ 41\%) \\
\text{aeration (non}-\text{capillary) porosity } (\mathrm{f_a}) &= \left(1 - \frac{\text{bulk density}, \rho_b}{\text{particle density}, \rho_s}\right) - \theta_v \\
&= \left(1 - \frac{1.57\ g\ cm^{-3}}{2.65\ g\ cm^{-3}}\right) - \frac{0.18\ cm^3 \text{water}}{cm^3 \text{soil}} \\
&= 0.23\ (or\ 23\%) \\
\text{water}-\text{filled (capillary) porosity } (\mathrm{f_w}) &= \frac{\text{volumetric water content}, \theta_v}{\text{total porosity, } \mathrm{f_t}} \\
&= \frac{0.18\ \ cm^3 \text{water } cm^{-3} \text{soil}}{0.41\ \ cm^3 \text{water } cm^{-3} \text{soil}} \\
&= 0.439\ (or\ 43.9\%)
\end{aligned}$$

22. How much water is needed to wet a 15 *cm* deep dry soil to the equivalent of 36 % water by volume?

$$\text{volumetric water content } (\theta_v) \times \text{soil depth} = \text{amount water needed}$$
$$0.36\ cm \text{ water } cm^{-1} \text{soil} \times 15\ \ cm = 5.4\ cm\ (2.13\ in)$$

23. A soil column 90 *cm* long has a volumetric water content of 0.12 $cm^3\ cm^{-3}$. Determine how much water is needed to bring the volumetric water content up to 0.30 $cm^3\ cm^{-3}$.

$$\begin{aligned}
\text{current volume (depth) of water} &= 90\ cm \times 0.12\ cm^3 cm^{-3} \\
&= 10.8\ cm \\
\text{required volume (depth) of water} &= 90\ cm \times 0.30\ cm^3 cm^{-3} \\
&= 27\ cm \\
\text{additional volume (depth) of water needed} &= 27\ cm - 10.8\ cm \\
&= 16.2\ cm
\end{aligned}$$

24. A soil currently has a volumetric water content of 0.06 (or 6 %) with an average field capacity of 0.28 (or 28 %). If the soil receives a 0.75 *in* (1.9 *cm*) rainfall, how deep will the rain wet the soil?

$$\text{water penetration} = \frac{\text{depth water applied}}{\text{difference in } \theta_v}$$
$$= \frac{1.9\ cm}{(0.28 - 0.06)}$$
$$= 8.6\ cm\ (or\ 3.4\ in)$$

25. A 100 *g* sample of moist soil has a water content of 0.10 $g\ g^{-1}$ (designated as $soil_1$), (a) how many *ml* water are currently in the soil, and (b) how many *ml* of water are needed to increase the soil water content to 0.15 $g\ g^{-1}$ (designated as $soil_2$).

(a)
$$\text{gravimetric water content } (\theta_g) \text{ of soil}_1 = \frac{\text{soil}_1 \text{ wet wt} - \text{soil}_1 \text{ dry wt}}{\text{soil}_1 \text{dry wt}}$$
$$0.10 = \frac{100\ g - \text{soil}_1 \text{dry wt}}{\text{soil}_1 \text{dry wt}}$$
$$(0.10 \times \text{soil}_1 \text{dry wt}) = 100\ g - \text{soil dry wt}$$
$$(1.0 \times \text{soil}_1 \text{dry wt}) + (0.10 \times \text{soil}_1 \text{dry wt}) = 100\ g$$
$$1.10 \times \text{soil}_1 \text{dry wt} = 100\ g$$
$$\text{soil}_1 \text{dry wt} = \frac{100\ g}{1.10}$$
$$= 90.9\ g$$
$$\text{water content } (ml) \text{ currently in soil} = 100\ g - 90.9\ g$$
$$= 9.1\ g (\text{or } 9.1\ ml)$$

(b)
$$\text{gravimetric water content } (\theta_g) \text{ of soil}_2 = \frac{\text{soil}_2 \text{wet wt} - \text{soil}_2 \text{dry wt}}{\text{soil}_2 \text{dry wt}}$$
$$0.15 = \frac{\text{soil}_2 \text{wet wt}_2 - 90.9\ g}{90.9\ g}$$
$$(0.15 \times 90.9\ g) = \text{soil}_2 \text{wet wt}_2 - 90.9\ g$$
$$13.6\ g = \text{soil}_2 \text{wet wt}_2 - 90.9\ g$$
$$\text{soil}_2 \text{wet wt} = 104.5\ g$$
$$\text{water weight } (g) \text{ needed} = \text{soil}_2 \text{wet wt} - \text{soil}_1 \text{wet wt}$$
$$= 104.5\ g - 90.9\ g$$
$$= 13.6\ g$$
$$\text{water volume } (ml) \text{ needed} = 13.6\ g \times 1\ ml\ g^{-1}$$
$$= 13.6\ ml$$

Therefore, 13.6 *ml* water would be needed to increase the soil water content from 0.10 to 0.15 $g\ g^{-1}$.

26. Consider a soil core of equal diameter and length of 8 *cm*. Water is allowed to flow vertically downward under a constant head of 4 *cm*. The constant rate of water collected from bottom of core was 1 $cm\ min^{-1}$. Calculate the quantity of water passing through the soil core.

$$Q = \frac{KA \times (L + H)}{L}$$

where:
Q = quantity of water (cm^3) passing through the soil core
$K = K_{sat}$ = hydraulic conductivity ($cm\ s^{-1}$) = 1 $cm\ min^{-1}$
A = cross-sectional area (cm^2) of the soil core = 50.25 cm^2
L = length (*cm*) of the soil core = 8 *cm*
H = hydraulic head above soil core
dH = total head (*cm*) of the water imposed on the core, L + H (length of soil column + height of water) = 12 *cm*

$$\begin{aligned} Q &= \frac{KA \times (L + H)}{L} \\ &= \frac{(1\ cm\ min^{-1} \times 50.24\ cm^2) \times (4\ cm + 8\ cm)}{8\ cm} \\ &= 75.36\ cm^3 min^{-1} \times 60\ min\ h^{-1} \\ &= 4{,}521.6\ cm^3 h^{-1} \end{aligned}$$

27. A sand being considered for a sports field was placed in a 10 *cm* column with a cross-sectional area of 25 cm^2 with a water head imposed above the column at 5 *cm*. If the quantity of water measured through the column is 1,335 $cm^3\ h^{-1}$, what is its K_{sat} value?

$$\begin{aligned} Q &= \frac{KA \times (L + H)}{L} \\ K_{sat} &= \frac{Q}{AT} \times \frac{L}{dH} \\ &= \frac{1{,}335\ cm^3 h^{-1}}{25\ cm^2} \times \frac{10\ cm}{15\ cm} \\ &= 35.6\ cm\ h^{-1} \\ &= 14\ in\ h^{-1} \end{aligned}$$

28. A football field has an initial volumetric water content of 0.10 with a volumetric water content (θ_v) at field capacity of 0.30. (1) How much water is needed to bring the top 125 *cm* of the field up to field capacity? (2) In theory, what depth of soil will be brought up to field capacity with a 10 *cm* rain event?

1.

$$\begin{aligned} \text{required volume (depth) of water} &= 125\ cm \times 0.30\ cm^3cm^{-3} \\ &= 37.5\ cm \\ \text{currently volume (depth) of water} &= 125\ cm \times 0.1\ cm^3cm^{-3} \\ &= 12.5\ cm \\ \text{additional volume (depth) of water needed} &= 37.5\ cm - 12.5\ cm \\ &= 25.0\ cm \end{aligned}$$

A more direct way to calculate step 1 involves :

$$\begin{aligned} (\theta_{v\ final} - \theta_{v\ initial}) \times \text{soil depth} &= (0.30\ cm^3cm^{-3} - 0.10\ cm^3cm^{-3}) \\ &\quad \times 125\ cm \\ &= 25.0\ cm \end{aligned}$$

2.

$$\begin{aligned} \text{water penetration} &= \frac{\text{depth water applied}}{\text{difference in } \theta_v \text{of two soils}} \\ &= \frac{10\ cm}{(0.3 - 0.1\ cm^3cm^{-3})} \\ &= 50\ cm \end{aligned}$$

Another means to determine the soil depth brought to field capacity from a 10 *cm* rain event is cross multiply the known depth of 125 *cm* soil brought to field capacity with 25 *cm* rain.

$$\frac{25\ cm\ \text{water}}{125\ cm\ \text{soil}} = \frac{10\ cm\ \text{water}}{\text{X}}$$

$$\text{X} = 50\ cm$$

29. If a sports field has an average rooting depth of 10 *cm* and a soil moisture release curve indicates field capacity is at a volumetric water content (θ_v) of 23.2 % and the wilting point is at a θ_v of 12.2 %, how much moisture is available to the grass from the soil?.

$$\begin{aligned} \text{total water available} &= \text{soil depth} \times \text{available water content } cm^3cm^{-3} \\ &= \text{soil depth} \times (\text{field capacity} - \text{permanent wilting point}) \\ &= 10\ cm \times (23.2 - 12.2\%) \\ &= 1.1\ cm\ (or\ 0.43 - in) \end{aligned}$$

30. A circular golf green 32.5 *ft* (9.76 *m*) in diameter and 12 *in* (30 *cm*) deep with a bulk density of 1.4 $g\ cm^{-3}$ using a TDR probe, recorded wilting point at an average 11 % volumetric water content (θv) and field capacity at 33 %. If the current θ_v of the soil is 15 %, how many gallons of water are necessary to reach field capacity?

1.

$$\begin{aligned} \text{area of golf green} &= \text{r}^2(\pi) \\ &= (32.5\ ft \div 2)^2 \times 3.14 \\ &= 829\ ft^2 \end{aligned}$$

2.

$$\begin{aligned} \text{equivalent depth of water, } D_e@\ 15\%\theta_v &= 30\ cm\ \text{depth} \times 0.15\ cm^3cm^{-3} \\ &= 4.6\ cm \end{aligned}$$

3. $\text{equivalent depth of water, D}_e\text{@ }33\%\theta_v = 30\ cm\ \text{depth} = 0.33\ cm^3cm^{-3}$
$$= 10.1\ cm$$

4. $$\text{additional depth of water needed} = 10.1\ cm - 4.6\ cm$$
$$= 5.5\ cm\ (or\ 2.16\ in)$$

5. $$\text{additional volume of water needed} = 2.16\ in \times \frac{ft}{12\ in} \times \frac{gal}{0.134\ ft^3} \times \frac{829\ ft^2}{\text{area}}$$
$$= 1108\ gal\ (4194\ L)$$

or

$$= 2.16\ in \times \frac{ac}{43,560 ft^2} \times \frac{829\ ft^2}{\text{area}} \times \frac{27,154\ gal}{ac - in}$$
$$= 1116\ gal\ (4225\ L)$$

31. From the previous question, the wilting point is at 11 % volumetric water content (θ_v) and the forecast calls for an additional 3 % soil moisture volume will be lost via ET. If the superintendent wishes to increases the soil volumetric water content (θ_v) to 16 % to account for this loss, how many gallons are necessary to achieve this?

1. $\text{equivalent depth of water, D}_e\text{@ }11\%\theta_v = 30\ cm\ \text{depth} \times 0.11\ cm^3cm^{-3}$
$$= 3.3\ cm$$

2. $\text{equivalent depth of water, D}_e\text{@ }16\%\theta_v = 30\ cm\ \text{depth} \times 0.16\ cm^3cm^{-3}$
$$= 4.8\ cm$$

3. $\text{additional depth of water needed} = 4.8\ cm - 3.3\ cm$
$$= 1.5\ cm\ (or\ 0.59 - in)$$

4. $$\text{additional volume of water needed} = 0.59\ in \times \frac{ft}{12\ in} \times \frac{gal}{0.134\ ft^3} \times \frac{829\ ft^2}{\text{area}}$$
$$= 304\ gal\ (1,151\ L)$$

or

$$= 0.59\ in \times \frac{ac}{43,560 ft^2} \times \frac{829\ ft^2}{\text{area}} \times \frac{27,154\ gal}{ac - in}$$
$$= 305\ gal\ (1,154\ L)$$

Chapter 2
Soil Drainage

Water management is the primary key to success for most commercial turfgrass facilities. Soil serves as the storehouse for water used for plant growth that must be readily available to satisfy the demand created by transpiration. Being able to apply water when needed (irrigation) and being able to expediently remove excess water (drainage) ensures good plant growth and prevents prolonged delay in play. Improper or inadequate drainage is the most common agronomic problem cited by golf course superintendents and sports field managers (Fig. 2.1a, b). As with many topics in turfgrass management, drainage is a subject widely misunderstood, full of myths, nonscientifically-based practices, and unproven materials and products.

All too often, the concepts, machines, and technology used to design and construct roads are used to build turf facilities. In most cases this is a serious mistake, as the exacting requirements and internal drainage needs for turf sites are much different and more precise than for roads.

2.1 Drainage Methods

Two primary forms of drainage are utilized in turfgrass facilities—surface and subsurface.

1. In surface drainage, land surfaces are reshaped, sloped, and smoothed as needed to eliminate ponding and to induce gravitational flow overland to an outlet (Fig. 2.2a). Diverting and excluding water from an area often involves diversion ditches, swales, and floodways (Fig. 2.2b).
2. With subsurface drainage, soils may be modified to induce surface water infiltration and percolation through the rootzone to buried drains that collect and transport excess soil water to an outlet (Fig. 2.3). The drop in pressure (or water potential) due to outlet discharge induces excess soil water flow into the drains.

L.B. McCarty et al., *Applied Soil Physical Properties, Drainage, and Irrigation Strategies*, DOI 10.1007/978-3-319-24226-2_2

Fig. 2.1 Improper or inadequate drainage is a very common agronomic problem cited by sports field managers (*left*) and golf course superintendents (*right*)

Fig. 2.2 A combination of surface and subsurface drainage systems are needed for high profile turf venues which must play regardless of weather conditions. *Left* illustrates a "crowned" sports field with sideline drains which capture surface runoff while *right* demonstrates surface contouring to redirect excessive surface runoff away from a golf green

Subsurface drainage may also involve interceptor drains oriented perpendicular to the direction of groundwater flow.

A combination of surface and subsurface drainage is often required to quickly remove water from the soil surface to minimize delays in play, avoid excessive compaction, and allow maintenance practices to continue (Fig. 2.4).

Surface Drainage

Surface drainage is often a missing component in the design of modern golf courses and sports fields. Traditionally, sports fields were raised (crowned) in the center to encourage surface drainage. More recently, soccer fields, for example, have almost totally gone to "flat" surfaces, as have many football fields. Some of the major problems of poor playability and performance of these facilities are caused by

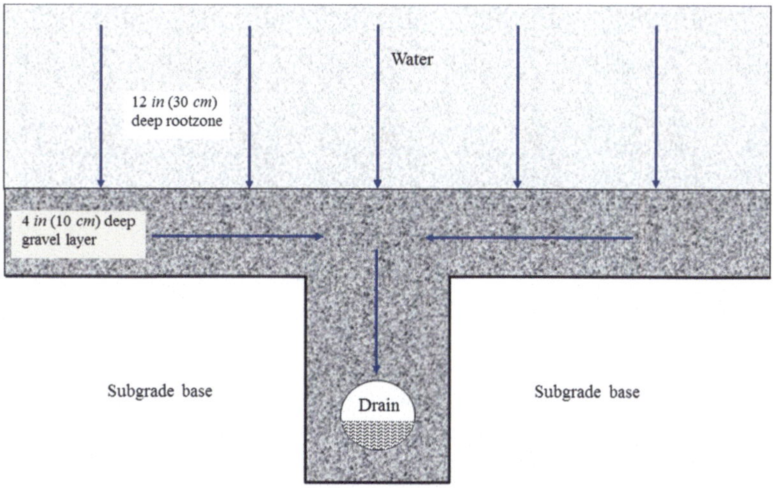

Fig. 2.3 Subsurface drainage often involves installing drainage lines at appropriate depths and spacing to remove excess soil moisture. In addition, rootzone modification to facilitate water movement through it is often performed on higher profile turf areas such as golf greens and sports fields

Fig. 2.4 Improper or inadequate drainage and flood prevention still plague the commercial turfgrass business

insufficient surface drainage, especially when the rootzone has poor internal drainage properties. Almost all long-term successful turfgrass facilities have adequate surface slope (grade) to remove excess surface water. Surface drainage uses the potential energy existing due to elevation change to provide a hydraulic gradient.

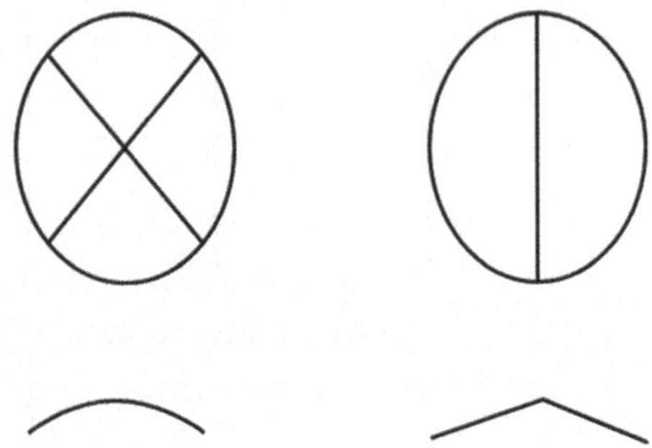

Fig. 2.5 Surface drainage design for golf courses often involves domed shaped surface with appropriate breaks to facilitate drainage. The *left figure* illustrates a 4-way surface domed contouring to remove surface water while the *right* one demonstrates a 2-way ridge contouring

The surface drainage system creates a water-free surface by moving surface water to an outlet at a lower elevation. For native soil constructed (or push-up) facilities characterized by low infiltration and poor internal drainage from high silt and clay content of the soil, surface drainage represents the only effective method for removal of excess surface water. Several designs are available to help facilitate surface drainage (Fig. 2.5).

Runoff occurs when the rate of precipitation or irrigation exceeds the soil infiltration rate (the rate water can enter a soil). The infiltration rate is dependent on the permeability of groundcover and on two soil parameters: soil structure and soil texture. Infiltration into heavier textured soils, such as clay, will be slower than infiltration into lighter soils, such as sandy soil. Soils with a low moisture content have higher infiltration rates that continue until the point of saturation is reached, the rate of water entry then begins to slow.

As water enters the soil, pores (large and small) near the soil surface fill first. When pores become full, gravity begins to move water downward. Water on the soil surface will puddle (or pond) if the water application rate exceeds the amount of water gravity can pull further down the profile. Once soil saturation is reached in shallow golf green or sports field profiles, the rate of water entering the soil is dependent on the rate the subsoil can remove it. If water sits or ponds on the surface, the whole topsoil is saturated. This is most common in surface depressions and on flat surfaces. If play commences while soil is saturated, the moisture acts like a lubricant allowing the soil particles to slide closer together, causing compaction. Turf plants and roots are easily damaged when soils are saturated (Fig. 2.6). In addition, saturated soils contain less oxygen, thus encouraging anaerobic conditions that lead to root loss and possible buildup of toxic gases such as carbon dioxide and methane, as well as substances such as iron and aluminum oxides, the chief causes of black layer.

A major advantage of good surface drainage is the capability to remove large volumes of water. This capability is especially important during heavy rainfall events as a 1 *in* rainfall across 1 *ac* equals 27,154 *gal* (25 *mm* over 0.40 *ha* equals 102,870 *L*).

Fig. 2.6 Excessive soil moisture acts like a lubricate allowing soils to be damaged when saturated and exposed to uncontrolled traffic

Slopes

The slope at which a particular surface should be constructed is determined by several variables. Slopes up to 3 % (1:33) are acceptable for soils with poor infiltration rates (Fig. 2.7). In competitive sports, players and coaches often feel slopes greater than 3 % affect ball roll and play. A minimum of 1 % slope (1:100) is almost always necessary for proper surface drainage, except with extensively modified rootzones and subsurface drainage such as USGA or California-style constructed greens or sports fields. For these modern greens, the surface slope surrounding the cup should typically be no more than 3 % for bermudagrass or ryegrass or no more than 2 % for bentgrass greens to prevent putting speeds from becoming excessive. For most non-modified soils, a 1.5 % (1:66) to 2.5 % (1:40) slope is usually adequate.

The following equation calculates the velocity of water across a bare surface as influenced by the surface slope and depth of ponded water or rainfall:

$$\mathrm{V} = 0.35 \times D^{0.67} \times \mathrm{S}^{0.5}$$

where:
V = Velocity ($in\ s^{-1}$)
D = water depth (*in*)
S = slope (decimal)

Note: The constant 0.35 includes conversion factors valid only for units shown.

Fig. 2.7 Surface slopes are generally between 1 and 3 %. This greatly enhances surface water drainage without compromising play

Examples

1. What runoff velocity would a 1 *in* (25 *mm*) rainfall onto saturated soil with a 1 % slope yield?

$$\begin{aligned} V &= 0.35 \times (1)^{0.67} \times (0.01)^{0.5} \\ &= 0.035\, in\, s^{-1} (0.09\, cm\, s^{-1}) \text{ water movement over a bare surface} \end{aligned}$$

2. A similar rainfall on a 2 % slope would yield:

$$\begin{aligned} V &= 0.35 \times (1)^{0.67} \times (0.02)^{0.5} \\ &= 0.049\, in\, s^{-1} (0.12\, cm\, s^{-1}) \text{ water movement over a bare surface} \end{aligned}$$

3. On a 3 % slope, velocity increases to:

$$\begin{aligned} V &= 0.35 \times (1)^{0.67} \times (0.03)^{0.5} \\ &= 0.06\, in\, s^{-1} (0.15\, cm\, s^{-1}) \text{ water movement over a bare surface} \end{aligned}$$

These examples demonstrate the large amount of surface water drainage provided by properly designed and constructed slopes. Insufficient slope means water must be drained through soil infiltration, which can be too slow to be efficient.

Fig. 2.8 Wet areas (or seeps) often develop when slopes flatten or when varying textural soils meet each other

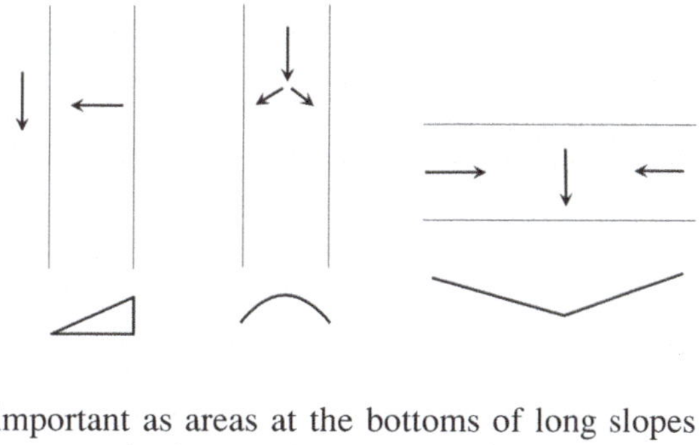

Fig. 2.9 Surface drainage of golf courses should be directed away from the fairway center when traffic is heaviest. Shown are three designs to facilitate surface drainage either as a slope (*left*), dome (*center*), or mowable center slopes (*right*)

The length of slope becomes important as areas at the bottoms of long slopes remain wet for longer periods than areas further up the slope; thus, they become subject to wear and compaction. Such areas are often found at the intersection of surface drainage from the fairway and front of golf greens (Fig. 2.8). This type of damage also often occurs in front of soccer and football goals. Golf course fairways should be designed so surface drainage is toward the outside edges of the fairway, rather than down the slope toward the green (Fig. 2.9). A maximum practical distance for surface drainage is approximately 150 *ft* (46 *m*). A minimum slope for adequate grassed surface drainage is 2 to 3 %.

Subsurface Drainage

Subsurface drainage involves water movement through a soil profile and often includes the installation of subsurface drains to remove excess water that can create

Fig. 2.10 Subsurface drainage most often involves drain lines to facilitate excessive water removal

undesirable (i.e., saturated) growing conditions (Fig. 2.10). Water available to plants is held in soil by capillarity, while excess water flows by gravity into drains. This lowers the groundwater level below the rootzone of plants. The movement of water into drains for turf facilities is influenced primarily by:

1. **Soil permeability**—this includes soil horizontal and vertical water permeability.
2. **Drain spacing**—this is often determined using Hooghoudt's equation.
3. **Depth of drain**—drain depth and spacing are interrelated. As the depth of the drain increases, generally so does the optimum spacing distance between drain lines.
4. **Drain size**—more correctly, the ability of the drain to lower the water potential sufficiently to promote water movement to and out of the drain.

Soil Modification to Improve Permeability

Soil modification to enhance internal soil moisture percolation is a common practice in the turfgrass industry. However, several misconceptions exist regarding soil modification to improve permeability. One such misconception is manifested in the practice of applying a 2 to 6 *in* (5 to 15 *cm*) layer of sand over a native soil with little or no surface slope provided and no subsurface drain lines installed. This is often referred to as the "bathtub" effect where the finer-textured native soil will not adequately drain and the coarse-textured sand holds water like a bathtub (Fig. 2.11). Heavy rainfall then causes saturation of the added sand layer and surface water accumulates, causing poor playing conditions. This is why most heavy use turf areas need 10 to 12 *in* (25 to 31 *cm*) of modified topsoil and properly spaced drain lines to lower this excess surface moisture further down in the soil profile (refer to Chap. 3 to determine appropriate sand depths). The drains act similar to a drain in a bathtub, providing a means of water removal.

Fig. 2.11 "Bathtub" effect of inadequate surface and subsurface drainage due to underlying clayey soils without sufficient drainage and outlets for the amount of rainfall received

Another misconception is that an inch (~2.5 *cm*) or so of a coarse sand, such as a river bottom sand, can be tilled into the top 3 to 6 *in* (7.6 to 15 *cm*) of native soil to enhance internal percolation. Unfortunately, this practice is rarely successful. First, a uniform, medium to medium-coarse sand that has consistent particle size should be used. River bottom sand often has a wide range of particle sizes; this variety in particle size allows smaller silt and clay particles to become dispersed among the larger sand particles, effectively reducing the pore space for water to percolate. Similarly, adding sand to native soil, which often has a high degree of silt and/or clay, often "clogs" these larger internal sand pores, again reducing internal percolation. Lastly, trying to uniformly "mix" the surface applied sand with the underlying soil is virtually impossible with a tractor-mounted roto-tiller. These machines will not provide the blended soil mix desired (Fig. 2.12). Proper mixing requires "off-site" machine blending.

Table 2.1 demonstrates the results of blending high-quality (USGA specified) sand into a native Cecil clay soil. The sand:clay blend was performed "off-site" in a laboratory, providing a very uniform distribution of sand and soil in the various ratios. As shown in Table 2.1, adding just 10 % clay soil to this sand reduced its hydraulic conductivity by almost 85 % (from 58 to 9 $in\ h^{-1}$, 148 to 23 $cm\ h^{-1}$). Conductivity values quickly dropped as the clay soil content increased; for example, with a 50:50 blend, the hydraulic conductivity was less than 0.2 $in\ h^{-1}$ (0.5 $cm\ h^{-1}$), totally unacceptable by today's standards. Furthermore, adding 20 % sand to the soil reduced drainage more than 50 % compared to straight

Fig. 2.12 A soil profile where an organic source was placed on the soil surface and roto-tilled in creating an uneven rootzone

Table 2.1 Hydraulic conductivity of a USGA medium sand and a Cecil clay soil, alone and in various combinations

	Hydraulic conductivity (K_{sat})	
Sand:soil ratio	($in\ h^{-1}$)	($cm\ h^{-1}$)
0:100	0.07	0.18
10:90	0.05	0.13
20:80	0.03	0.06
30:70	0.09	0.22
40:60	0.13	0.33
50:50	0.15	0.39
60:40	0.19	0.47
70:30	1.89	4.80
80:20	3.24	8.23
90:10	9.01	22.89
100:0	58.1	147.6

(100 %) soil. This again represents small soil particles "clogging" the larger pores between sand particles.

The following equation provides a guideline for using a suitable sand with a soil of known mechanical composition to create a rootzone with the desired drainage rate:

$$|A| = \frac{[R-B]}{[C-R]} \times 100$$

where:

A = weight of sand to add to 100 weight units of the original soil. *Note*: this is an absolute value regardless of their positive or negative signs.
B = percent of original soil in the desired particle-size range (e.g., 0.125 to 0.5 *mm*).
C = percent of desired particle-size range (e.g., 0.125 to 0.5 *mm*) in the sand used as an amendment.
R = percent of desired particle-size range (e.g., 0.125 to 0.5 *mm*) sand in the final mix.

Example
Assume the following particle-size distribution (%) and bulk density values are found in the sand and soil sources listed below. Determine volume and weight of sand to be added to the soil to achieve a K_{sat} value of 9 $in\ h^{-1}$ (23 $cm\ h^{-1}$) based on values in Tables 2.1 and 2.2.

Table 2.2 Calculated values of various v/v ratios of sand to soil from known particle-size distribution and bulk density values

Soil type	Percent particle-size distribution (*mm*)							Bulk density ($g\ cm^{-3}$)
	2–1	1–0.5	0.5–0.25	0.25–0.125	0.125–0.05	0.05–0.002	<0.002	
Known values								
Sand	3	32	44	21	0	0	0	1.65
Soil	3	17	15	20	5	29	11	1.35
Calculated values of various sand:soil ratios								
1:1	3	24.5	29.5	20.5	2.5	14.5	5.5	1.50
2:1	3	27	34	21	1.7	9.7	3.7	1.55
3:1	3	28	37	21	1.3	7.3	2.8	1.58
9:1	3	30.5	41	21	0.5	2.9	1.1	1.62

If a 9 $in\ h^{-1}$ (23 $cm\ h^{-1}$) percolation rate is desired for this sand:soil rootzone, the R value would be 90 % as determined from Table 2.1 in the desired particle-size range of 0.125 to 0.5 *mm*. The values of B and C would be determined by adding the known values in the columns for 0.5 to 0.25 *mm* and 0.25 to 0.125 *mm* particle size for the soil (B) and sand (C) from Table 2.2.

$$|A| = \frac{[90 - 35]}{[65 - 90]} \times 100$$
$$= 220$$

Therefore, 220 *tons* of sand per 100 *tons* of soil would be required to raise the percentage of soil particles between 0.125 and 0.5 *mm* to 90 % in the final mix.

Note: Values generated are absolute values regardless of their positive or negative signs, as the actual calculation value in the previous example is −220.

If mixed on a volume basis (such as with off-site blending) instead of a weight basis, one must find the volumetric ratio of sand to soil using the equation: volume = mass/density. The bulk density of sand in this example is 1.65 $g\ cm^{-3}$ and soil is 1.35 $g\ cm^{-3}$, giving:

$$\text{Volume (ratio)} = \frac{V_{sand}}{V_{soil}} = \frac{M_{sand}/\rho_{b\ sand}}{M_{soil}/\rho_{b\ soil}}$$
$$= \frac{220/1.65}{100/1.35} = \frac{220 \times 1.35}{100 \times 1.65}$$
$$= 1.8$$

Therefore, 1.8 unit volumes of this particular sand are needed per one unit volume of this soil to achieve the desired ratio of 220 *tons* of sand per 100 *tons* of soil.

Calculating Volume to Volume (V/V) Ratios

If one wishes to determine the outcome of mixing sand with topsoil on a volume to volume (v/v) ratio basis, the following calculations can be performed for a sand to soil ratio mix.

$$\text{New percent particle size} = \frac{[\text{sand fraction}\,\% \times \text{ratio sand}] + [\text{soil fraction}\,\% \times \text{ratio soil}]}{\text{total sand} + \text{soil ratio}}$$

Example
Calculate the new percent particle size in the 0.5 to 0.25 *mm* range from the sand/soil ratio listed in Table 2.2 in a 3:1 ratio.

$$\text{New percent particle size} = \frac{[\text{sand fraction}\,\% \times \text{ratio sand}] + [\text{soil fraction}\,\% \times \text{ratio soil}]}{\text{total sand} + \text{soil ratio}}$$
$$= \frac{[44 \times 3] + [15 \times 1]}{3 + 1}$$
$$= 37\,\%$$

The following example demonstrates how to determine the new particle-size distribution obtained by tilling a known volume of sand into native soil.

Example
1,500 *tons* (1,814 *metric tons*) of sand with a bulk density of 1.65 $g\ cm^{-3}$ is tilled into the top 5 *in* (13 *cm*) of native soil 1.7 *ac* (0.7 *ha*) in area. Calculate the predicted new particle-size distribution percentages and bulk densities.

First, determine the depth of 1,500 *tons* of sand over the 1.7 *ac*:

$$\frac{1500\,ton}{1.7\,ac} \times \frac{2000\,lb}{ton} \times \frac{454\,g}{lb} \times \frac{cm^3}{1.65\,g} \times \frac{ac}{43{,}560\,ft^2} \times \frac{ft^2}{929\,cm^2} \times \frac{in}{2.54\,cm} = 4.7\,in\,\text{deep}$$

Since 4.7 *in* (12 *cm*) of sand in depth is to be tilled into the top 5 *in* (13 *cm*) of soil, it is possible to approximate the new particle size distribution percentages and bulk density using a 1:1 ratio as presented in Table 2.2.

Although this equation helps predict projected particle-size distribution and bulk density values of two known sand/soil sources, it cannot be reliably used to predict hydraulic conductivity (or "perc") rates. For example, with the same sand and soil from the previous example (Table 2.1) in a 1:1 ratio, the following calculations could be performed to attempt to predict a percolation rate for the mix.

$$\begin{aligned} \text{predicted percolation rate} &= \frac{[\text{perc. rate sand} \times \text{ratio sand}] + [\text{perc. rate soil} \times \text{ratio soil}]}{\text{total sand} + \text{soil ratio}} \\ &= \frac{[58 \times 1] + [0.07 \times 1]}{2} \\ &= 29\,in\,h^{-1}\,(74\,cm\,h^{-1}) \end{aligned}$$

However, when actual samples are mixed in a 1:1 ratio, the percolation rate is only $0.15\,in\ h^{-1}$ ($0.4\,cm\ h^{-1}$) (Table 2.1). The small amount of fine-textured clay in the soil mix is sufficient to "clog" the pores in the sand, thus reducing the actual percolation. This demonstrates the importance of actually measuring particle-size distribution, bulk density, and hydraulic conductivity (percolation rate) of the various soil/sand mix being considered as well as the ratios of each.

Table 2.3 reflects the percent (by volume) change when a known amount of amendment is mixed into a soil. For example, if a contractor places 2 *in* (5 *cm*) of an amendment on the existing soil surface and roto-tills this 6 *in* (15 *cm*) deep, the theoretical percent volume this added amendment occupies is 25 %.

Table 2.3 Amount (*in*) of surface-applied amendment converted into percent volume

Amount (*in*) amendment mixed into soil	Depth of roto-till (*in*)	percent by volume change
0.5	2	20
0.5	3	14
0.5	4	11
0.5	6	8
1	2	33
1	3	25
1	4	20
1	6	14
1.5	2	43
1.5	3	33
1.5	4	27
1.5	6	20
2	2	50
2	3	40
2	4	33
2	6	25

Multiply inch (*in*) by 2.54 to obtain centimeters (*cm*)

Lateral Soil Water Movement

Lateral (sideways) water movement in a soil is influenced or restricted by three factors:

1. Depth (hydraulic head) of the saturated free-water zone in the topsoil.
2. Hydraulic conductivity of the rootzone soil.
3. Slope of the subgrade or base.

As a soil absorbs more water into its pores, a saturated zone develops and reaches the subsoil base. Until this saturated zone reaches the subsoil base and a buildup of "free water" occurs atop this much less permeable layer, little water will move laterally (sideways). This saturated zone of free water is the only water moved horizontally by resultant forces due to the vertical force of gravity. The smaller the soil pore space, the slower water will move laterally.

Lateral water movement ceases when the free-water zone is removed. This occurs even if the capillary fringe is still saturated. Hooghoudt's equation (discussed later) is used to calculate the rate at which the saturated free-water zone of the topsoil will drain at the midpoint between two drains (the slowest draining point).

Lateral water movement in soil is generally limited in distance and time. However, gravity is able to "pull" water down a sloped base (subgrade). The steeper the subgrade slope, the greater the effect of gravity. Generally, water will move laterally (sideways) along the subgrade's surface in direct proportion to the subgrade's slope. For example, if the slope is 2 % (1:50), water will move laterally 2 % (or one-fiftieth) as fast as it will move downward. If the rate of downward movement of water in a soil is $15\,in\,h^{-1}$ ($38.1\,cm\,h^{-1}$), the maximum rate at which water would move laterally due to the same head would be 2 % of $15\,in\,h^{-1} = (0.02 \times 15\,in\,h^{-1}) = 0.3\,in\,h^{-1}$ ($0.76\,cm\,h^{-1}$).

In addition, for water to continuously drain (move) down the subgrade, water must be removed from the end or low point of the subgrade with drains or ditches. If this water is not removed, an equilibrium will be reached, often resulting in ponding of water on the surface and excessively wet conditions at the end of a slope or against an impermeable obstruction such as a wall. Such conditions frequently occur when water drains down banks or hills onto a flatter playing surface (Fig. 2.13). In this case, disposal of water from the bottom of the slope can be achieved by: (1) installing a cutoff surface drain at the top of the hill to collect water before it reaches the hill (Fig. 2.14); (2) constructing a terrace to move the water gently across or around the perimeter of the hill; or (3) most commonly, by placing an interceptor drain near the bottom of the hill (discussed earlier).

Fig. 2.13 A wet "seep" that develops when sloping soil meets flatter elevations

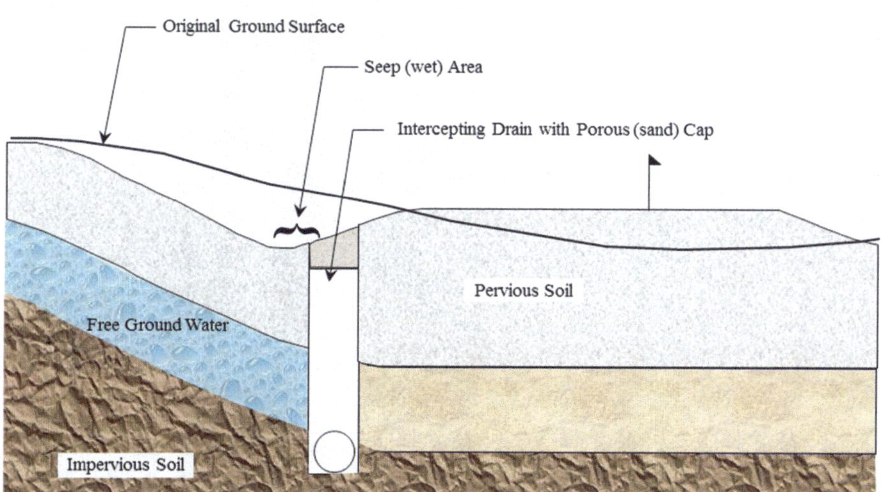

Fig. 2.14 Placing an intercepting drain in a seep (wet) area to promote drainage

Drain Lines

Subsurface drain lines are designed to function as open channels, meaning the water flow through the pipe is from the influence of gravity due to the slope or grade of the pipe, not from pressure pushing water through the pipe (Fig. 2.15). If the subsurface drainage pipe tries to convey more water than it was designed for, it will first fill to capacity and then become pressurized along some portion of its length. When perforated pipes become pressurized, water tries to escape through the inlet holes of them. This pressure on the water in the drain line trench can create flow back into the surrounding soil, causing the soil to become saturated. This can cause the saturated soil to begin to act like a fluid and flow, thus making it prone to high levels of erosion. When under extreme pressure, drainage pipes can erode out of the trenches, requiring extensive repair. This is especially true near the outlet of a long run of pipe at a steeper slope. This pipe pressure can also create a floating 'lens' of water between the turf and soil. The turf and thatch layer essentially floats off the soil surface, creating play and maintenance problems. The key to avoid this problem is to properly size the drainage pipe, taking into account the expected water flows, so the pipe can function as an open channel.

Two parameters largely determine the rate at which water is removed by a drain; (1) depth and (2) spacing of drain lines. In addition, the slope of drain lines in the

Fig. 2.15 Four inch *in* perforated tile placed into the subgrade and back-filled with gravel to help facilitate subsurface drainage

trenches also affect drainage capacity. Generally, the deeper the drain lines and closer their spacing, the quicker and more effectively soil moisture is removed. However, water cannot enter a drain any faster than the soil around it can conduct water into it. Optimum depth and spacing are directly related to the permeability of the soil. Since golf greens have a relatively shallow rootzone (~1 *ft*, 30 *cm*) of highly permeable soil (sand), and need to quickly and completely remove surface water so play can resume, their optimum drain spacings are much narrower (closer) than most unmodified soil situations.

As mentioned, the closer drain lines are together, the faster a profile will drain. Also, as the free-water depth in the soil profile decreases, so too does the gravitational gradient. A deeper topsoil has a greater storage space in the profile for the free-water zone. Therefore, in shallower soils, the rate of drainage and soil water storage capacity decrease and drains need to be spaced closer together. Golf green drainage lines should be spaced so water will not have to travel more than 10 *ft* (3 *m*) to reach any individual line. If the golf course is situated on an area with a high water table, it may be necessary to place larger drainage lines deeper into the subgrade to lower the water table and handle the increased internal flow of water.

Calculating drainage line spacing can be done with a modification of **Hooghoudt's equation**. In Hooghoudt's equation, the drain discharge is assumed to equal the incoming rainfall or irrigation, and the water table midway between drains is maintained at a steady height above the drain level. Water enters the soil more rapidly nearer to the drains than midway between them. The equation takes into account both horizontal flow and radial flow caused by the convergence of flow lines over the drains (Fig. 2.16). In shallow topsoil, widely spaced drains only remove water from a very small area immediately adjacent to the drains and do not adequately drain the topsoil between them.

$$K_{sat} = \frac{S^2 v}{4h^2}$$

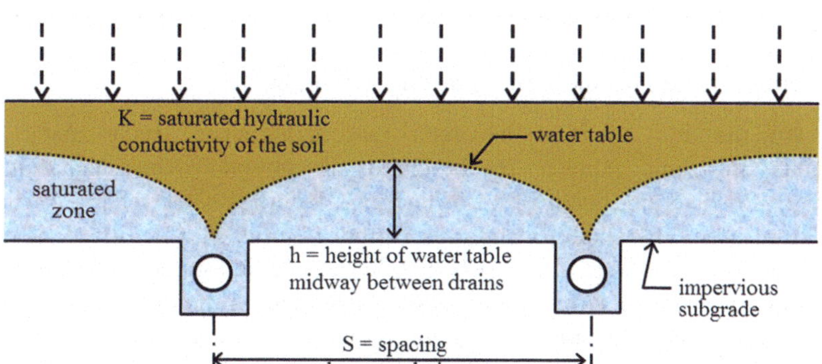

Fig. 2.16 Hooghoudt's equation and components used to determine proper spacing of drainage lines. *S* is the distance between drain line spacing, *v* is the amount of rainfall or irrigation applied, and *h* is the height or depth of the saturated zone with free water, also known as the water table. Hydraulic conductivity of the soil is another variable needed to calculate drain tile spacing. Hooghoudt's equation calculates drainage at the slowest draining point over the total soil surface in consideration

Hooghoudt's Equation

$$S = \sqrt{\frac{4Kh^2}{v}} \quad \text{or} \quad S^2 = \frac{4Kh^2}{v}$$

where:
S = Drain line spacing (*in*); the units used for h must be the same as those used for S.
$K = K_{sat}$ = Saturated hydraulic conductivity ($in\ h^{-1}$) of the soil.
h = Height of the (saturated) free-water zone midway between the two drains (*in*).
v = Drain discharge rate, assumed to equal irrigation or rainfall rate ($in\ h^{-1}$).

Normally, the anticipated maximum rainfall or irrigation event rate is used here.

Since S and h are squared in Hooghoudt's equation, varying them will change the drainage rate by the square of the magnitude of drains distance apart or height, respectively. In other words, if drain spacing (S) is halved, or depth of the saturated zone (h) is doubled, the effective drainage rate of drains increases fourfold. Conversely, shallower topsoil and wider-spaced drains decrease soil water drainage rates exponentially.

Two calculations are suggested. The first involves the assumption of a worst-case scenario where the free-water zone extends to the surface or the total depth of the topsoil resulting in ponding. The other calculation is performed with the free-water zone lowered by 2 or 3 *in* (5 or 7.6 *cm*). This gives an indication of how quickly water can be removed from the top of the profile. Often this removal is slow if drains alone are being relied on.

Several points are illustrated by Hooghoudt's equation:

1. As the allowable free water (water table) depth (h) decreases (i.e., the shallower the topsoil), the gravitational gradient decreases, resulting in decreased drainage, and the closer drain lines need to be spaced. Conversely, the deeper the topsoil, the greater the storage space in the profile for the free-water zone, and the further apart drain lines may be spaced.
2. The closer the drains, the faster a profile will drain.

Note: Hooghoudt's equation becomes inaccurate when drain spacing (S) approaches the same value as the height of the saturated free-water zone (h). However, this does not occur often in turfgrass facilities since relatively shallow topsoils are used. Hooghoudt's equation cannot be used in a two-tier soil profile with a sand rootzone over a gravel bed (USGA specified green). The gravel bed allows rapid vertical rootzone drainage and movement to the drain pipes, and soil water movement is most influenced by the saturated hydraulic conductivity of the rootzone. For Hooghoudt's equation to be accurate, the soil must be uniform in hydraulic conductivity and must have an impervious layer located below the soil and the drain.

Example

If the hydraulic conductivity of a loam soil is 12 *in* h^{-1} (30 *cm* h^{-1}), the height from the drain line to the soil surface is 18 *in* (46 *cm*), and the design rainfall event is 1 *in* h^{-1} (2.5 *cm* h^{-1}), determine the drain line spacing required to prevent ponding.

$$S = \sqrt{\frac{4Kh^2}{v}} = \sqrt{\frac{4(12\,in\,h^{-1})\,(18\,in)^2}{1\,in\,h^{-1}}} = 125\,in\ (10.4\,ft\ \mathrm{or}\ 3.2\,m)$$

Calculating Necessary Soil Hydraulic Conductivity

Hooghoudt's equation also can be rearranged to calculate the desired hydraulic conductivity (percolation) for a given drain line spacing:

$$K = \frac{S^2 v}{4h^2}$$

Example

1. If an area has a proposed drain spacing of 10 *ft* (120 *in* or 30 *cm*) between drain lines, a 10 *in* (25.4 *cm*) deep rootzone above the drains, a 1 *in* h^{-1} (2.54 *cm* h^{-1}) anticipated rainfall rate, and the free-water zone extends to the surface, as might occur after prolonged rain, determine the necessary hydraulic conductivity (*in* h^{-1}) of the soil.

$$K = \frac{S^2 v}{4h^2} = \frac{(120\,in)^2 \times (1\,in\,h^{-1})}{4 \times (10\,in)^2} = 36\,in\,h^{-1}\left(91\,cm\,h^{-1}\right)$$

2. If the same area had a 12 *in* (1 *ft* or 30 *cm*) deep rootzone instead of 10 *in* (25 *cm*), what would be the necessary hydraulic conductivity of the soil?

$$K = \frac{S^2 v}{4h^2} = \frac{(120\,in)^2 \times (1\,in\,h^{-1})}{4 \times (12\,in)^2} = 25\,in\,h^{-1}\left(63.5\,cm\,h^{-1}\right)$$

3. If the same 10 *in* (25 *cm*) rootzone area had an anticipated maximum rainfall of 0.5 *in* h^{-1} (1.3 *cm*) instead of 1 *in* h^{-1} (2.5 *cm* h^{-1}), what would be the necessary hydraulic conductivity of the soil?

$$K = \frac{S^2 v}{4h^2} = \frac{(120\,in)^2 \times (0.5\,in\,h^{-1})}{4 \times (10\,in)^2} = 18\,in\,h^{-1}\left(46\,cm\,h^{-1}\right)$$

Calculating Drainage Rates

Hooghoudt's equation can also be rearranged to calculate the drainage rate between subsoil drains.

$$v = \frac{4Kh^2}{S^2}$$

where: v = drainage rate ($in\ h^{-1}$) of the saturated free-water zone at the midpoint between drains.

Example

1. A soil has a hydraulic conductivity of $2\,in\ h^{-1}$ ($5\,cm\ h^{-1}$), a saturated depth midway between the drains of $10\,in$ ($25\,cm$), and drains spaced $10\,ft$ ($120\,in$ or $3\,m$) apart. Determine the drainage rate at the midpoint between the drain lines.

$$v = \frac{4Kh^2}{S^2} = \frac{4 \times (2\,in\,h^{-1}) \times (10\,in)^2}{(120\,in)^2} = 0.055\,in\,h^{-1}\,(0.14\,cm\,h^{-1})$$

2. If the soil's depth in the above example is increased to $12\,in$ ($0.3\,m$), determine the new drainage rate at the midpoint between the drain tiles.

$$v = \frac{4Kh^2}{S^2} = \frac{4 \times (2\,in\,h^{-1}) \times (12\,in)^2}{(120\,in)^2} = 0.08\,in\,h^{-1}\,(0.2\,cm\,h^{-1})$$

3. If the same soil in question 2 has drain tile spaced at $15\,ft$ ($180\,in$ or $4.5\,m$) instead of $10\,ft$ ($3\,m$), what will be the resulting drainage rate?

$$v = \frac{4Kh^2}{S^2} = \frac{4 \times (2\,in\,h^{-1}) \times (12\,in)^2}{(180\,in)^2} = 0.036\,in\,h^{-1}\,(0.09\,cm\,h^{-1})$$

   Increasing tile spacing from 10 to $15\,ft$ (3–$4.5\,m$) decreases the drainage rate from 0.08 to $0.036\,in\ h^{-1}$ (0.2–$0.09\,cm\ h^{-1}$).

4. Now determine the drainage rate for the above example if tile lines are spaced $5\,ft$ ($60\,in$ or $1.5\,m$) apart.

$$v = \frac{4Kh^2}{S^2} = \frac{4 \times (2\,in\,h^{-1}) \times (12\,in)^2}{(60\,in)^2} = 0.32\,in\,h^{-1}\,(0.8\,cm\,h^{-1})$$

   Decreasing tile spacing from 10 to $5\,ft$ (3–$1.5\,m$) increases the drainage rate from 0.08 to $0.32\,in\ h^{-1}$ (0.2–$0.8\,cm\ h^{-1}$).

These examples illustrate that, the closer the drain tiles or deeper the saturated rootzone, the faster a profile will drain. Specifically, if the drain spacings are halved, drainage increases fourfold. Similarly, as soil depth is doubled, drainage increases fourfold.

Determining Drain Line Discharge Rates

If the length of the drain line is known (Fig. 2.17), then the total amount of water expected to drain from a particular area following a known amount of rainfall or irrigation can be determined from the following equation, modified from Darcy's and Hooghoudt's equations:

$$Q = \frac{2Kh^2w}{S}$$

where:
Q = discharge rate of water from drain line ($in^3\, h^{-1}$),
$K = K_{sat}$ = saturated hydraulic conductivity ($in\ h^{-1}$),
h = height of saturated free water zone midway between drains (in),
w = length of the drain line (in),
S = drain line spacing (in).

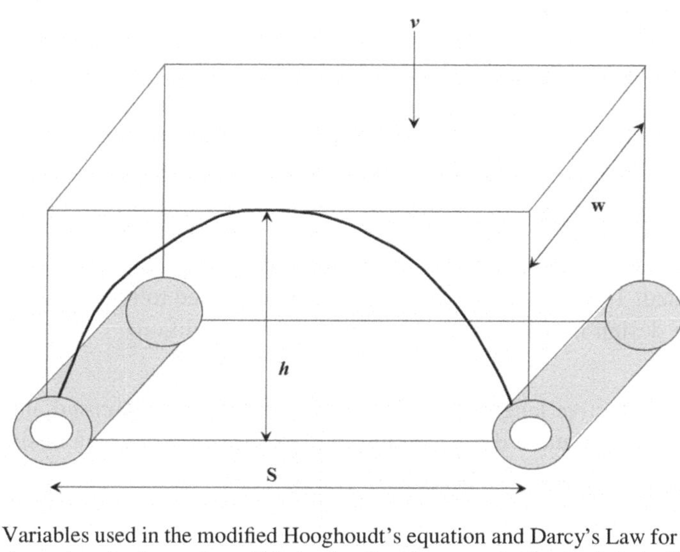

Fig. 2.17 Variables used in the modified Hooghoudt's equation and Darcy's Law for determining the total volume (area) of a section of drained soil and appropriately sized drain lines. *S* is the distance between drain lines, *v* is the amount of rainfall or irrigation applied, and *h* is the height or depth of the saturated zone of free water, also known as the water table; *w* is the width (length) of the drain line

Example
Determine the volume of water flowing from an area with a drain spacing of 10 *ft* (120 *in* or 3 *m*), drain lines of 12.5 *ft* (150 *in* or 3.8 *m*) length, and a rootzone hydraulic conductivity of 16 $in\ h^{-1}$ (41 $cm\ h^{-1}$), with the saturated zone midway between the drain lines at the surface of a 10 *in* (2 *cm*) rootzone. (Water has a volume of 0.00434 $gal\ in^{-3}$, 1 $ml\ cm^{-3}$.)

$$Q = \frac{2Kh^2w}{S} = \frac{2 \times \left(16\,in\,h^{-1}\right) \times (10\,in)^2 \times (150\,in)}{(120\,in)}$$
$$= 4000\,in^3\,h^{-1} \times 0.00434\,gal\,in^{-3}$$
$$= 17.4\,gal\,h^{-1}$$

Therefore, drain lines should be selected that can remove at least 18 $gal\ h^{-1}$ (68 $L\ h^{-1}$).

Hooghoudt's equation can be used here to calculate the rainfall or irrigation rate expected to cause the saturated zone to rise to the surface, in this example, the rootzone depth of 10 *in* (25 *cm*).

$$v = \frac{4Kh^2}{S^2} = \frac{4 \times \left(16\,in\,h^{-1}\right) \times (10\,in)^2}{(120\,in)^2} = 0.44\,in\,h^{-1}\left(0.17\,cm\,h^{-1}\right)$$

Determining Drain Size and Length

Sizing drain pipe for a particular area requires a considerable amount of information, including proposed drain depth, slope, width, length, and spacing; average rainfall event (inflow rates); soil type; area to be drained; and surface slope. First, the amount of water to drain following a rainfall event needs to be determined. Hooghoudt's equation is only valid for drainage rates (equal to rainfall rates) where the saturated zone does not rise above the rootzone surface. A more conservative design for effective pipe length can be performed by assuming the entire rootzone to be saturated. This results in a drainage system designed to remove water at rates equal to the design rainfall rates, even during flood conditions.

Example
Calculate the effective length of (a) 2 *in* (5 *cm*) and (b) 4 *in* (10 *cm*) diameter drain pipe with 1 % slope and a drain spacing of 10 *ft* (3 *m*), following a design rainfall event of 2 $in\ h^{-1}$ (5 $cm\ h^{-1}$) [1 ft^3 (0.028 m^3) = 7.5 *gal* (28 *L*)].

First, the rate of water every foot (0.3 *m*) of trench should collect is calculated as:

$$10\,ft \times 1\,ft \times \frac{2\,in}{h} \times \frac{1\,ft}{12\,in} = 1.7\,ft^3\,h^{-1}$$

$$\frac{1.7\,ft^3}{hr} \times \frac{7.5\,gal}{ft^3} =\sim 13\,gal\,h^{-1} \text{ or } 0.21\,gal\,min^{-1}\left(0.79\,L\,min^{-1}\right)$$

a. The maximum drainage rate the pipe can handle is 7.9 $gal\ min^{-1}$ (30 $L\ min^{-1}$) based on the manufacturer's specification for 2 *in* (5 *cm*) pipe with 1 % slope). Therefore, the 2 *in* (5 *cm*) pipe's effective length can be calculated as:

$$\frac{7.9\,gal}{min} \times \frac{min^1\,ft^1}{0.21\,gal} = 38\,ft\,(11.4\,m)$$

A collector (lateral) drain would be needed after a maximum 2 *in* (5 *cm*) pipe run of 38 *ft* (11.4 *m*).

b. Per the manufacturer's specifications, the 4 *in* (10 *cm*) pipe has a maximum flow rate on a 1 % slope of 0.85 $gal\ s^{-1}$ (or 51 $gal\ min^{-1}$). Therefore, the 4 *in* (10 *cm*) pipe's effective length can be calculated as:

$$\frac{51\,gal}{min} \times \frac{min^1\,ft^1}{0.21\,gal} = 243\,ft\,(74\,m)$$

A 4 *in* (10 *cm*) drain pipe is the current standard for most golf greens. These are more than sufficient to handle most rainfall events. Lateral lines are typically increased to 6 *in* (15 *cm*) in diameter to handle the total output of a draining green.

Drain Line Types

A wide array of drain line types, sizes, and configurations are available (Fig. 2.18). A common misconception is that all of these products work equally well. Unless the drainage line provides lower water potential than the surrounding soil and the pipe is laid on grade in the bottom of the trench, water will not efficiently enter nor move down it. A suitable outlet is also needed to remove drained water.

Corrugated Pipe

In the past, drainage lines were built from agricultural clay tile, concrete, or flexible corrugated plastic. Today 4 *in* (10 *cm*) diameter corrugated high-density polyethylene (HDPE) pipe with perforations (slits or holes) is the industry standard, due to its ease of installation and relatively low cost (Fig. 2.19a, b). The perforations in HDPE pipe are laid facing downward on a bed of gravel to prevent clogging by downward-migrating soil particles. Drain lines with smooth inner walls provide more efficient water removal than lines with corrugated inner walls.

Water moves down through the medium and stops at the bottom of the trench. As the water level rises in the trench, it moves into the pipe through slits or holes fabricated by the manufacturer (Fig. 2.20) and then moves down the pipe.

Fig. 2.18 Various sized and types of drainage pipes and other apparatuses used in the turfgrass industry

Fig. 2.19 Corrugated tile pipe is widely used for golf course drainage (*left*). A smooth inner-walled pipe should be considered for increased drainage efficiency. Slits in pipes where water enters (*right*)

An alternative to corrugated HDPE pipe is rigid polyvinyl chloride (PVC) pipe with two rows of holes drilled adjacent to each other over the length of the pipe for drainage water to enter. PVC pipe is generally more expensive but is able to withstand heavier traffic pressure than flexible HDPE pipe.

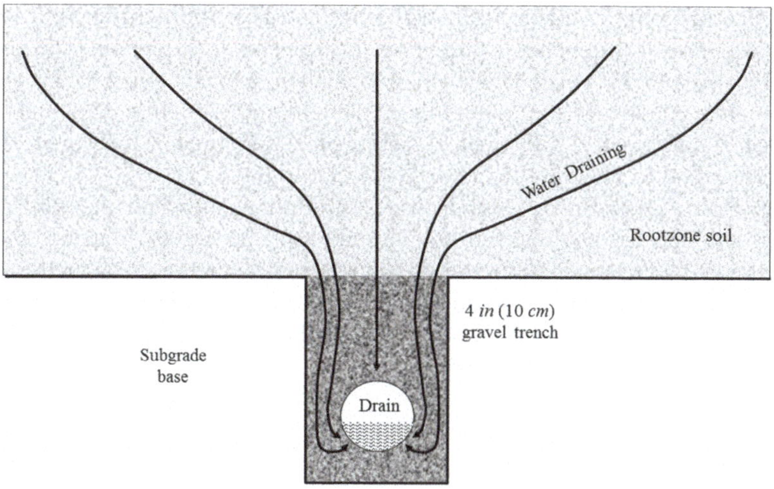

Fig. 2.20 Water enters drain lines from the bottom of trenches, as this is the point of lowest water potential

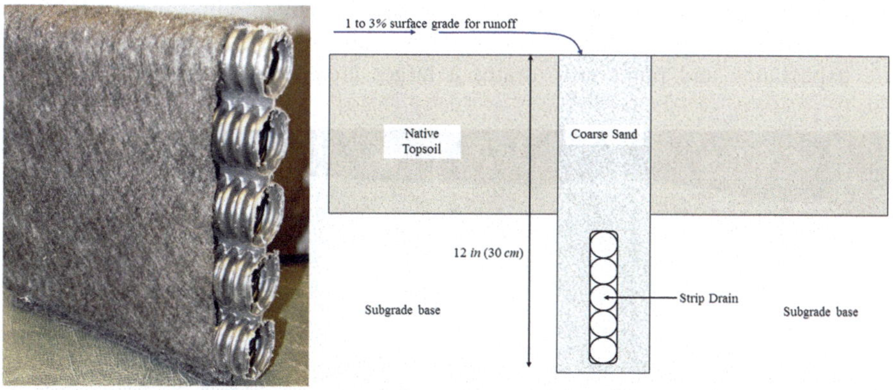

Fig. 2.21 Strip drains (*left*) placed vertically in a narrow trench of sand with pipes embedded to remove surface water (*right*)

Strip Drains

Strip drains exist where a narrow (2 to 4 *in* wide, 5 to 10 *cm*) trench is dug, a narrow (1 to 2 *in* wide, 2.5 to 5 *cm*) perforated drain (or "strip") sleeve is installed, and trenches are backfilled with sand (Fig. 2.21). The drain types used include cloth-wrapped, waffle- or honeycomb-shaped drain sleeves, vertically stacked small diameter pipes (composite drains), or other similar narrow sleeved material. Drains are placed at the bottom of the trench and extend about halfway to the soil surface. Water will enter the lowest pipe of a composite drain first. Once this pipe is filled, water will enter and flow through the next lowest, and so on. Narrower trenches are used for this stacked pipe, reducing the costs of trenching and fill material. However, the smaller diameter pipe means greater surface area contacting the water; thus, there is more friction loss and lower water-carrying capacity. This makes stacked pipe less efficient for water removal and more prone to clogging than single larger tile lines.

Strip drains allow an increase in water infiltration rates into the soil surface without complete renovation of the native soil profile. However, in many cases the trenches filled with gravel and sand on grade actually work as "dry wells" where they lower the water table, removing surface water. The drain lines are not directly involved in this water removal. Drain lines must be installed deep enough to avoid disruption or displacement by heavy equipment.

Slit Drains

Slit drains are essentially narrow trenches (slits) dug into a soil and filled with permeable medium to facilitate drainage. A modification of the slit drain is a **French drain**, which consists of a trench 4 to 8 *in* (10 to 20 *cm*) wide dug on a 1 to 3 % slope, with a drainage pipe laid at its bottom and backfilled with sand (Fig. 2.22). This drainage system provides a wider trench than strip drains, which extends its life expectancy and potentially drains a larger area. Additional fill material is

Fig. 2.22 Installation of slit drains (also called French drains) in a fairway where a 4 to 8 *in* (10 to 20 *cm*) wide trench is backfilled with sand. A drain tile may also be placed at the bottom of these to facilitate water removal

Fig. 2.23 Turf desiccation and/or nutrient deficiency at the surface of slit drains composed of sand that are prone to drying and leaching. Cool-season grasses grown in heavy soils during warm, dry weather are especially susceptible

necessary for this system and more surface area disruption occurs. However, many facilities readily have access to the trenching equipment and the wider trench is usually easier to work with. In heavy soils, the wider trench filled with sand easily desiccates, shows nutrient deficiencies, and may become more susceptible to low-temperature damage (Fig. 2.23).

The advantages of strip and slit drains include installation with minimal surface disturbance and the need for less labor, as these installations are largely mechanized. They also provide drainage to poorly designed and constructed facilities at an attractive cost compared to complete soil profile renovation (Fig. 2.24). However, due to the complex design of the drains, personnel turnover in management often leads to eventual disturbances of the drain's integrity as new turfgrass managers are not as aware of the positioning of these drains or their required upkeep. Sufficient surface slopes must be present for lateral water movement to these drains. Gravel and possibly small pipe in the bottom of the slits also are needed to expedite water removal from the site. Traffic from play, soil migration, erosion, and aerification can also cause glazing of silt and/or clay over the sand, reducing water infiltration. Topdressing with finer sand or soil used to fill the trench can also clog pore spaces and disrupt water infiltration. Heavy equipment can crush or disturb the integrity of the drain. The smaller diameter (1 *in*, 2.5 *cm*) pipe used in a strip drain is also prone to eventual clogging by downward-migrating soil particles. These drains typically require closer spacings than traditional lateral drains; thus more pipe junctions are needed.

Fig. 2.24 Narrow slit drains backfilled with sand to reduce surface moisture. This helps lower the water table until the slits fill with water

Due to these inherent limitations, strip or slit drains should be viewed as a technique for removing the symptoms of waterlogging rather than as a technique for curing the problem. They also have a finite life span and often are a means of providing a temporary fix that will have to be performed repeatedly to succeed continually. They generally are not viewed as a substitute for sound surface and subsurface drainage planning and installation, which should occur prior to construction.

Filter Cloth

Nylon-netted filter drainage sleeves are also available, with the filter cloth wrapped around perforated pipe to prevent soil particles from impeding water flow into drainage lines. There is much debate over the probability of these nets/sleeves clogging over time from soil particle movement toward the drain. If excessive (>5 to10 %) silt and clay are present in the topsoil, these drainage sleeves may clog. In this situation, filter cloth should be considered to line the drainage ditch but should not be physically wrapped around the individual drain lines. It is also believed this cloth can become clogged from the bio-products of algae and other organisms that may colonize the perpetually wet cloth.

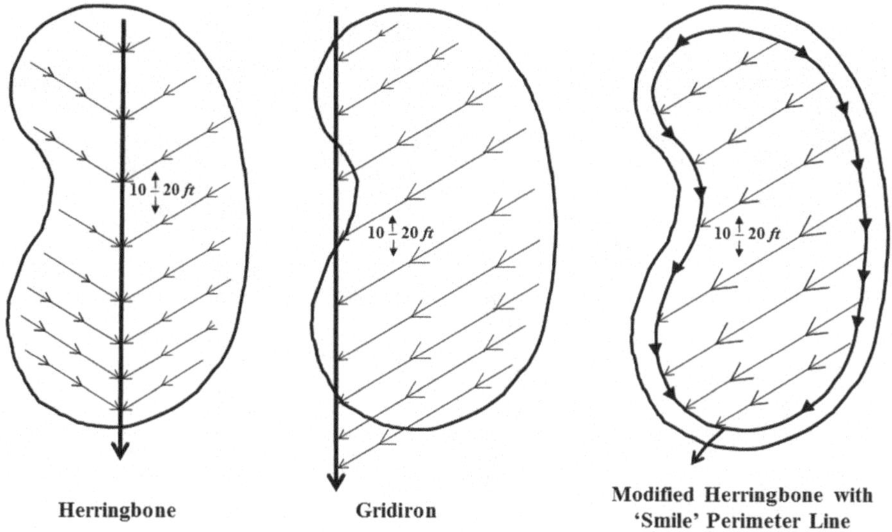

Fig. 2.25 Commonly used patterns to drain turf areas with drainage tile include: (*left*) herringbone; (*center*) gridiron, and (*right*) a modified herringbone pattern with a perimeter "smile" line to facilitate draining edges

Drainage Line Patterns

Typically, a gridiron or herringbone pattern is used for drainage line arrangement (Fig. 2.25). The drainage pattern should be designed so drain lines are placed nearly perpendicular to the slope and rotated downhill as required to drain. However, any pattern is acceptable as long as each line has a continuous downward slope. Water in golf greens should not have to travel more than 10 *ft* (3 *m*) to a drainage line. An additional lateral drain line is placed at the furthest downslope location of the green, adjacent to the perimeter of the green. This perimeter drain (referred to as a 'smile' drain) helps avoid wet areas where the modified greens sand meets native soil.

Drainage Line Trenches

Trenches in which golf green drainage lines are to be laid should be cut a minimum of 6 to 8 *in* (15 to 20 *cm*) in depth into the subgrade and 5 to 6 *in* (12.7 to 15 *cm*) in width (Fig. 2.26). In native soil, 3 to 4 *ft* (0.9 to 1.2 *m*) deep drain lines are sufficient. Lines less than 2 *ft* (0.6 *m*) deep become subject to damage or disruption by heavy machinery or excessive traffic.

The bottom of the trench should be a minimum of 2 *in* (5 *cm*) wider than the outside diameter of the pipe. Trenches up to 12 *in* (30.5 *cm*) wide have been utilized. However, more gravel is needed to fill the wider trenches, which increases cost. Normally, a drainage line trench should be no more than twice the width of the

Fig. 2.26 Trenches in golf green subsoil with drainage tile and being backfilled with gravel

drain pipe. A 5 to 6 *in* (12.7 to 15 *cm*) wide "U"-shaped trench will allow for a 0.5 to 1 *in* (12.7 to 25.4 *mm*) bed of gravel to be placed around (below, above, and on either side of) a 4 *in* (10 *cm*) diameter drain line to reduce washing of subgrade soil into the drain line. The soil displaced by digging the trench should be removed or placed between drainage lines to provide a slight slope toward the trench and then compacted.

Prior to digging trenches, the area should be surveyed. Proposed trench lines should be staked and labeled with the desired depth of cut. Drain lines should not be placed any deeper than necessary to obtain the desired slope. Trenches should have a minimum downward slope of 0.5 % (1 *ft* of drop for every 200 *ft*, 0.3 *m* per 60 *m*) and a maximum slope of 4 % (1 *ft* of drop for every 25 *ft*, 0.3 *m* per 7.5 *m*). Slopes of 1 to 2 % (1 *ft* of drop for every 100 *ft*, 0.3 *m* per 30 *m*, to 1 *ft* of drop for every 50 *ft*, 0.3 *m* per 15 *m*, respectively) are ideal. Drain lines with slopes of less than 0.5 % are difficult to properly grade, install, and maintain due to the slight elevation changes and slow flow rates. Drain lines with slopes greater than 4 % will lose lateral drainage capability. Steeper slopes also require greater elevation changes within the drain line and a lower outlet point.

When establishing the subgrade of a drain system, it is best to start at the outlet and establish the grade of the main collector line. After establishing this main line grade, the grade of each lateral can be determined. Care must be taken to ensure the drainage trench and drain lines always slope downward to avoid any entrapment or collection of water along the drainage lines. If a section of pipe is lower than the section closer to the outlet, water will pond in the lower section. This causes any sediment in the water to settle and collect in the bottom of the pipe, eventually clogging (or slowing) drainage. Grades of all main and lateral drainage lines should be checked with a level prior to backfilling.

2.2 Putting Greens

Putting green rootzones are formulated to drain quickly and allow play to be resumed shortly after heavy rain or irrigation. However, installation of a well-designed drainage system is critical for water removal from the subgrade, especially if the native soil is a clay or has an impermeable layer. Without drainage, the green could remain excessively wet and unplayable for several days after heavy rain.

Subgrade

Final subgrade contours should closely reflect the contours of the surface. Consequently, successful green construction starts with a properly planned and constructed subgrade (Fig. 2.27). Internal drainage follows the contours of the

Fig. 2.27 Subgrade of golf greens should be within 1 in (2.5 cm) of the eventual surface grade to facilitate more even drainage and soil moisture. Otherwise, shallow soils tend to stay saturated while deeper soil remain excessively dry

subgrade. Under normal circumstances, subgrade contours should not be sloped exclusively toward the front of the green since this will cause the front edge to be extremely wet. A soggy turf exposed to concentrated foot traffic quickly becomes worn and thin. It is better to have the green's slopes draining away from high traffic areas and also from any side facing the cart path's entrance and exit.

Depending on the green design and elevation of the site, the subgrade will be built into the existing grade or cut into the subsoil. If the grade is to be cut into the subsoil, the stripped topsoil may be stockpiled for future construction, such as mounds adjacent to the green, or distributed over the fairway and rough. Usually, greens built into the existing grade are elevated, requiring outside fill material for the subgrade. Heavier soils, such as clays, are desirable for the subgrade since these are easily compacted to form a firm base that does not readily shift or settle. In either case, the subgrade must be compacted to prevent future settling that might create depressions or pockets of poor drainage or, in the event of a higher grade, droughty areas. This is accomplished with a power-driven vertical compactor (modified jack-hammer), a vibratory plate, or with a water-filled mechanical roller operated in several directions across the subgrade.

The subgrade for a USGA specification green should be constructed 16 *in* (41 *cm*) below the planned surface, and should look like the finished green, but at a lower elevation. Contours of the subgrade should match those of the surface to within a tolerance of 1 *in* (2.5 *cm*). The gravel layer must conform to the finished surface grade even if the subgrade does not. Initial shaping of subgrade contours involves placement of fixed grade stakes that are referenced to a permanent bench mark. The grading equipment operator then follows these pre-marked stakes to the depths indicated. Once the initial grade is established, it should be re-surveyed and then inspected by the architect to ensure the settled contour elevations match original specifications (Fig. 2.28).

A uniform subgrade, or uniform depth of green, is critical since soil and water physics that dictate the amount of water retained in a soil profile are inversely proportional to its depth. This means the deeper a soil profile, the less water the top surface will hold. Uneven soil profile depths will have areas that remain excessively dry (high spots) while others will remain soggy (low spots). This greatly increases costs later as the superintendent struggles to maintain uniform soil moisture, usually by using extensive hand watering.

The finished subgrade should be smooth, free of any pockets, rocks, or tire tracks, and firm enough to support construction equipment to prevent settling later. Any plants growing in the subgrade should be removed or killed before applying gravel or sand layers.

Gravel Size and Shape

In USGA specification profiles, the height of the perched water table is also determined by the contact area between the gravel and the sand above it. As the gravel size decreases, contact with the sand above increases and a shallow perched

Fig. 2.28 Checking the integrity of settled drain lines to ensure continuous fall to facilitate proper drainage

water table develops; more water is able to flow downward across these contacts. In addition, if the gravel particle shape becomes flatter and narrower, it is able to pack closer together, lie more horizontally, and thus create a larger surface area in contact with the sand. Gravel more round in shape will have only a small point of contact with the sand and less water will flow downward across these contacts, creating a higher (or deeper) perched water table. The USGA has developed specific guidelines for matching gravel size to rootzone sand mix texture. These guidelines include factors for bridging, permeability and uniformity. Proper gravel sizing is discussed further in Chap. 3.

Drainage Systems

Drain Line Outlets

The first task in drainage installation is locating an adequate outlet area for the water. Typically, drain lines are routed into nearby ditches, ponds, retention areas, larger drain lines, existing French drains in the fairways, or nearby out-of-play grass

Fig. 2.29 Sump and pump to removed water from a drainage system

areas. Discharge lines are normally non-perforated pipe and should be laid across, rather than down, a steep slope to reduce the flow rate from the green. In some cases, a suitable discharge area may not be readily available and a sump and pump may be required. The sump may be formed with several concrete rings placed on top of each other and enclosed with a lid. A low-lift pump is installed inside the sump with float-activated switching so the water level may be controlled within specified limits (Fig. 2.29). Once a predetermined level of water is drained into the sump, the water is then pumped up to an appropriate discharge area. Sumps should be located away from the green and in areas receiving little traffic. Avoid directing the main drain line from the green into adjacent sandtraps, as washouts will be common. It is also a good idea to cover the main drain line outlet with a screen to prevent animals from entering the line.

Drain Spacing

Drain lines should be spaced 10 to 20 *ft* (3 to 6 *m*) apart. If the golf green is in an area with a high water table, it may be necessary to place larger drain lines deeper into the subgrade to lower the water table and handle the increased drainage. Specific drain line spacing can be calculated using Hooghoudt's equation as discussed earlier in this chapter, based on rainfall intensity, rootzone hydraulic conductivity, and rootzone depth.

Fig. 2.30 A herringbone drainage pattern is commonly used for golf green and sports fields

Drain Layout Design

Typically, drainage lines are installed diagonally to the grade in a gridiron or herringbone pattern (Fig. 2.30). However, any arrangement is acceptable as long as each line has a continuous downward slope and water does not have to travel more than $10\,ft$ ($3\,m$) to a drain line. Greens with slopes greater than $2\,\%$ or having surface water run-off from higher surroundings should have an interceptor drain line that rings the perimeter of the green, especially in the front or lowest areas.

Herringbone designs are generally the most popular, and are well-suited for irregularly shaped or relatively large turf areas due to the numerous lateral drain lines. However, herringbone systems are complicated to install and the pipes may be difficult to locate once installed. If slit drainage is needed later, cutting the slits at 90° angles to the lateral lines becomes difficult.

Drain Line Types

In the past, drain lines were fashioned of agricultural clay tile or concrete. Today, 2 to 4 *in* (5 to 10 *cm*) diameter corrugated, flexible, plastic pipe with slits is widely used because it is easy to install and inexpensive. The slits in the plastic pipe should always be placed face-down on the gravel bed to prevent clogging of drain lines with soil migrating downward from the rootzone. Nylon drain sleeves that wrap around the line are available. However, if silt and clay exist in the rootzone, these may plug the filters and ultimately restrict drainage. Another popular design is to

Fig. 2.31 Flat panel tile being used instead of the traditional round perforated tile. Flat panel tile is laid on the subgrade and not imbedded into it saving trenching and spoil disposal costs

place a fabric along the perimeter of the tile ditch, fill to grade with gravel, and place the edges of the fabric over the drainage ditch. Other pipe or tile designs are also available; however, little research exists on the total benefits of these.

An alternative design involves using flat drainage pipe instead of the traditional round pipe (Fig. 2.31). The flat pipe is laid directly on the subgrade base and is not cut into the subgrade as with round pipe. Pea gravel is then placed around the flat pipe. The flat pipe still must be on a downward grade to facilitate drainage. This technique is cheaper as drainage ditches are not needed and less gravel is required to surround the flat pipe. Limited research suggests this pipe design is beneficial; however, use of this system is a new technique, and this construction design has not been proven for all situations and environmental conditions.

Drain Line Installation

Drain lines are laid in trenches dug into the subgrade 6 to 8 *in* (15 to 20 *cm*) deep and 6 to 8 *in* (15 to 20 *cm*) wide. Wider trenches are sometimes used, but this means more gravel and higher costs are required to fill the trench. Normally, the trench width and depth should be no greater than twice the diameter of the drain line. Soil (or spoil) dug from the trenches should be removed or spread between the drain lines and then compacted to provide a slight crown. A 1 *in* (2.5 *cm*) bed of pea gravel should be placed in the bottom of the trenches before the drain line is laid. Once drain tile is installed, the trenches should be filled with gravel. Care should be taken not to contaminate the gravel with surrounding native soil or drainage may be sacrificed.

Fig. 2.32 Grade stakes to clearly mark proper depths of various components of layered systems

Slopes

Before excavation, drainage trenches should be surveyed and staked with the desired depth of cut clearly marked (Fig. 2.32). Drains should be placed only as deep as necessary to obtain the desired slope. Stakes should be marked to give drain lines a minimum downward slope of 0.5 % (or $1\,ft\ 200\,ft^{-1}$, $0.3\,m\ 60\,m^{-1}$), an ideal slope of 1 to 2 % (or $1\,ft\ 100\,ft^{-1}$ to $1\,ft\ 50\,ft^{-1}$, $0.3\,m$ 30 to $15\,m^{-1}$), and a maximum slope of 3 to 4 % (or $1\,ft\ 33\,ft^{-1}$ to $1\,ft\ 25\,ft^{-1}$, $0.3\,m$ 9.9 to $7.5\,m^{-1}$). Care must be taken to ensure the trench and drain line always slope downward so pockets of standing water do not develop. These lines should be placed diagonally to the slope of the green and not at right angles. All main and lateral lines should be double-checked with a level prior to backfilling to ensure the grade provides the desired drainage. Joints connecting drain lines should be covered with tape, asphalt paper, fiberglass composition, plastic spacers, or covers to prevent gravel and sand from entering the line.

It is recommended that the main drain line has its upper end extended to the soil surface and capped (Fig. 2.33). If this line becomes clogged with soil in the future, the cap can be removed and the line periodically flushed. This greatly extends the useful life of the drainage system and reduces the need to disturb the playing surface to clean the lines.

Fig. 2.33 Main tile line extended to the soil surface and capped to allow future clean-out if clogging is suspected

2.3 Fairways

Subsurface Drainage Design

Surface drainage, as discussed previously, is the first and quickest means of removing excess surface water. However, in areas that do not surface drain sufficiently, subsurface drainage is often used to lower the water table below the soil surface and avoid waterlogged conditions. Subsurface pipe drain lines can be installed as either singular or composite systems. A singular system consists of an array of individual drain lines, each emptying into an outlet. Composite systems consist of laterals connected to a common main line (Fig. 2.34).

Similar to golf greens, fairway subsurface drainage design can have a variety of patterns such as a gridiron, herringbone, or random. Gridiron and herringbone patterns are used to drain larger areas while random drains are used when small localized areas need drainage while areas in between are satisfactory drained. A gridiron system is often used to drain an area with a uniform slope in one direction while a herringbone system is generally best used to drain an area with a swale near the center.

With each design, the main drainage lines should generally follow natural valleys and be perpendicular to the contours. Lateral drain lines are generally laid across the slope with a gentle downward grade of 0.5 to 2.0 %. These drains

Fig. 2.34 Subsurface drainage being installed on a golf course fairway. Drainage ditches are back-filled with gravel prior to soil placement

intercept subsurface interflow that generally moves perpendicular to the contours. Lateral drains should maintain a sufficient uniform grade while keeping the laterals at a consistent depth from the soil surface. The laterals lines typically are from 2 to 2.5 *ft* (0.6–0.8 *m*) deep. Spacing varies from as little as 10 *ft* (3.3 *m*) on less permeable soils such as clays and silt loams to as much as 30 *ft* (9 *m*) on highly permeable sandy soils. Hooghoudt's equation, as discussed earlier in this chapter, can be used to determine the drain tile spacing or hydraulic conductivity needed for a particular drain spacing design. Modifications of Hooghoudt's equation are available for designing fairway subsurface drainage systems where a drainage coefficient is used to estimate water loss from a soil profile and is then multiplied by the area and converting it to the desired units. This provides the outflow volume of drainage which allows one to then choose the appropriately sized pipe to carry this flow using a drainage pipe capacity chart.

Interceptor Drains

Surface drainage from areas adjacent to golf course fairways, such as parking lots, hills, or adjacent fairways, often becomes problematic (Fig. 2.35). Water that

Fig. 2.35 Unrestricted runoff from adjacent wooded property onto a turf area

infiltrates into the soil can either continue to move downward to eventually recharge the groundwater or can move laterally through the soil down a hill, this is referred to as interflow. Interflow is the major source of water for stream and pond recharge during periods between rains and slows considerably near the bottom of a hill. Wetter soils near the hillside base often occur and result in a seep. Attempts to drain seeps by installing subsurface drainage typically fail since the source of the seeping water remains unchecked. Usually this water is easily collected by installing surface cutoff (or interceptor) drains to collect the water at the bottom (or "toe") of slopes, prior to entering the playing surface, or by diversion using surface terraces (or swales) (Fig. 2.36). Interceptor drains consists of a gravel- or coarse sand-filled trench cut along the contour and perpendicular to the overland flow.

Sloping water tables are found in slightly rolling, hilly, or mountainous areas. The free groundwater in these areas will flow in the direction of the slope, usually along an underlying impervious soil layer. Precipitation on the soil surface percolates downward until it encounters this impervious layer and then flows laterally over this layer. The most likely place for a water table (seep) to appear at the soil surface is near the intersection of a steep slope and a flatter slope (Fig. 2.37). This is a common problem on golf courses, such as when the surrounding land area meets an elevated green. Wet seep areas are also common on approaches where the fairway slopes downhill toward the green, which is slightly elevated. Here the approach may be wet from irrigation water being retained in the green base material, and a seep may be caused in the same approach area from a surfacing water table on the fairway side. Interceptor drains are placed in these situations where the free groundwater of the hill meets the flat area to intercept the water flowing on the slowly permeable subsoil layer before it appears on the soil surface.

Fig. 2.36 Using a slit drain to intercept unrestricted runoff from adjacent property. Swales and terraces also are often used to redirect this runoff

Determining placement of an interceptor drain can best be performed by digging test holes or miniature wells (called **piezometers**) when most of the surrounding area is dry enough to use, but the seep area is still wet. Piezometers are small-diameter pipes driven into the subsoil so no leakage occurs around the pipes and water entrance is only from the open bottom. This indicates hydrostatic pressure of groundwater at the specific point in the soil. The piezometers should extend, in a grid pattern, upslope from the seep area to a depth of 2 to 3 ft (0.6 to 0.9 m). By observing the water level in the piezometer holes 24 h after being dug, the depth to the water table or water flowing over the impervious layer in the ground can be determined. Once this occurs, a trench should be dug to approximately 2.5 ft (0.75 m) deep to extend below the water table. To facilitate drainage, the trench should be backfilled to the depth of the water table with gravel. If the water table intersects the soil surface, additional drains may be necessary. If not, additional interceptor drains may be needed further down the slope.

The bottom of drain trenches should be uniform in slope to prevent depressions and should have a minimum slope of 2 % (1:50) if a pipe is not placed at the bottom of the trench. Placement of a pipe in the trench allows grades down to 0.5 % (1:200). This allows quick removal of surface water, and helps prevent ponding, wheel depressions, and trash accumulation. Mowable drains or graded drains are ideal to minimize maintenance requirements and to facilitate play.

Fig. 2.37 The most likely place for a water table (seep) to appear at the soil surface is near the intersection of a steep slope and a flatter slope. This occurs due to surface and subsurface moisture accumulation at this junction

Springs

Springs are weak points in the soil strata where groundwater is under sufficient pressure to allow surfacing of the water. Springs are drained by placing a perforated drain pipe directly in the actual spring head to a depth of about 2.5 *ft* (0.76 *m*) and about 5 to 10 *ft* (1.5 to 3.0 *m*) beyond it and filling with gravel to facilitate water entry into the drain. In some instances it is possible to collect spring water for irrigation purposes.

Outlets

Water intercepted by surface and subsurface drainage requires a suitable outlet to discharge its flow, typically into channels, streams, or lakes. If the outlet is inadequate, the effectiveness of the entire drainage system can be reduced. Outlets types include the classic outlet or extension of the subsurface drainage pipe to the discharge location, pumped outlets, siphon outlets, dry wells and subsurface reservoirs, and wetlands (Fig. 2.38).

Fig. 2.38 Collection point for several main lateral drainage lines. The collected water is then removed or redistributed away from the property

With classic outlets, the location of the drainage pipe outlet must be at the low point of the drainage system. Efficient drainage system design requires identifying the outlet location for an area and then extending the drainage system array upslope from this location. An adequate slope must also occur along the entire run of the system, along with adequate soil cover as a protection from crushing, and excessively deep excavations should be avoided.

The drain outlet often is the weakest portion of a drainage system since it is exposed and subject to damage or clogging. To prevent this, extending the tile or plastic tubing directly to the discharge point should be avoided. A section of non-perforated plastic or metal pipe 10 to 15 *ft* (3 to 4.5 *m*) in length is used to carry the water from the point where sufficient soil cover is available to the discharge to avoid crushing that may occur if insufficient cover is present to protect the pipe. A concrete collar is placed at this intersection of pipes to prevent pipe displacement. The outlet pipe should be the same size or larger than the main discharge line at the collar and should discharge at least 1 *ft* (0.3 *m*) above the normal water level in the receiving waterway. If flooding periodically occurs, the outlet pipe should be equipped with a flood gate to prevent water backing up into the pipe. The outlet pipe should be covered with a wire mesh to prevent animals from entering it.

Sometimes a pump and siphon outlet is necessary if a gravity outlet is unavailable or the area to be drained is completely contained with a large depressed area. A pumped outlet consists of an automatically controlled pump with float switches set to start and stop levels, placed within a small sump to provide some degree of active water storage. A siphon outlet is when the entire drainage system is located in a depression and a sump contains a non-perforated, 2 *in* (5 *cm*) siphon tube leading to a remote discharge location. As long as the entrance and exit of the siphon pipe remains under-water, the tube can convey water across higher elevations than the location of the sump or relief point. These systems work best for relatively flat areas and should be connected to an irrigation line so it can be primed and occasionally flushed.

Dry wells are holes dug into the ground at the end of a drain line that are used to receive normal drainage water from relatively small areas. They are used when discharge locations are too far to trench and pipe. Dry wells are usually buried beneath the soil surface and covered with turf or other material to hide them. Stormwater wetlands are constructed systems designed to mitigate downstream impacts of stormwater quantity and quality by temporarily storing drainage waters in shallow pools and marshes. Drainage design specialists consider these and other options when planning stormwater and normal surface and subsurface drainage systems.

Sand Capping

Sand capping can be the most reasonable means of "drying" a fairway located in perpetually wet (low) area without installing expensive sump and pump systems (Fig. 2.39). However, on most courses, unless sound soil science is applied to the situation, unsatisfactory results may occur.

Sand-capping increases the depth of growing medium, thus increasing the depth from the soil surface to the water table, and reducing surface puddling and wet conditions. These benefits may or may not be realized for several reasons. As the depth of the soil profile increases, the gravitational pull on the water throughout the profile above increases, thereby decreasing the soil water content, and consequently increasing the storage capacity for rain water. However, water flow may be so slow in the original soil below the sand-cap that vertical drainage in the sand-cap zone may also be too slow, especially if the sand-cap is not of sufficient thickness. Refer to Chap. 3 for more information on using soil moisture retention curves to determine sand capping depths for particular soils and situations.

2.4 Sports Fields

Water drains or exits a field in four major ways: (1) evaporation; (2) surface runoff; (3) internal rootzone drainage; and eventually, (4) percolation or other movement out of the rootzone profile, preferably, through an underground drainage network.

Fig. 2.39 Sand capping a perpetually wet site to raise the turf above the naturally occurring high water table

Three types of soil profiles are currently used for sports fields in most areas which include one or more of these drainage means.

Soil Profiles

Native Soil Fields

These fields use existing soils and depend primarily on surface drainage to remove excess water. The advantages of native soil fields include: (1) they hold adequate nutrients and have a high water holding capacity, thus, require less fertilizer and water; (2) they provide good stability, shear strength and traction; and, (3) they are less expensive to construct as soil is on-site. Costs depend on how much surface grading is performed and if drain tile is installed (Fig. 2.40).

Disadvantages of native soil fields include: (1) most provide inadequate internal drainage, as these fields depend on a crown for surface drainage, thus, may compact easily; (2) due to the heavy nature of many native soils, internal drainage of the playing area is inefficient during heavy rainfall; (3) perimeter drain lines are needed to move surface run-off; and, (4) they are prone to surface rutting, puddling, and tracking unless aggressive maintenance is performed.

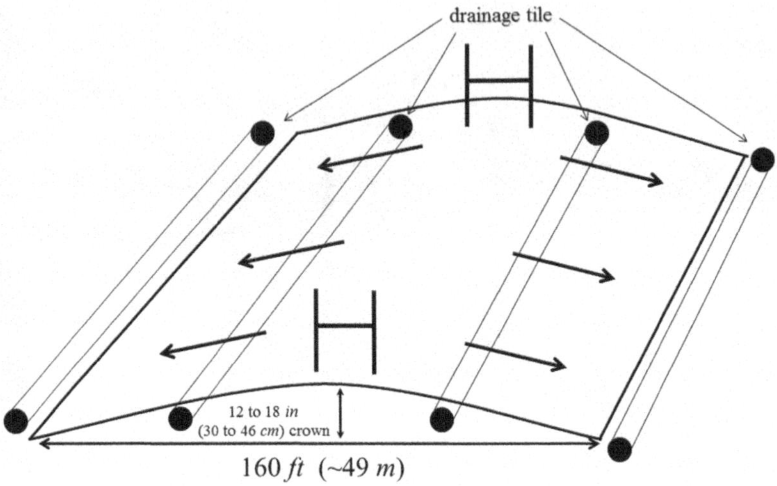

Fig. 2.40 Surface drainage with appropriately spaced drainage tile should be a standard design component of most sports fields, especially those that cannot afford rootzone modification

Modified Soil Fields

These are native soil-based fields modified by topical addition and roto-tilling in of sand. Performance depends on various proportions of sand and soil and the relative particle size distribution of each.

Advantages of modified soil fields include: (1) they are less expensive to build and maintain than sand fields; and, (2) they may have better drainage than native soil fields.

Disadvantages include: (1) their drainage still may be limited, and like native soil fields, they must still depend heavily on surface crowning; (2) they need irrigation and semi-aggressive fertilization; and, (3) their proper construction is difficult to achieve.

Often, with modified soil fields, lower budgeted fields have 1 to 4 *in* (2.5 to 10 *cm*) of sand placed on the existing soil surface and then roto-tilled in the top 4 to 6 *in* (10 to 15 *cm*). As Fig. 2.41 indicates, this procedure is often more deleterious than beneficial as the small particles of the existing soil will "clog" the pore spaces created by the much larger sand particles. For example, 10 % clay was added to a sand, which reduced its hydraulic conductivity by almost 85 % (from 58 to 9 $in\ h^{-1}$, 147 to 23 $cm\ h^{-1}$). Conductivity values quickly dropped as the clay soil content increased, for example, with a 50:50 blend, the hydraulic conductivity was less than 0.2 $in\ h^{-1}$ (5 $mm\ h^{-1}$), unacceptable by today's standards. Furthermore, adding 20 % sand to soil reduced drainage more than 50 % compared to straight (100 %) soil. Significant increases in drainage and aeration properties are not normally seen until sand volumes are greater than 80 %.

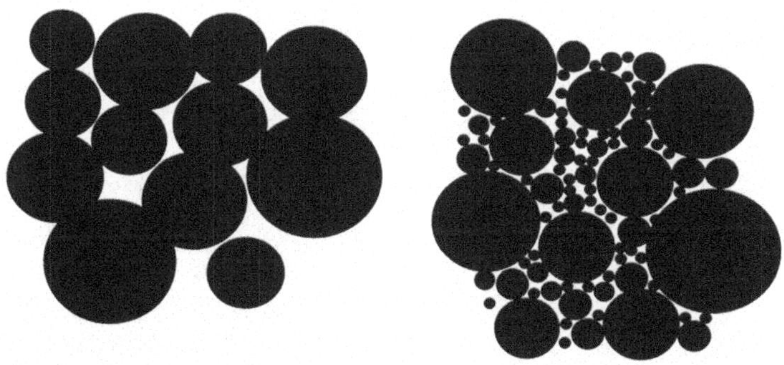

Fig. 2.41 Modified soil fields typically have a layer of sand applied on the surface and then roto-till it in. This rarely improves internal soil drainage as excessive fine soil particles (silt and clay) usually "clog" the pores between larger sand particles (*right*) and is not normally recommended

Sand-capped fields are a modified soil construction method where a 3 to 6 *in* (7.6 to 15 *cm*) sand layer is "capped" over a native soil and not roto-tilled into the subgrade. The advantages and disadvantages of these fields are similar to where the sand is mixed with the native soil. However, this construction technique can pose problems when subgrade has been compacted and does not allow for drainage. Water will rapidly move through the sand "cap" and not penetrate the compacted subsoil, creating a "bath tub" effect. The field then holds too much water, too close to the playing surface leading to wet conditions and ultimately thin turfgrass. Many of these problems can be minimized by deep tillage (4 to 8 *in*, 10 to 20 *cm*) of the subgrade prior to adding the sand cap, and not re-compacting the tilled area prior to "capping" the surface with sand. The addition of drain tile is still necessary with sand-capped fields for expedient water removal. Refer to Chap. 3 for more information on determining appropriate depths of sand capping for a particular site, soil, and sand source.

Sand-Based Fields

These rely on 80 to 100 % sand rootzones plus 0 to 20 % native soil or other amendment (Fig. 2.42). Sand-based fields are essentially flat, not heavily crowned, and have high infiltration rates. Internal drainage needs to be designed to move large amounts of water away quickly. Selecting the proper, uniform sand particle size is the key.

Advantages of sand-based fields include: (1) they provide the best internal drainage of the three designs; (2) minimum crown is needed, since internal drainage is high; and, (3) minimum soil compaction occurs as properly sized sand has a greater resistance to soil compaction compared with silty or clayey soils.

Disadvantages of sand-based fields include: (1) they require increased irrigation and fertility compared to native soils as sands have less cation exchange and water holding capacities; (2) they can be subject to layering problems as only a 1/8 *in*

Fig. 2.42 A sand-based sport field rootzone used when fields are built essentially flat. This is necessary to facilitate subsurface drainage since a surface crown is absent

(3.2 *mm*) thick layer of dissimilar soil can interfere with drainage; (3) they are usually more expensive to build as sand typically has to be trucked in; (4) expert management is needed, due to sand holding nutrients and moisture poorly; (5) increased organic matter buildup due to excessive nitrogen needed to provide satisfactory playing conditions and/or less soil organisms present in inert sand fields which normally decompose organic matter for a food source, and, (6) decreased surface stability often occurs early in the life of the field, typically this is less problematic in the second year. Stabilization products may be incorporated to reduce shearing and tearing and allow for better grass growth, and recuperation, i.e., mats, carpets, fabrics, fragments of interlocking mesh, grids, fibers, and fibers sown into the rootzone.

Within the sand rootzone profile, two main drainage systems are currently used. The most proven is one with a 12 *in* (30 *cm*) layer of rootzone mix overlying a 4 *in* (10 *cm*) layer of "pea" gravel with 4 *in* (10 *cm*) drain tiles embedded in the subsoil (Fig. 2.43). This provides optimum drainage when heavy rain necessitates prompt water removal and allows the "flattest" surface in terms of minimum crown. The gravel layer, however, helps retain enough soil moisture in the rootzone to prevent constantly dry soil often experienced with pure sand rootzones and no gravel layer.

The second popular profile deletes the 4 *in* (10 *cm*) gravel layer leaving 12 *in* (30 *cm*) of pure sand rootzone along with the embedded drain lines. Pure sand is not as effective at removing soil water as a sand/gravel rootzone, since the moisture has to traverse the soil profile laterally to a drain line before it is removed. Research indicates for sand-based fields to equal the time necessary to drain compared to fields with a 4 *in* (10 *cm*) gravel layer, an increase in percolation rate of 20 $in\ h^{-1}$

Fig. 2.43 A sand-based sports field with drainage lines imbedded into the subgrade to facilitate subsurface water removal

($51\ cm\ h^{-1}$) is necessary in the rootzone sand. This is because in a gravel layer field, water is essentially drained vertically when it encounters this layer, typically 12 *in* (30 *cm*) deep. However, when the gravel layer is absent, water must move down and across the soil profile and encounter a drain line before it is removed. To overcome the drainage issues in a field without the gravel layer, designs may include a deeper soil profile (i.e., 14 to 16 *in* deep, 36 to 41 *cm*) or closer tile spacing (i.e., 10 *ft*, 3 *m*). Sand-based fields often stay drier than the first two field types, but require more irrigation and fertilization.

Sports field managers typically have only one opportunity to build or renovate a facility. Careful attention to expected use and quality weighted against maintenance budgets should be considered during this process.

Football Fields

Minimum Drainage Requirements

For many high school and local municipal fields, adequate surface contouring is the most effective and economical means of providing surface drainage. If insufficient

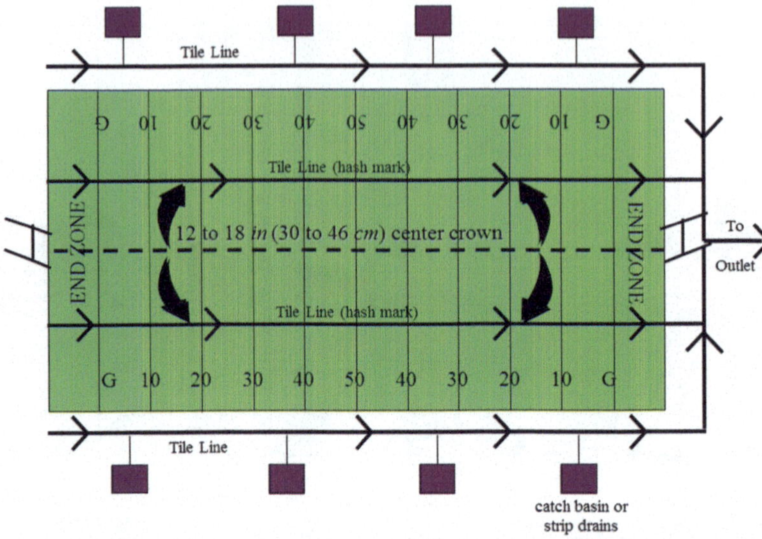

Fig. 2.44 Optimum surface drainage of lower-budgeted sports fields consisting of a 12 to 18 *in* (30–46 *cm*) crown with a pair of drainage tile imbedded along the hash marks of football fields and a pair parallel to the sidelines. Surface catch basins or strip drains help remove surface from the playing surface, sidelines, and water draining from spectator stands

sloping of the surface occurs, water will stand (puddle), saturating the soil causing compaction and damage by traffic. To provide surface drainage, high school or similar football fields should have a 12 *in* (30 *cm*) crown for sandy soils and 18 *in* (46 *cm*) for clay soils from center to the sideline, or a 1 to 2 % slope (Fig. 2.44). The slope at the sideline may be reduced, but the area should not be flat. Surface water movement away from the high-traffic sideline areas where players stand is important.

A minimum of four drainage lines should be installed, one running parallel to the center crown, typically down each hash mark, and the other set just off the field along each sideline. Drain lines are usually 6 *in* (15 *cm*) wide and 12 to 36 *in* (30 to 91 *cm*) deep. Two *inch* (5 *cm*) of gravel is placed at the bottom of the lines, a 4 *in* (10 *cm*) perforated drain pipe laid on top of this gravel layer and "pea" gravel (industry designation '789', ¼ to ½ *in*, 6.4 to 12.7 *mm* diameter) or very coarse sand is used to fill the trench to grade. Sometimes the drainage trench is lined with a geotextile fabric to prevent clogging of the drainage system.

The drainage tile should not be laid within 4 *in* (10 *cm*) of the surface, to prevent future aerification practices from disrupting the integrity of the drainage system. The pipe should be laid on a continuous ½ to 1 % downward slope (3 to 6 *in* drop in 50 *ft*, 7.6 to 15 *cm* in 15 *m*) and should be connected at the ends to allow for water to drain away from the field. Surface catch basins or surface strip drains should also be installed between the playing field and both sideline stands (Fig. 2.45a). These intercept surface drainage from the field as well as water draining from the spectator stands. At least 3 (preferably 4 or more) catch basins

Fig. 2.45 Traditionally used surface catch basis to capture excessive surface water (*left*); strip surface drains to capture excessive surface water (*right*)

or strip drains should be considered for each side of the field. For safety purposes, catch basins should be located no closer than 10 *ft* (preferably 15 *ft*, 3 to 4.6 *m*) from the playing surface. Strip surface drains are becoming more popular as a replacement for catch basins (Fig. 2.45). Strips are less noticeable and less likely to cause injury, plus their length offers more intercepting surface area to facilitate drainage.

To provide a higher level of drainage, an additional crown can be installed starting at about each 20 *yd* (18 *m*) line and sloped at 1 to 2 % toward the end zones (Fig. 2.46). Variations of this exist; one is commonly referred to as the "turtle-back" design while the other is a "hip-roof" drainage design (Fig. 2.47). These provide additional surface drainage without significantly altering the field's playing characteristics. Also, these designs allow for a flatter field, yet provide some surface drainage. Multiple field designs rely mainly on surface drainage with appropriately placed drain outlets (Fig. 2.48a, b, c).

Optimum Drainage

High profile fields used for college and professional sporting events require optimum drainage so play can commence on schedule. This involves replacing the existing rootzone soil with an appropriate sand blended with an organic source and/or loamy soil as mentioned previously. A series of parallel drainage tile lines spaced 10 to 20 *ft* (3 to 6 *m*) apart running the length of the field should be used (Fig. 2.49). The shallower the rootzone, the closer the drain lines should be. The amended rootzone should have an initial infiltration rate between 6 and 16 $in\ h^{-1}$ (15 to 41 $cm\ h^{-1}$). If an amended sand profile is used, then the center field crown can be reduced to approximately 6 to 10 *in* (1 to 25 *cm*). This field profile requires increased maintenance inputs such as fertilizer and water, but will provide optimum drainage and playing conditions. If maintained properly, the field should have a **minimum** life expectancy of 20 years. This design is strongly recommended for those who demand the highest quality fields and best assurance against poor drainage.

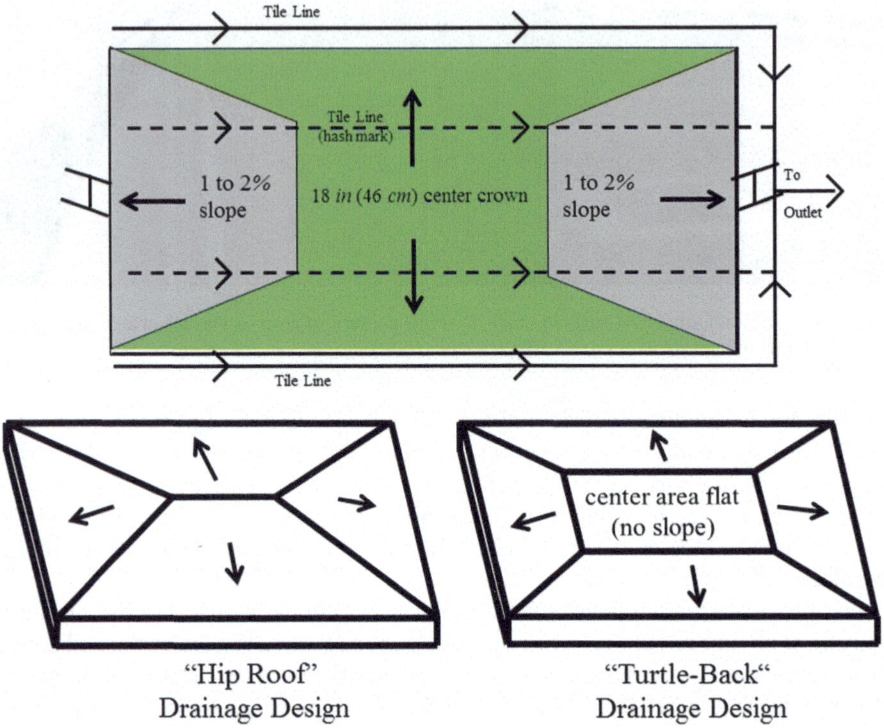

Fig. 2.46 Two popular surface slope designs for sports fields where surface water drains from 'hip roof' or 'turtle-back' slopes

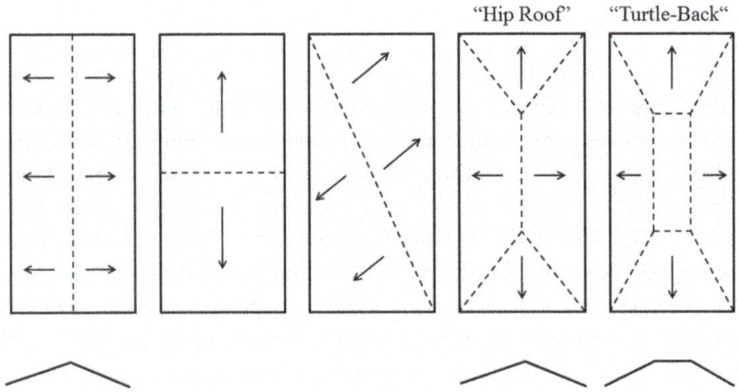

Fig. 2.47 Additional possible surface slope designs for sports fields

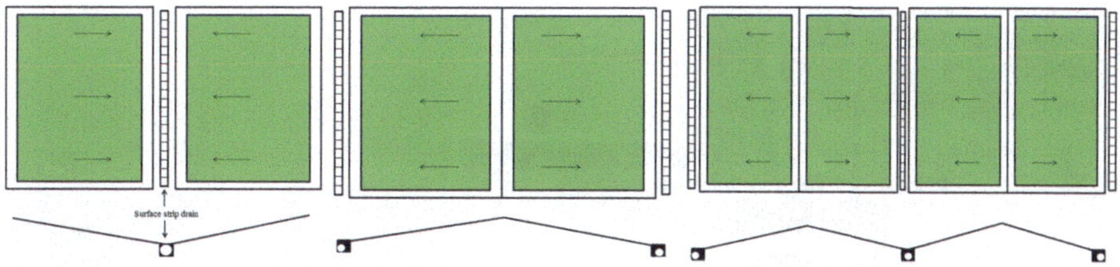

Fig. 2.48 Popular surface slopes and drainage locations for multiple fields. Two fields sloped towards each other with drainage between them (*left*); Two fields sloped away from each other with drainage on their outer perimeter (*center*); Four fields with the inner two sloped towards each other with drainage between them and the outer two sloped away with perimeter drainage (*right*)

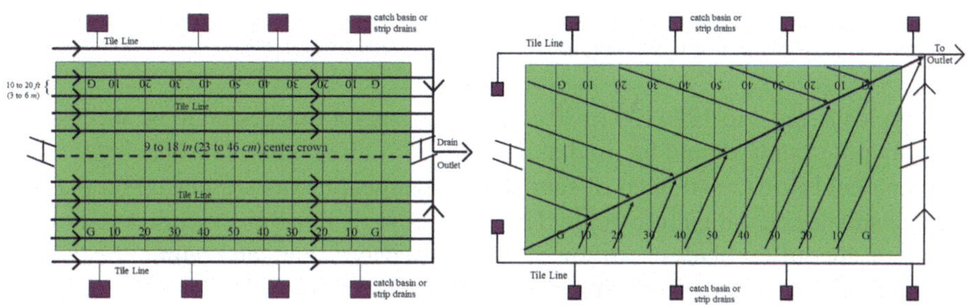

Fig. 2.49 Extensive subsurface drainage line use for fields with relatively flat surfaces and/or greater (quicker) drainage is needed so sporting events can be completed in a timely manner. Parallel drainage line design (*left*); herringbone design (*right*)

Additional Drainage Designs

Numerous alternative sports field designs are available, varying in sophistication and costs. For example, suction pumps can be connected to the drainage outlet points to enhance water removal. Although successes have occurred with these systems, most fields utilizing suction pumps rarely have over a 5-year life expectancy. Other designs regulate drainage by raising or lowering the field's water table. These designs are expensive to build, complicated to operate, and have had agronomic issues with shallow turf rooting and surface algae invasion.

Some fields have used 10 *in* (25 *cm*) of sand instead of 12 *in* (30 *cm*) for the rootzone mix depth. This saves about 17 % of cost for rootzone material. If routine topdressing is performed, the field will likely gain 2 *in* (5 *cm*) of depth over the first 5 years or so. If this design is chosen, it is advisable to use a faster drainage rate and place the subsurface drains closer together, i.e., 10 *ft* to no more than 15 *ft* apart (3–4.6 *m*).

Similarly, 6 *in* of sand rootzone have been used instead of 12 or 10 *in* (30 or 25 *cm*). This shallower depth restricts the used of "deep-tine" aerification as the

Fig. 2.50 Installing slit drains (also referred to as "cell system") where a narrow trench is backfilled with sand to remove excess surface moisture (*left*); A 1 to 2 *in* diameter (2.5 to 5 *cm*) pipe placed at the bottom of the trench to facilitate water removal (*right*)

tines typically are 8 to 12 *in* (20 to 30 *cm*) long. Also, for this design to be successful, high draining sands should be used along with 8 *ft* (2.6 *m*) drain line spacing.

Some fields are constructed of sand and then amendments are placed on the soil surface and mixed with a roto-tiller. This procedure then has an amended soil to the depth of the tiller blades. Although less expensive than amending the whole rootzone, differential drainage and turf quality often results.

The "cell" or "grid" design incorporates a very sophisticated series of small drainage lines, crisscrossing to forms "cells" or "grids" (Fig. 2.50) This system does not mix sand into the existing soil. Instead, the drainage grid consists of a cross matrix of 3 *in* (7.6 *cm*) wide trenches. Drains are spaced 5 to 10 *ft* (1.5 to 3 *m*) apart and are filled to the surface with sand. Sometimes small drains are placed at the bottom of these cells to facilitate water removal. Although successful if designed and built correctly, the cell system is expensive and often has a short life expectancy, due to narrow trenches which easily clog or collapse, and the high level of knowledge and experience required for the increased technology and maintenance. This experience and knowledge is often lost as field managers change jobs or as team management and coaching personnel change.

Fields requiring frequent resodding also introduce various types of soils which generally reduce the effectiveness of these and other systems. This is further amplified as new field managers use a different topdressing materials than the soil used to construct the field. These real-life situations can pose significant problems and should be considered closely during the design planning phase.

If an alternative design is used, then one should have limited expectation of field performance. These are not rapidly draining fields but should absorb small rain showers and provide better growing conditions than no modification. However, they should not be expected to rapidly drain during heavy rainfall and typically require additional aerification and have shorter life expectancies than a sand-based rootzone facility.

Fig. 2.51 Soccer fields are constructed with little to no surface slopes which may influence sideline shots. In such instances, subsurface drainage becomes more important means to removing excessive soil moisture

Soccer Fields

Soccer fields tend to be flatter to interfere less with crosses (side shots) (Fig. 2.51). A 6 to 12 *in* (15 to 30 *cm*) crown should be planned, with the higher crown height for native clay soils (Fig. 2.52). Due to the greater width of a soccer field compared to a football field, these crown heights result in a slightly lower surface slope. For those desiring soccer fields with a flatter crown, subsurface drainage, including use of a sand-based rootzone and drain lines similar to optimum draining football fields should be considered.

Baseball and Softball Fields

Most of the water that falls on the skinned area on a baseball or softball infield should be removed by surface runoff (Fig. 2.53). To facilitate this, the skinned area should have at least a 1 % fall from front to back (Fig. 2.54). Baseball fields have the pitcher's mound as the high point (10 *in* or 25 *cm* above home plate), and slope towards the sidelines and outfields. Infields should have a 1 % slope or an 8 *in* (20 *cm*) fall from the bottom of the pitcher's mound to beyond the baseline. The outfields should slope 1 to 2 % from the infield skinned area toward the warning track. Minimally, drain lines should be placed just off (i.e., 5 to 10 *ft*, 1.5 to 3 *m*) the playing surface around the perimeter of the entire infield. Drain lines installed under the infield skinned area are usually ineffective as the high clay content prevents expedient drainage. Additional drain lines should be considered along the outside of

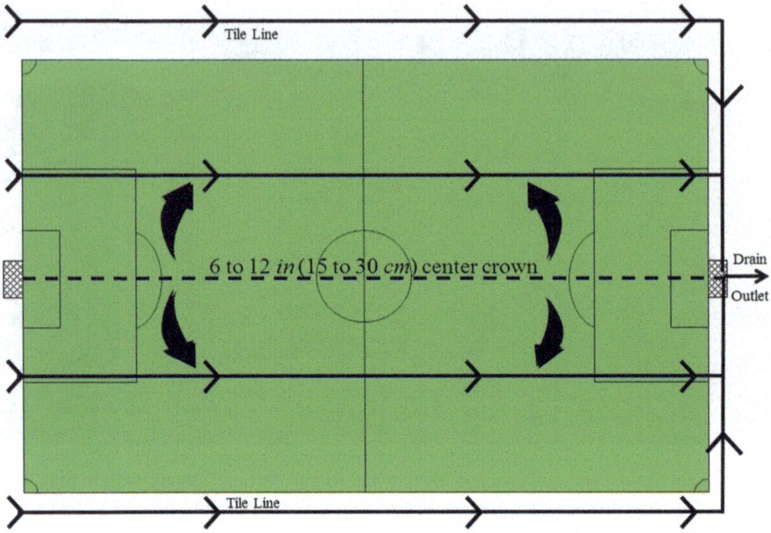

Fig. 2.52 Soccer field design utilizing a slight surface slope (crown) of 6 to 12 *in* (15 to 30 *cm*) and four parallel strategically imbedded drainage lines. More extravagant drainage systems are needed for professional fields where little to no surface slope is allowed. These often have more complicated and extensive drainage patterns and along with the use of a sand-based rootzone

Fig. 2.53 Surface drainage is necessary for highly compacted, clay content surfaces such as softball and baseball infields

the foul lines (or "hip" area) and on the inside of the warning track (Fig. 2.55). A 1 to 2 % slope should be utilized in the outfield to allow drainage toward the drain lines and a series of culverts (catch basins) should also be placed in the outfield for surface drainage and as outlet points for mechanically absorbed water. Higher

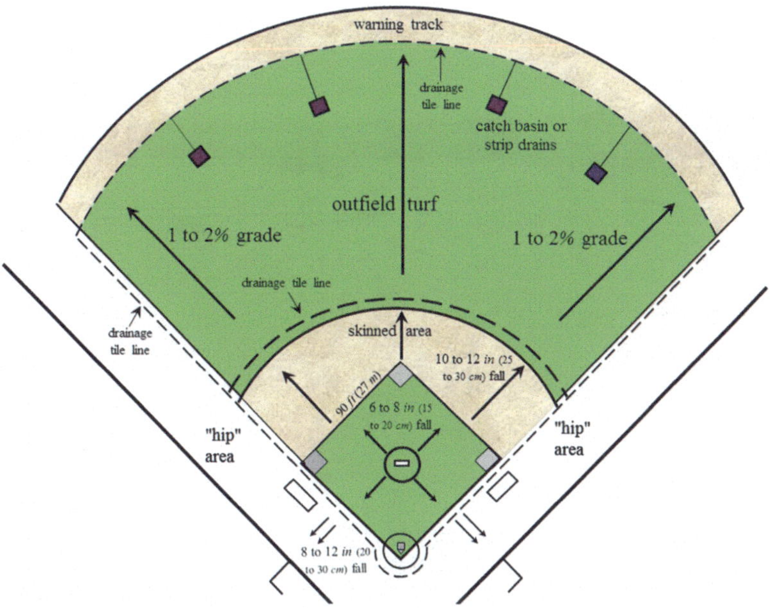

Fig. 2.54 Baseball drainage design incorporating surface grade and moisture capturing drainage tile and catch basins or strip drains

Fig. 2.55 Installing a perimeter drainage line adjacent to the warning track of a baseball field

Fig. 2.56 More extravagant drainage systems along with a sand-based rootzone commonly used on collegiate and professional baseball fields

profile fields are often constructed similarly to golf greens in terms of completely modifying the rootzone and installing sophisticated drainage systems (Fig. 2.56).

For smaller fields such as little league or softball fields, an alternative drainage design is often used where a center crown is utilized, slicing the field in half from the catcher's back drop through home plate, pitchers' mound, second base, and into centerfield (Fig. 2.57). A 1 to 2 % slope is installed away from this center crown towards the first and third base lines. A drain line should be installed on the perimeter of the playing surface just outside the field's foul lines. Outfield surface catch basis can also be installed to help remove surface moisture. This design works well for smaller fields since the water does not have to drain excessive distances and this design is initially easier and cheaper to install.

Baseball Infield Rootzones

Skinned baseball infield soils are modified to provide drainage and playability. Different percentages and combinations of soil, sand, and clay are used (Fig. 2.58). Silt and clay plus water are the binding agents that hold soil together. Most infields consist of 50 to 75 % sand with the remaining 25 to 50 % equally split between a local soil source and calcined clay. A combination of 60 % sand, 20 % silt and 20 % clay (i.e., a sandy clay loam to sandy loam) is often used. The silt and clay give the

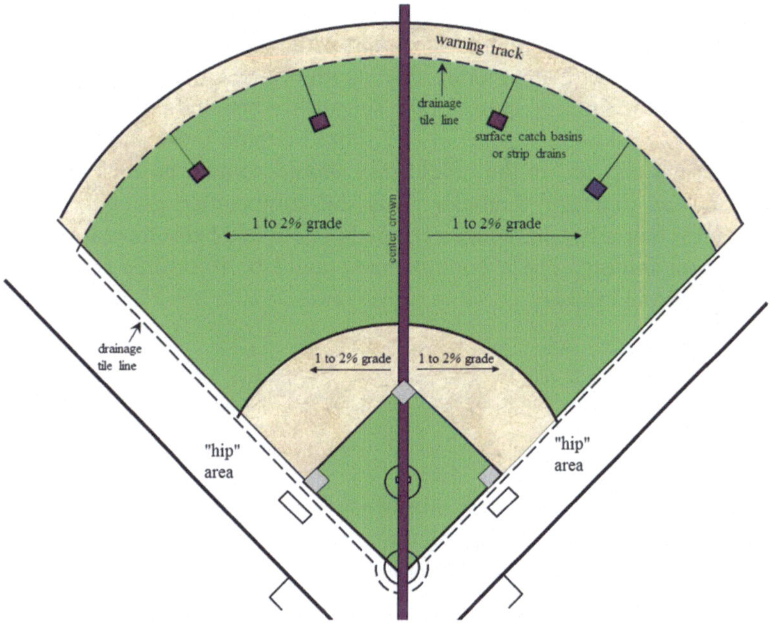

Fig. 2.57 Simpler and cheaper drainage for smaller softball or baseball fields where a center crown extends from home plate, pitcher's mound, through the outfield. Perimeter drainage tile and surface catch basins or strip drains are also used to help facilitate surface drainage

Fig. 2.58 Typical soil of a baseball skinned infield composed of native and calcined clay plus sand

mix firmness. Soils with higher sand content often become too loose, will not pack tightly, causing low spots in high traffic areas, while soils with excessive silt or clay become hard, compacted, and muddy. The depth of this infield mix is 3 to 6 *in* (7.6 to 15 *cm*) with a liner placed between the subsurface bed and infield rootzone mix. The top ¼ to ½ *in* (0.64 to 1.3 *cm*) of soil should remain loose and hold moisture. Ideally, the sand and soil in the infield mix should contain no rocks and pebbles greater than ¼ *in* (0.64 *cm*) diameter. This soil composition gives the infield the consistency for ball roll with reduced erratic bounces and helps increase field safety with consistent footing. These components should be roto-tilled into the infield rootzone to prevent crusting.

2.5 Questions

1. Drainage involves surface and subsurface water removal. Discuss the use of each, where they are most appropriate and advantages and disadvantages of using them separately or in combination.

   In surface drainage, land surfaces are reshaped, sloped, and smoothed as needed to eliminate ponding and to induce gravitational flow overland to an outlet. Diverting and excluding water from an area often involves diversion ditches, swales, and floodways.

   With subsurface drainage, soils may be modified to induce surface water infiltration and percolation through the rootzone to buried drains that collect and transport excess soil water to an outlet. The drop in pressure (or water potential) due to outlet discharge induces excess soil water flow into the drains. Subsurface drainage may also involve interceptor drains oriented perpendicular to the direction of groundwater flow.

2. When depending on surface drainage, the following equation can be used to calculate the velocity of water across a bare surface as influenced by the surface slope and depth of ponded water or rainfall amount.

$$\mathrm{V} = 0.35 \times D^{0.67} \times \mathrm{S}^{0.5}$$

   where:
   V = velocity ($in\ s^{-1}$)
   D = water depth (in)
   S = slope (decimal)

   Calculate the amount of water moving across a 1.5 % slope with a 1 *in* (2.5 *cm*) rainfall event.

$$V = 0.35 \times (1)^{0.67} \times (0.015)^{0.5}$$
$$= 0.043 in\, s^{-1} (0.11 cm\, s^{-1}) \textit{ of water movement over a bare surface}$$

3. Explain why placing a layer of sand over a native soil and then roto-tilled in usually is unsuccessful in achieving better internal soil drainage.

   The smaller native soil particles typically "clog" the pores between the larger sand particles.

4. Determine the necessary depths at 15 % non-capillary porosity of the following sand with a bulk density of 1.52 $g\ cm^{-3}$, K_{sat} of 0.56 $m\ h^{-1}$, and anticipated rainfall extreme of 0.75 $in\ h^{-1}$ with and without drain tile installation.

Tension *cm*	Total Porosity $cm^3\ cm^{-3}$	θ_v $cm^3\ cm^{-3}$
0	0.427	0.427
10	0.427	0.425
20	0.427	0.410
30	0.427	0.333
40	0.427	0.195
50	0.427	0.185
60	0.427	0.160

$$air\text{-}filled\ porosity\,(cm^3cm^{-3}) = total\ porosity\,(cm^3cm^{-3}) - \theta_v(cm^3cm^{-3})$$
$$water\text{-}filled\ porosity\,(\%) = volumetric\ water\ content\,(\theta_v) \div total\,soil\ porosity\,cm^3cm^{-3}$$
$$air\text{-}filled\ porosity\,(\%) = 100 - water\text{-}filled\ porosity\,(\%)$$

Tension, *cm*	Total porosity, $cm^3\ cm^{-3}$	θ_v, $cm^3\ cm^{-3}$	Air-filled porosity, $cm^3\ cm^{-3}$	Water-filled porosity, %	Air-filled porosity, %
0	0.427	0.427	*0.000*	*100*	*0*
10	0.427	0.425	*0.002*	*99*	*1*
20	0.427	0.410	*0.017*	*96*	*4*
30	0.427	0.333	*0.094*	*78*	*22*
40	0.427	0.195	*0.232*	*46*	*54*
50	0.427	0.185	*0.242*	*43*	*57*
60	0.427	0.160	*0.267*	*37*	*63*

Soil depth at 15 % aeration (capillary) porosity is between 30 and 40 cm (say 35 cm or ~14 in).

Use this depth to calculate drain line spacing using Hooghoudt's Equation: $K_{sat} = 0.56\ m\ h^{-1}$, *~ 22 in* h^{-1}

$$S = \sqrt{\frac{4Kh^2}{v}} = \sqrt{\frac{4\left(22\,in\,h^{-1}\right)\left(14\,in\right)^2}{0.75\,in\,h^{-1}}} = 151.6\,in\,(12.6\,ft\,or\,3.9\,m)$$

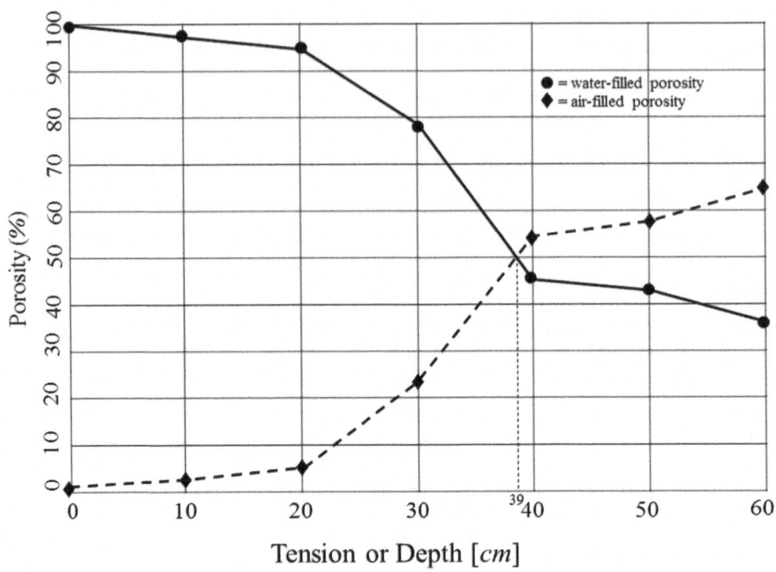

Fig. 2.59 From question 4, an approximate depth of rootzone needed in the absence of drainage line can be estimated by graphing water-filled and air-filled porosity and measuring the depth (or tension) where these two lines meet (refer to Chap. 3). In this example, 39 *cm* (15 *in*) is the depth of rootzone needed to provide sufficient air-filled porosity in the absence of drain lines

*If drain lines are not being installed, then the soil depth at intersection of the capillary and aeration porosity (Chap. 3) is used. Water-filled porosity (%) and air-filled porosity (%) is graphed at the various tensions (*Fig. 2.59*). In this example, this intersection is approximately 39 cm (or ~15 in).*

5. A football field is 100 *yd* long (91 *m*) and 53.3 *yd* (49 *m*) wide, with a 10 *in* (25 *cm*) deep rootzone and a hydraulic conductivity of 7 $in\ h^{-1}$ (18 $cm\ h^{-1}$). Drain lines run along each sideline. For a 1.5 *in* rainfall, what would the drain line discharge rates be (water has a volume of 0.00434 $gal\ in^{-3}$, 1 $ml\ cm^{-3}$)?

$$Q = \frac{2Kh^2w}{S} = \frac{2 \times (7\ in\ h^{-1}) \times (10\ in)^2 \times (3{,}600\ in)}{(1920\ in)}$$
$$= 2{,}625\ in^3h^{-1} \times 0.00434\ gal\ in^{-3}$$
$$= 11.4\ gal\ h^{-1}$$

Therefore, drain lines should be selected that can remove at least 12 gal h^{-1}*(45 L* h^{-1}*).*

6. a. For a 1.5 $in\ h^{-1}$ (3.8 $cm\ h^{-1}$) rainfall, determine the effective length of 4 *in* (100 *mm*) diameter corrugated drain pipe with smooth interior with 1 % slope and drain line spacing of 25 *ft* (7.6 *m*). The manufacturer's given discharge

rate for 4 *in* diameter corrugated drain pipe with a smooth interior on a 1 % slope is 0.17 $ft^3\,s^{-1}$ ($1\,ft^3 = 7.5\,gal$)

Each foot (0.3 m) of trench should collect:

$$Q = 25\,ft \times 1\,ft\,(trench) \times \frac{1.5\,in}{h} \times \frac{1\,ft}{12\,in} = \frac{3.1\,ft^3}{h}$$

Determine the maximum effective length the 4 in pipe.

$$\frac{0.17\,ft^3}{s} \times \frac{h\ linear\ ft^1}{3.1\,ft^3} \times \frac{60\,min}{hr} \times \frac{60\,s}{min} = 197\ linear\ ft$$

b. If a 6 *in* (15 *cm*) diameter drain pipe is used instead of the 4 *in* (10 *cm*) diameter pipe, determine the maximum length it can have (per manufacturer's specifications, a 6 *in* (15 *cm*) diameter pipe on 1 % slope has a maximum discharge rate of 0.49 $ft^3\,s^{-1}$ or 0.014 $m^3\,s^{-1}$).

$$\frac{0.49\,ft^3}{s} \times \frac{h\ linear\ ft^1}{3.1\,ft^3} \times \frac{60\,min}{h} \times \frac{60\,s}{min} = 569\ linear\ ft\ \ or\ 173\,m$$

7. A stadium manager wishes to modify a field's soil mix to be more predominately sand. In order to save money, the manager still wants to use some portion of the native soil present in this blending process. The desired K_{sat} value is 6 $in\ h^{-1}$ (15 $cm\ h^{-1}$). Determine the amount (weight and volume) of sand that needs to be added to the soil to achieve the desired K_{sat} rate. Refer to Table 2.1 for hydraulic conductivity of a USGA medium sand combined with a Cecil clay soil at various combinations.

   Calculated values of various v/v ratios of sand to soil from known particle-size distribution and bulk density values.

Soil type	Percent particle-size distribution (*mm*) 2–1	1–0.5	0.5–0.25	0.25–0.125	0.125–0.05	0.05–0.002	<0.002	Bulk density (*g cm*$^{-3}$)
Known values								
Sand	2	23	45	25	5	0	0	1.59
Soil	5	15	18	22	15	15	10	1.40
Calculated values of various sand:soil ratios								
1:1	*3.5*	*19*	*31.5*	*23.5*	*10*	*7.5*	*5.0*	*1.495*
2:1	*3.0*	*20.3*	*36.0*	*24.0*	*8.3*	*5.0*	*3.3*	*1.53*
3:1	*2.8*	*21.0*	*38.3*	*24.3*	*7.5*	*3.8*	*2.5*	*1.54*
7:1	*2.4*	*22.0*	*41.6*	*24.6*	*6.3*	*1.8*	*1.3*	*1.57*
8:1	*2.3*	*22.1*	*42.0*	*24.7*	*6.1*	*1.7*	*1.1*	*1.57*
9:1	*2.3*	*22.2*	*42.3*	*24.7*	*4.8*	*1.5*	*1.0*	*1.57*

a. **Weight**. The following equation provides a guideline for using a suitable sand with a soil of known mechanical composition to create a rootzone with the desired drainage rate:

$$|A| = \frac{[R-B]}{[C-R]} \times 100$$

where:

A = weight of sand to add to 100 weight units of the original soil.

B = percent of original soil in the desired particle-size range (i.e., 0.125 to 0.5 *mm*).

C = percent of desired particle-size range (i.e., 0.125 to 0.5 *mm*) in the sand used as an amendment.

R = percent of desired particle-size range (i.e., 0.125 to 0.5 *mm*) sand in the final mix.

If a 6.0 in h^{-1}*(15 cm* h^{-1}*) percolation rate is desired for this sand:soil rootzone, the R value would be 85 % (in the desired particle-size range of 0.125* to *0.5 mm) as extrapolated from* Table 2.2.

$$|A| = \frac{(85 - 40)}{(70 - 85)} \times 100 = 300$$

Therefore, 300 tons of sand per 100 tons of soil would be required to raise the percentage of soil particles between 0.125 and 0.5 mm to 85 % in the final mix.

b. **Volume**. If mixed on a volume basis (such as with off-site blending) instead of a weight basis, one must find the volumetric ratio of sand to soil using the equation: volume = mass/density. The bulk density of sand in this example is 1.59 $g\ cm^{-3}$ and soil is 1.40 $g\ cm^{-3}$, giving:

$$\text{Volume (ratio)} = \frac{V_{sand}}{V_{soil}} = \frac{M_{sand}/\rho_{b\ sand}}{M_{soil}/\rho_{b\ soil}} = \frac{300/1.59}{100/1.40} = 2.64$$

Therefore, 2.64 unit volumes of this particular sand are needed per one unit volume of this soil to achieve the desired ratio of 300 tons of sand per 100 tons of soil corrected for their respective bulk density values. This indicates the enormous amount of sand necessary to add to a soil to increase the final mix's K_{sat}*value.*

8. Calculate the new percent particle size in the 0.5 to 0.25 *mm* range from the sand/soil ratio listed in the previous example in an 8.5:1 ratio.

$$\text{New percent particle size} = \frac{[\text{sand fraction}\,\% \times \text{ratio sand}] + [\text{soil fraction}\,\% \times \text{ratio soil}]}{\text{total sand} + \text{soil ratio}}$$

$$= \frac{(45 \times 8.5) + (18 \times 1)}{8.5 + 1}$$

$$= 42\%$$

Therefore, approximately 42 % of the sand to soil mix at a 8.5:1 ratio would be in the 0.5–0.25 mm diameter range.

9. Determine the new particle-size distribution percentages and bulk densities obtained by tilling 300 *tons* of sand (bulk density of 1.65 $g\ cm^{-3}$) into the top 6 *in* of soil (bulk density of 1.40 $g\ cm^{-3}$) 2 *ac* in area.

   First, determine the depth of 300 tons of sand over the 2 ac:

$$\frac{300\ ton}{2\ ac} \times \frac{2000\ lb}{ton} \times \frac{454\ g}{lb} \times \frac{cm^3}{1.65\ g} \times \frac{ac}{43{,}560\ ft^2} \times \frac{ft^2}{929\ cm^2} \times \frac{in}{2.54\ cm}$$

$$= 0.8\ in\ deep$$

This ratio (0.8 in:6 in) is the same as a 1:7.5 ratio. Therefore:

Soil type	Percent particle-size distribution (mm)							Bulk density, ($g\ cm^{-3}$)
	2–1	1–0.5	0.5–0.25	0.25–0.125	0.125–0.05	0.05–0.002	<0.002	
Known values								
Sand	2	23	45	25	5	0	0	1.59
Soil	5	15	18	22	15	15	10	1.40
Calculated values of various sand:soil ratios								
1:7.5	*4.6*	*15.9*	*21.2*	*22.4*	*13.8*	*13.2*	*8.8*	*1.42*

Chapter 3
Rootzone Selection

Proper selection and construction of rootzone media is one of the most important, yet often least considered steps in any turfgrass construction project. Rootzones are "out of sight and out of mind" until excessive rainfall and/or soil compaction occurs, thereby delaying or cancelling events or becoming unsafe for participants. A common construction mistake is to place a couple of inches of sand on the existing soil and roto-tilling this in. Rarely, if ever, does this approach provide the desired results. Proper rootzone selection not only provides desirable drainage when necessary, but also retains sufficient moisture and nutrients for normal agronomic growth, resists compaction, and provides necessary soil aeration. Facilities must decide initially if they need optimum rootzones to provide these qualifications with minimum long-term agronomic and drainage problems or if they are willing to accept closings or delayed openings due to weather conditions.

3.1 Golf Putting Greens

Golf greens typically experience heavy use throughout the year. Although putting greens only represent approximately 2 % of the total course area, 50 % of the game is actually played on them. This concentrated traffic combined with daily mowing and other management practices involving machinery almost guarantees a problem with soil compaction, especially if the greens are constructed with improper soils or drain inadequately (Fig. 3.1).

Profiles

The modern putting green consists of 2 to 4 distinct components or layers, including (from top to bottom) the rootzone medium, choker sand layer (optional), gravel

L.B. McCarty et al., *Applied Soil Physical Properties, Drainage, and Irrigation Strategies*, DOI 10.1007/978-3-319-24226-2_3

Fig. 3.1 Concentrated traffic combined with daily mowing and other management practices involving machinery almost guarantees a problem with soil compaction, especially if the greens are poorly constructed or drain inadequately

layer, and drain lines. The rootzone medium is the finest textured, the choker layer (if used) is intermediate-textured, and the gravel layer is the coarsest-textured component. This profile creates a **perched water table** (or zone of saturation) at the lower level of the finer-textured layer, since water does not move (or percolate) readily from the small pores of the finer-textured rootzone layer into the large pores of the coarser layer unless the finer layer is saturated with water to some depth (Fig. 1.17). An example of this principle involves placing a saturated sponge on top of a bed of gravel, coarse sand, or another material. The water will stay in the sponge due to the differential particle size between it and the coarser material beneath it (this is called **granular discontinuity**). However, if additional water is added to the sponge, the water's weight will eventually break the tension between the two materials and water will start flowing. This allows sand which normally drains excessively to be used successfully as a rootzone. There are several successful putting green construction systems, each using some or all of these components.

USGA Specifications

The best known and most widely used system is a tiered or layered system used by the United States Golf Association (USGA) (Fig. 3.2). In this system, 12 to 14 *in* (30 to 36 *cm*) rootzone medium overlays an optional 2 to 4 *in* (5 to 10 *cm*) coarse sand layer (choker), which in turn covers a 4 *in* (10 *cm*) layer of gravel. Drainage is provided by drain lines cut into subsoil at 15 to 20 *ft* (4.5 to 6 *m*) spacings.

Fig. 3.2 The best known and most widely used system of golf green construction is a tiered or layered system used by the United States Golf Association (USGA). In this system, 12 to 14 *in* (30 to 36 *cm*) of rootzone medium overlays an optional 2 to 4 *in* (5 to 10 *cm*) coarse sand layer (choker), which in turn covers a 4 *in* (10 *cm*) layer of gravel

The gravel blanket helps move water rapidly to the drainage lines and out of the green while the choker layer prevents migration of fine sand into the gravel layer. The physical textural difference between the gravel and rootzone mix creates a capillary break (or "perched water table"), where water will not move freely into the gravel unless the rootzone mix above it is saturated, like the sponge example discussed previously.

USGA greens, if constructed properly, have a history of providing many years of satisfactory service. However, appropriate sands and gravel may be difficult and expensive to obtain, and the expertise and care required in construction and maintenance are demanding.

Hybrid Greens

A modification to the USGA system allows the intermediate choker layer to be eliminated (Fig. 3.3). This deletion depends on meeting very specific criteria, as determined by laboratory analyses, for the rootzone medium and the gravel. If the gravel is too coarse or rootzone medium too fine, problems may arise when sand from the rootzone migrates into and clogs the coarse gravel layer.

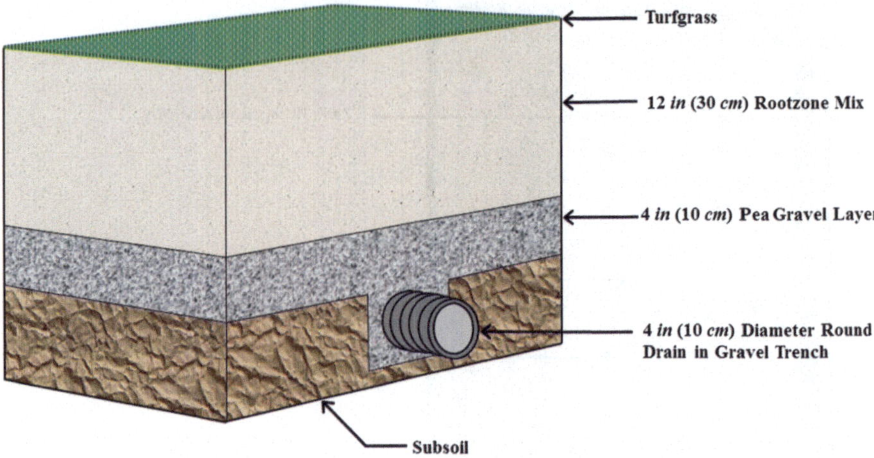

Fig. 3.3 A modification to the USGA system allows the intermediate choker layer to be eliminated. This deletion depends on very specific criteria, as determined by laboratory analyses, are met by the rootzone medium and the gravel

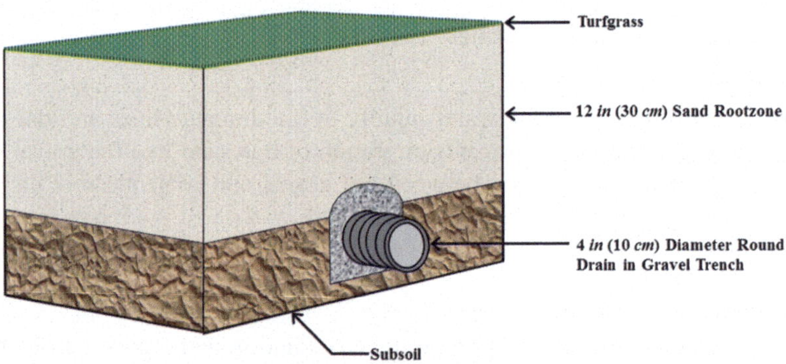

Fig. 3.4 A one-tier profile with simplified construction standards for pure sand golf greens (often referred to as the "California Method"). It consists of 12 to 14 *in* (30 to 36 *cm*) of appropriate rootzone sand overlaying the native soil. Drain lines are trenched into the subgrade and backfilled with gravel

Sand Greens

Figure 3.4 depicts a one-tier profile with simplified construction standards for pure sand golf greens (often referred to as the "California Method"). It consists of 12 to 14 *in* (30 to 36 *cm*) of appropriate rootzone sand overlaying the native soil. Drain lines are trenched into the subgrade and backfilled with gravel. Unlike the previous two profiles, the 4 *in* (10 *cm*) gravel layer is deleted, as is the 2 to 4 *in* (5 to 10 *cm*) choker layer, and only pure sand is used as the rootzone medium. This type of green is simple and relatively inexpensive to construct. It can perform satisfactory if the native soil underlying the rootzone is either impermeable or a layer of plastic (e.g., 6 mil polyethylene) is placed on the subsoil before adding the rootzone mix to prevent the downward movement of water. However, if the native soil readily drains, moisture will be sucked out of the rootzone medium and the green will be extremely droughty and difficult to manage. Research also indicates for sand greens to drain at rates comparable to two-tier greens, sand infiltration and percolation rates (designated when combined as saturated hydraulic conductivity, or K_{sat}) must be at least $20\,in\ h^{-1}$ ($51\ cm\ h^{-1}$) greater than K_{sat} rates for sand in two-tier greens. Drainage line spacing should also be based on the permeability of the rootzone sand, average rainfall rate, and the amount of water to be removed or retained. This drain spacing is much more critical for water removal than in a two-tier profile; therefore, a qualified laboratory should be consulted to make this determination.

Native Soil Greens

Despite the advances in putting green construction, a high interest remains in building native topsoil-containing greens, mainly for financial reasons (Fig. 3.5). In temperate areas, for example, where play is very seasonal, it is difficult to justify the considerably increased cost of a modified soil profile green. If native soil is used, however, one must recognize these greens will not drain as well internally; thus, sufficient (i.e., 1 to 3 %) surface drainage must be included in the design. Soil compaction is the other major

Fig. 3.5 Despite the advances of putting green construction, a high interest remains in building native topsoil-containing greens, mainly for financial reasons. If native soil is used, however, one must recognize these will not drain as well internally; thus, sufficient (i.e., 1 to 3 %) surface drainage must be included in the design. Also, soil compaction is the other major danger of using most native soils

potential problem when using most native soils. Compaction is minimized during construction by keeping large, heavy machines off the greens. Once established, compaction typically requires additional coring (aerification) and topdressing to help combat drainage problems. Advantages of native soil greens include (1) they are cheaper to build, (2) they hold water and nutrients much more efficiently than sand-based soils, and (3) they are less likely to have drastic changes in soil chemical properties (e.g., soil pH). However, be prepared to accept some risk of failure with these, especially during periods of excessive rainfall since these greens vary from a proven standard. Visit similar examples of the type of construction and materials being considered. No substitute for proper materials and construction methods exists.

Rootzone Mix Selection

The most common material traditionally used for rootzone construction was simply native soil. A bulldozer operator would "push up" the surrounding soil to a final grade, followed by grass planting. These greens performed adequately as long as traffic was light, adequate crowning to facilitate surface drainage, and the soil was not excessively wet during play.

As golf became more popular and courses received additional play, many of these greens declined or failed. They became seriously compacted, drained poorly, became algae infested, and were more susceptible to damaging outbreaks of disease, particularly *Pythium* and weeds such as *Poa annua*. An extensive survey of "push-up" greens revealed poorest turf was associated with heavier loams and clay soils, while healthiest turf was usually growing on sands or sandy soils. Today this seems obvious, since it is well-known sands resist compaction, maintain good drainage, and promote deep rooting, but at the time it was a revelation. This insight led to recommendations that putting greens should be constructed using sands or sandy soils as the primary ingredient.

Numerous refinements have been made to the sand-based rootzone over the past several decades, and modern recommendations for rootzone materials are considerably more specific and detailed. Because the success or failure of a putting green often hinges on the performance of the turfgrass root system, experts agree that choosing the rootzone mix is the most important decision when constructing golf greens.

Sand Sources

Minimal measurements necessary to evaluate potential components of a rootzone are:

1. Particle size analysis
2. Bulk density and porosity (total, capillary, and noncapillary)

3. Saturated hydraulic conductivity of an appropriately compacted rootzone sample; and
4. Soil moisture retention curves

Particle Size

The successful use of sand for rootzone construction depends primarily on three factors: (1) average particle size, (2) uniformity of particles, and (3) correct mixing of the sand with amendments. All sands are not created equal. Highly uniform sands are well-suited for constructing golf greens and sports fields while less-uniform sands are better for making concrete or providing a stable road bed.

Uniform sands are characterized as having most of the individual particles similar in size, which is termed a **narrow particle-size distribution** (Fig. 3.6). This is important since like-sized particles do not easily interpack, and result in good and stable soil porosity. By contrast, non-uniform sands have particles ranging from very coarse to very fine in size, and these can interpack and have higher bulk density. Intermediate-sized particles fill the spaces (pores) between the largest particles, smaller particles fill the spaces between the intermediate particles, and

Fig. 3.6 Uniform sands are characterized as having most of the individual particles similar in size, which is termed a narrow particle-size distribution. This is important since like-sized particles do not interpack, and result in good and stable porosity in the soil. By contrast, non-uniform sands have particles ranging from very coarse to very fine in size, and these can interpack and have higher bulk density. Intermediate-sized particles fill the spaces (pores) between the largest particles, smaller particles fill the spaces between the intermediate particles, and silt and clay can fill any remaining spaces. The net result is dense sand with reduced pore space, smaller average pores, and a tendency to compact. Shown are the various sand sizes as defined by the USDA with the coarsest being the *upper left* and finest *lower right*

silt and clay can fill any remaining spaces. The net result is dense sand with reduced pore space, smaller average pores, and a tendency to compact.

Soil physical properties are controlled or influenced by the size distribution of its particles. **In general, standard builder's sands used in construction or for concrete mixing are not suitable for rootzone construction unless they have been closely screened and sieved to remove unapproved particles.** Such sands are either too coarse, and remain droughty, or have a broad particle-size distribution, making the sand dense, hard, and impermeable. Builder's sands often contain small percentages of silt and clay, which will cause these sands to have poor infiltration and become very compacted. In fact, small amounts of silt and clay can affect the performance of even the most uniform sands. For example, the addition of only 4 % silt and clay to uniform fine sand has been shown to reduce the infiltration rate from 27 to 6 $in\ h^{-1}$ (69 to 15 $cm\ h^{-1}$) (Davis et al. 1990).

Chapter 1 lists the United States Department of Agriculture (USDA) particle-size classification for those materials of general interest for building the rootzone of desirable turf uses. Some sand companies provide particle-size distribution, but in many cases the analysis is based on engineering criteria, not the USDA sieve sizes.

Particle Size Analysis

The mineral fraction of soil is composed of sand, silt, and clay. The relative proportion of these in a soil determines its texture (Chap. 1). A particle-size analysis provides a general description of physical soil properties to soil scientists, and is the basis for assigning the textural class name (i.e., sand, sandy loam, clay) to the soil sample. Once the percentage of sand, silt, and clay has been determined, the specific textural class of the soil can be determined from the USDA's textural triangle (Chap. 1). If native soils are used for push-up type golf greens, they should fall in the sand or loamy sand textural classes. If a modern sand rootzone is being constructed, pure sand with minimum silt and clay should be used.

Sand Specification

The particle-size distribution should be determined for all sands being considered for the rootzone. Values from the analyses can then be compared to the specifications listed in Table 3.1, which summarizes recommendations from several different sources. Although experts may slightly disagree on precisely which sands are best suited for golf greens, three general recommendations should be met:

1. The sand should be free of silt and clay. If present, silt should not exceed 5 % and clay 3 % by volume. Larger amounts of silt and clay will reduce infiltration and percolation (Fig. 3.7). Riverbed sands or other sedimentary type soils or muck are often unacceptable due to their high clay or silt content or non-uniform sands.

Table 3.1 Suggested specifications for sandy soils used for turfgrass rootzones by various references

Textural name	Particle size (*mm*)	ASTM Mesh	Approximate K_{sat} *in* h^{-1} (*cm* h^{-1})	USGA		ISTRC		UC	ATRI	
				Recommended %						
Clay	<0.002	—	—	≤3	≤10	≤3	≤10	0–8	5–10	<30
Silt	0.002–0.05	—	<5 (<13)	≤5		≤5				
Very fine sand	0.05–0.15	270	5 (13)	<5		≤5			—	
Fine sand	0.15–0.25	140	18 (46)	≤20	—	10–15	65–85	82–100	60–70	
Medium sand	0.25–0.5	60	59 (150)	>35 (75 ideal)	≥60	>40				
Coarse sand	0.5–1.0	35	217 (551)	<45		15–25			10–30	
Very coarse sand	1.0–2.0	18	>217 (>551)	<7	≤10	—	≤10	0–10	<5	
Gravel	>2	10	—	<3		<3			0	

ASTM American Society for Testing and Materials
USGA United States Golf Association
ISTRC International Sports Turf Research Center. Note: ISTRC recommends an optimum 65 to 85 % sand particle in the coarse (0.5 *mm*) + medium (0.18 to 0.25 *mm*) range and a maximum of 20 % in the 0.18 *mm* (80 mesh) range
UC University of California (sand only)
ATRI Australian Turfgrass Research Institute

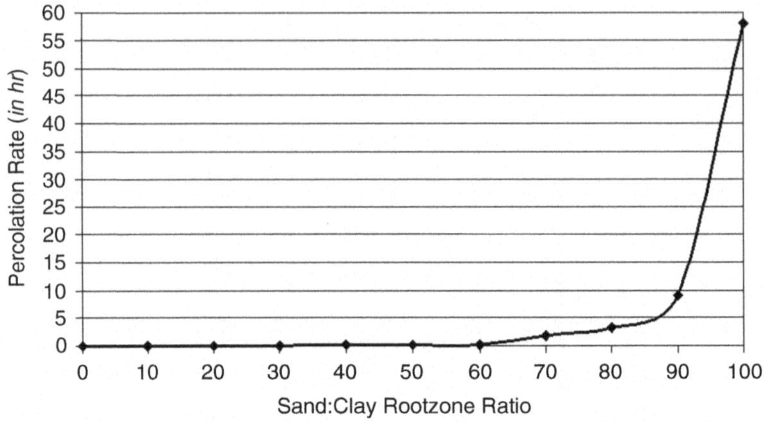

Fig. 3.7 Sand used to construct golf courses and sports fields should ideally be free of silt and clay. If present, silt should not exceed 5 % and clay should not exceed 3 % by volume as these reduce infiltration and percolation. Shown is an example of adding only 10 % clay reducing sand's percolation rates from 57 $in\ h^{-1}$ (145 $cm\ h^{-1}$) to less than 10 $in\ h^{-1}$ (35 $cm\ h^{-1}$)

Sometimes these sands are washed to remove the silt and clay, but all materials being considered should be tested before use.

2. The sand should be free of very coarse sand and gravel. If present, very coarse sand should not exceed 7 % and gravel 3 % by volume. If these limits are exceeded:

   (a) Large particles may cut or bruise the stolons/rhizomes of the turfgrass.
   (b) Large particles tend to accumulate at the soil surface, resulting in hard greens.
   (c) Large particles at the surface may dull mower blades.
   (d) Large particles make cup-setting and core aerification difficult.
   (e) The soil may not hold adequate water or nutrients.

3. The sand should have a particle-size distribution with the majority (>80 %) of the particles falling in the fine, medium, and coarse sand (0.1 to 1.0 *mm*) fractions. Within this range, the medium-sized particles (0.25 to 0.5 *mm*) should comprise at least 50 to 70 %.

Composition

Quartz silica sand is preferred for every golf course use (bunkers, greens, tees, and fairways) because it is very resistant to breakdown and retains its original shape. If possible, the chosen sand should contain 95 % or greater quartz silica. Manufactured (mechanically crushed) sands generally have poor quality and undesirable chemical content, while calcareous sands are soft, unstable, and have a high pH, which can reduce the availability of some micronutrients (e.g., Fe and Mg) to grass.

Sand Sieve Analysis

Sieves are used to separate sands into different fractions based on effective particle size. Effective particle size is the distance between wires in a square grid fabric of woven wire making up the bottom of a sieve through which particles with smaller effective diameters pass. The %-retained by each sieve is determined by dividing the weight of material on a given sieve by the total sample weight. The cumulative %-retained value is determined by summing the %-retained values from all sieves greater than or equal to the size class of interest. Finally, the cumulative %-passing values are determined by subtracting the corresponding cumulative %-retained value from 100 (cumulative %-passing = 100 – cumulative %-retained).

Example Determine the cumulative %-retained and cumulative %-passing for the following sand size distribution (*answers*).

Size class	Particle diameter (*mm*)	% Retained	Cumulative %-retained	Cumulative %-passing
Gravel	>2.0	1.5	*1.5*	*98.5*
Very Coarse Sand	1–2	4.0	*5.5*	*94.5*
Coarse Sand	0.5–1	34.5	*40.0*	*60.0*
Medium Sand	0.25 to 0.5	45.5	*85.5*	*14.5*
Fine Sand	0.1–0.25	12.0	*97.5*	*2.5*
Very Fine Sand	0.05–0.15	2.5	*100.0*	*0.0*

From either column, this sand sample consists primarily of medium and coarse particles (80 % of the sample falls within these size classes).

Tabular data can be used to draw a graph of the sand size distribution. A 3-cycle semi-log graph paper is often used with the particle diameter (mesh openings) as the x-axis and cumulative %-retained (or cumulative %-passing) as the y-axis values. Figure 3.8 is a graphical presentation of the data in the previous example. A sand sample with coarser particles than this sample would have a curve that shifts more to the right while a finer-graded sand would have a curve shifted to the left. When using graphical presentation of sieve data, a steep cumulative %-retained curve indicates a relatively uniform sand. A flattened curve indicates a non-uniform sand, where a large range of diameters are needed to go from low to high cumulative %-retained. The graphical presentation allows greater ease in assessing particle size distribution compared with tabular presentation and is useful when comparing multiple sands.

The D_x value of a sand refers to the sieve opening through which x% of the sand (by weight) will pass. The sieve opening equals the effective particle diameter of the largest grain which will pass through a sieve separating the finer (smaller) particles from coarser (larger) particles retained. The 'x' in a D_x value can be arbitrarily assigned from 1 to 99 to indicate a specific cumulative %-passing. Thus, a D_{10} value is the estimated sieve size where 10 % of the sand particles would pass through and 90 % of the particles would be retained. To determine a Dx value for a

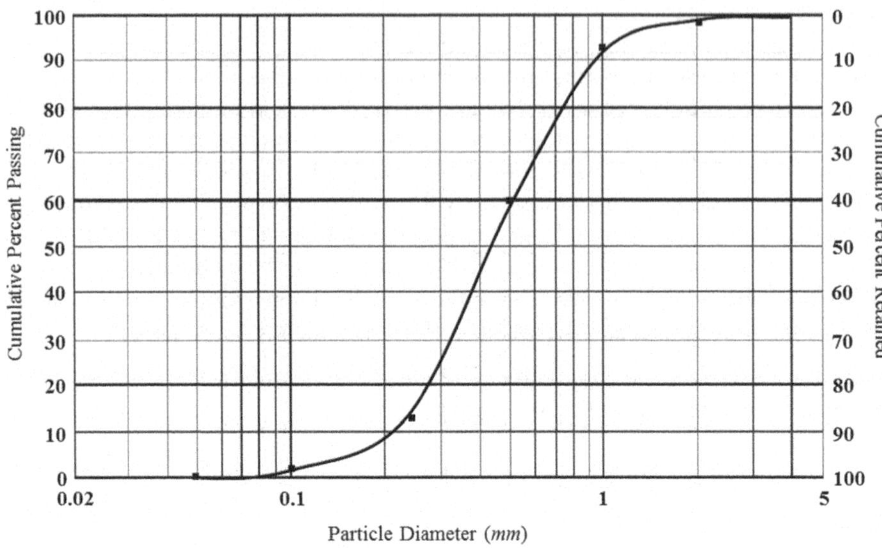

Fig. 3.8 Using 3-cycle semi-logarithmic paper to graph sieve size versus cumulative percentage passing or cumulative percentage retained for a sand sample example

specific 'x' value, not corresponding to an exact sieve size used in the analysis, cumulative %-passing values for several sieve sizes (particle diameters) are plotted on semi-log paper. The plotted points are connected to form a smooth curve. A line is then traced from the desired 'x' value (cumulative %-passing) on one axis to intersect the curve. From that point of intersection, a line is traced to the other axis to find the particle diameter (theoretical sieve size) for the particular 'D_x' value. For example, from Fig. 3.8, the D_{60} value is 0.5 *mm*.

Gradation Index

A gradation index (or coefficient of uniformity) describes the uniformity of the particle sizes, thus predicts the potential for particle interpacking. A larger coefficient of uniformity value indicates a wider range of particles sizes are present, increasing potential for compaction, while smaller values indicate a more uniform sand, less prone to compaction. Interpacking in sands with a large gradation index reduces total pore space, reducing hydraulic conductivity. A uniform sand has a large %-retained within a single or two adjacent size classes, while a sand having approximately equal %-retained in all size classes would be very non-uniform. Engineering terminology classifies uniform sand as being poorly-graded while non-uniform sand is referred to as well-graded.

Gradation indices utilize a D_x approach to assess sand uniformity. The gradation index expresses the ratio of larger to smaller particles. For example, a gradation index of D_{90}/D_{10} is determined by dividing the D_{90} value (the grain diameter where

90 % of the particles are smaller) by the D_{10} value (the grain diameter where only 10 % of the particles are smaller). From Fig. 3.8, the D_{90} value is approximately 0.95 *mm* while the D_{10} value is approximately 0.22 *mm*, thus D_{90}/D_{10} would yield 0.95/0.22 or 4.3. Other gradation indices have also been used such as D_{95}/D_5 or D_{60}/D_{10}. Gradation index value guidelines for sands used for golf and sports turf are available, including a D_{90}/D_{10} of ≤ 3.3 (Adams and Gibbs 1994); D_{95}/D_5 of 2 to 6 (Bingaman and Kohnke 1970); $D_{85}/D_{15} \leq 4$ (Stewart 1994); and a $D_{60}/D_{10} < 4.0$ (Blake 1980).

Fineness Modulus

Another calculated index used in the sand industry to quantify the particle sizes in a sample is the fineness modulus. This index is the summation of the cumulative %-retained on the openings of 0.15, 0.3, 0.6, 1.18, 2.36, and 4.75 *mm* or mesh numbers of 100, 50, 30, 16, 8, and 4 which is then divided by 100. Finer textured sands would have a smaller fineness modulus value while coarser sands would have a larger value. Blake (1980) recommends sands used for putting greens generally fall within a fineness modulus range of 1.7 to 2.5.

Example From the following sand particle size distribution, determine the Finenss Modulus and its acceptability according to Blake (1980) (*answers*).

Particle diameter (*mm*)	Mesh number	%-Retained	Cumulative %-retained
4.75	4	2.0	*2.0*
2.36	8	2.3	*4.3*
1.18	16	2.5	*6.8*
0.6	30	23.0	*29.8*
0.3	50	56.0	*85.8*
0.15	100	14.2	*100*
sum	—	100	*~229*

The sum of the cumulative %-retained equals 229 which, when divided by 100, equals 2.29, which is within the suggested range of 1.7 to 2.5.

Particle and Bulk Density

Particle density is defined as the mass (or weight) of dry soil per unit volume of the soil solids (excluding pores). If one could melt the soil into a solid mass of known volume, this would be its density. An average value of 2.65 $g\ cm^{-3}$ has been found for most mineral soils, and is the standard value used by soil scientists and soil laboratories in calculating other soil properties. A soil's overall particle density decreases if peat or inorganic amendments are added to the soil.

Bulk density, defined as the mass (or weight) of dry soil per actual volume of the soil, is the more important parameter for turf use. Bulk density, unlike particle density, includes the pore space volume as well as the volume of soil solids. Since most soils are about half solids and half pore space, bulk densities tend to be about half the particle density. The more compacted a soil is, the higher its bulk density. The preferred bulk density range for golf greens is between 1.35 and 1.55 $g\ cm^{-3}$, with a lower limit of 1.20 $g\ cm^{-3}$, an upper limit of 1.60 $g\ cm^{-3}$, and an optimum level of 1.40 $g\ cm^{-3}$. Incorporating organic matter, such as peat, is one means of reducing the bulk density of a rootzone.

Sand particle size and the uniformity of sands influence bulk density. Widely graded sands tend to pack tightly, producing higher bulk densities. More uniform sands tend to pack less (or not at all) resulting in lower bulk densities.

Soil Porosity

Soil porosity or **total pore space** is the fraction of soil volume not occupied by solid particles (Chap. 1). Porosity is important, since it is the pores between solid particles that hold both air and water. The arrangement and size of particles largely determines pore space. The optimum porosity range for golf greens is 35 to 55 % by volume, evenly divided between smaller capillary (water filled) and larger noncapillary (air filled) pore spaces (Table 3.2).

Smaller capillary pores largely determine the amount of water held by soil, while larger noncapillary pores determine air content. Noncapillary pores also control how quickly water and air move through a soil. If capillary pores predominate, moisture holding capacity of the soil will be high, but water and air movement may be limited. If noncapillary pores predominate, excessive drainage and high aeration result at the expense of adequate moisture holding capacity. As bulk density increases, total soil porosity decreases, and vice versa. For example, a rootzone with a bulk density of 1.55 $g\ cm^{-3}$ and a particle density of 2.65 $g\ cm^{-3}$ has a total porosity of 42 %. However, if the bulk density increases to 1.65 $g\ cm^{-3}$through compaction, total porosity would be reduced to 38 %.

Golf greens should have a capillary porosity between 15 and 25 % by volume, and noncapillary porosity between 15 and 30 %, with an ideal value between 18 and 25 % (Table 3.2). These values are based on laboratory analyses where rootzone samples have been compacted, then saturated, and allowed to drain for 24 h.

Table 3.2 Suggested porosity and solids of USGA and California golf green soils (Hummel 1998)

	Porosity			Solids
Reference	Capillary (water-filled) ($cm^3\ cm^{-3}$ or % by volume)	Noncapillary (air-filled) ($cm^3\ cm^{-3}$ or % by volume)	Total pores ($cm^3\ cm^{-3}$ or % by volume)	($cm^3\ cm^{-3}$ or % by volume)
USGA Greens	15 to 25	15 to 30	35 to 55	45 to 65
California Greens	10 to 20	15 to 30	35 to 55	45 to 65

The *minimum* noncapillary air-filled porosity that will support good turfgrass growth is between 10 and 15 %. Moisture content (capillary porosity following drainage) should fall between 12 and 25 % by volume, with 18 % being ideal.

Saturated Hydraulic Conductivity (K_{sat})

Even though the particle-size distribution of a sand falls within the ranges listed in Table 3.1, the sand could have unacceptable infiltration or water-retention values. Therefore, it is essential a soils lab perform a saturated hydraulic conductivity (designated as "K_{sat}") test on compacted sands before use. This will help eliminate questionable sands that might create drainage problems later. Saturated hydraulic conductivity values are not the same as infiltration or percolation ('perc') rates, which vary with moisture content. In fact, K_{sat} values are not rates at all, but are ratios relating flow of water through a soil (flux) to the gravitational force of water driving the flow (hydraulic gradient). Saturated hydraulic conductivity values allow soils to be compared with respect to their anticipated effects on a soil profile's infiltration and percolation rates. Refer to Chap. 1 on the procedure for measuring K_{sat}.

For most putting greens, initial K_{sat} values should be 10 to 15 $in\ h^{-1}$ (25 to 38 $cm\ h^{-1}$). This value will decline over time by approximately 33 %, but it should still be well above the minimum of 6 $in\ h^{-1}$ (15 $cm\ h^{-1}$). Bermudagrass, seashore paspalum, and zoysiagrass greens with adequate surface slope can have slightly lower initial K_{sat} values of 6 to 10 $in\ h^{-1}$ (15–25 $cm\ h^{-1}$). High or "accelerated" K_{sat} values of 12 to 14 $in\ h^{-1}$ (30 to 36 $cm\ h^{-1}$) may be appropriate for bentgrass/Poa courses at some locations to handle heavy rains or if irrigation water quality is poor or cool-season turfgrasses are being grown outside their range of adaptation. In this case, the sand should contain a minimum of 65 % coarse- (0.5 *mm*) and medium- (0.25 *mm*) sized particles. Even more desirable would be a sand having 75 % in the medium-sized, and the majority of the remaining 25 % as coarse-sized sand particles.

Rootzones with accelerated (i.e., >16 $in\ h^{-1}$, >41 $cm\ h^{-1}$) K_{sat} values often require a longer period of time for full maturity due to excessive moisture applied and lower cation (nutrient) exchange capacity. These also tend to remain firm longer after grow-in and are more difficult to manage due to their low water and nutrient holding capacities. As mentioned, sand (or California-style or one-tier system) greens also require K_{sat} values about 20 $in\ h^{-1}$ (51 $cm\ h^{-1}$) greater than a two-tier (USGA) green to remove similar amounts of water with K_{sat} values ranging from 15 to 50 $in\ h^{-1}$ (38 to 127 $cm\ h^{-1}$).

For courses desiring a slower infiltration rate, the particle distribution should include 75 % medium and up to 15 % fine (0.10 *mm*) sand. Even slower infiltration rates can be achieved by selecting a sand with a minimum of 65 % in the fine and medium sand classes. Why would a golf course select slower-draining rootzones? Generally, faster rootzone drainage rates indicate lower water and nutrient holding capacity. A slower-draining rootzone should hold more water, be less prone to drought and localized dry spots, and perhaps require less fertilizer. A slower-draining rootzone is best for arid regions where heavy rainfall is infrequent.

The values just discussed are meant as guidelines and not necessarily building specifications. Depending on location, it may be difficult to obtain sands that meet these rigid specifications. In these cases, look for a sand containing at least 80 % of the particles in the fine, medium, and coarse classes combined. Avoid sands containing large amounts of very coarse sand and gravel, since they will drain too quickly and have excessively low water and nutrient holding capacity. However, rootzone mixes dominated by smaller-sized particles (fine sand, very fine sand, silt, and clay) will hold too much water, have poor aeration, and be conducive to algae, moss, and soil diseases. These are sometimes referred to as "dirty" sands. Sometimes a sand pit will wash a "dirty" sand over a number 140 or 200 screen to remove the very fine sand, silt, and clay.

Soil Moisture Retention Curves

As discussed in Chap. 1, an additional evaluation criteria for sand:soil mixes being considered for golf greens, sand capping fairways, or sports field mixes is a soil moisture retention curve. These curves help predict soil moisture distribution throughout a soil's profile, based on water potential. This information can help in designing optimal rootzone depths for various construction methods. For example, most successful sand rootzones contain ≥ 0.15 but $< 0.35\, cm^3$ water per cm^3 soil (15 to 35 %) volumetric water. Soil moisture retention curves can predict a sand's ability to retain this range of moisture content in the rootzone. The curves also help predict future soil compaction and resulting moisture patterns within the soil profile.

Soil Amendments

A variety of organic and inorganic soil amendments are available to add to soils to improve their physical and chemical properties. The most commonly used ones will be discussed. However, additional local sources are also often available.

Organic Soil Amendments

The addition of a well-decomposed organic amendment enhances soil structure by improving soil aggregation, nutrient retention, and the water-holding capacity of sands. Organic amendments commonly used in rootzone mixes include peats and composts such as those listed in Table 3.3. These improve germination and establishment by increasing moisture and nutrient holding capacity. However, peats are not all the same, they vary in their botanical origins and quality.

Another common belief is the thatch/mat layer that typically develops several years after turf establishment will substitute for adding organic amendments during

Table 3.3 Comparison of soil organic amendments used to modify golf green rootzones

Soil amendment	pH	Cation exchange capacity	Water-holding capacity	Durability (years)
Peat humus	Acid	Good	Good	5+
Reed-sedge peat	Acid	Good	Good	4 to 5
Peat moss	Acid	Fair	Excellent	1 to 3
Rice hulls	Acid	Fair	Poor	1 to 3
Ground fir bark	Acid	Fair	Fair	5
Lignified wood waste	Acid	Poor to fair	Good	8+
Sawdust	Acid	Fair to good	Fair to good	1+
Sphagnum moss peat	Acid	Good	Excellent	1 to 3
Yard waste compost[a]	Basic	Fair	Good	1 to 3
Biosolids compost[a]	Basic	Poor to fair	Good	1 to 5
Mushroom compost[a]	Basic	Fair	Good	1 to 3

[a]Salts can be extensive with some composted materials

construction. Thatch, however, is not considered as organic matter until it is well decomposed. Thatch will not return substantial nutrients back to the soil and can actually tie them up.

Peat

Peat is a generic term for partially decomposed plant material formed in bogs under cool and moist conditions and exceeding 75 % organic matter by weight (Fig. 3.9). Higher quality peats generally exceed 85 % organic matter by weight. Peat is the most commonly used organic amendment for putting green soil mixes. Most commercial peats are mined in Canada, North Dakota, Minnesota, and Europe. Since they are derived from different plant materials decomposing under different environmental conditions, these products vary considerably in pH, water retention, organic content, ash and fiber content, and level of decomposition. Peats are broadly classified into moss peat, reed-sedge peat, and peat humus. Moss peats are composed of sphagnum, hypnum, and other mosses. Reed-sedge peat is formed from reeds, sedges, cattails, marsh grasses, and other plants. Peat humus is decomposed to the point where the original plant materials are not recognizable. Fibrous peats are preferred over sedimentary and woody-type peats. Peats used to modify sands should be high in organic content and low in ash. Amounts of peat used range from 5 to 20 % by volume (Fig. 3.10). Since peat is generally a nonrenewable resource, many local organic sources are being considered as peat substitutes. For example, muck soils have 25 to 75 % organic matter by weight. Stability, hydraulic conductivity and soil moisture retention tests should be performed on the resulting soil mixes containing local organic amendments prior to use. Table 3.3 lists specific characteristics of some commonly used organic materials for modifying rootzone mixtures.

Fig. 3.9 Peat is a generic term for partially decomposed plant material formed in bogs under cool and moist conditions. It is the most commonly used organic amendment for putting green soil mixes

Fig. 3.10 Peats used to modify sands should be high in organic content and low in ash. Shown is the result of a typical 85 % sand amended with 15 % peat by volume

Inorganic Soil Amendments

Several inorganic amendments (Table 3.4) are marketed as rootzone amendments, and may merit consideration if they are readily available, meet K_{sat} specifications, and are affordable. These are mined highly porous materials that have been processed and sized (Fig. 3.11). Inorganic soil amendments do not promote microbial growth the way organic materials do, but may improve either water-holding capacity or aeration of the sand due to their high internal pore space. Since they do not decompose, inorganic amendments usually persist. Also, inorganic soil amendments tend to displace sand on a 1:1 basis while organic amendments do not. Less total material is needed when using inorganic amendments compared to organic. For example, if a 90:10 sand to amendment ratio is desired, 100 yd^3 (77 m^3) of organic amendment is necessary for each 1,000 yd^3 (765 m^3 of sand, since organic

Table 3.4 Comparison of miscellaneous inorganic soil amendments used in golf green construction

Soil amendment	pH	Cation exchange capacity	Water-holding capacity	Durability (years)
Porous ceramics	—	—	—	—
• Calcined clay	Slightly acidic	Moderate	High	10+
• Calcined diatomaceous earth	Slightly acidic	Moderate	High	10+
Clinoptilolite/zeolite	Neutral	High	High	10+
Perlite	Neutral	Low	Moderate	10+
Pumice	Neutral	Low	Low	10+
Vermiculite	Neutral	Poor	Low to fair	10+
Colloidal phosphate	Neutral	Good	Good	10+

Fig. 3.11 A number of inorganic amendments are marketed as rootzone amendments, and may merit consideration if they are readily available, meet the infiltration and percolation specifications, and are affordable. Shown is a mined highly porous materials that has been processed and sized

amendments do not appreciably displace (add to the total volume or bulk of) the sand. Inorganic amendments, however, do displace the sand; thus, only 900 yd^3 (689 m^3) of sand are needed with the 100 yd^3 (77 m^3) of inorganic amendment to achieve a total volume of 1000 yd^3 (765 m^3) of a 90:10 mixture.

Some unstable inorganic materials, however, may crush into finer particles when subjected to compaction. Inorganic amendments also may retain and release water and retain nutrients less efficiently than organic amendments; they are more difficult to grow-in and are more costly than most organic amended sands (Bigelow et al. 2004; Waltz et al. 2003). A ranking of inorganic amendments according to turfgrass establishment from best to worst is:

$$\text{peat} > \text{clinoptilolite zeolite} = \text{porous ceramic} > \text{crystalline silica} = 100\,\%\ \text{sand}$$

Gravel and Coarse Sand Layers

After drains are installed, the packed subgrade of the golf green should be covered with 4 *in* (10 *cm*) of "pea" (1/4 to 3/8 *in* diameter, 6.4–9.5 *mm*) gravel conforming to the proposed final surface grade of the green to a tolerance of ±1 *in* (2.5 *cm*) (Fig. 3.12). The gravel layer serves several purposes:

1. It is very porous and allows water to rapidly move laterally to the drain lines.
2. It separates the subgrade from the rootzone and prevents the subgrade soil from extracting water from the rootzone.

Fig. 3.12 In golf green and sports field construction, after the drains are installed, the packed subgrade can be covered with 4 *in* (10 *cm*) of "pea" (1/4 to 3/8 *in* diameter, 6.4–9.5 *mm*) gravel conforming to the proposed final surface grade of the green to a tolerance of ±1 *in* (2.5 *cm*)

Fig. 3.13 Successful greens and sports fields have been constructed without the 4 *in* (10 *cm*) pea gravel layer. However, drainage trenches should be backfilled with gravel

3. It impedes salt movement from the subsoil into the rootzone. Near coastal areas, where the water table may be contaminated by seawater intrusion, salts can move to the soil surface during periods of hot, dry weather.
4. It helps prevent an excessively wet rootzone due to a rising water table.
5. The interface between the gravel and the sand or rootzone mix above temporarily creates the perched water table which increases the water-holding capacity of the rootzone mix.

Successful greens have been constructed without the 4 *in* (10 *cm*) pea gravel layer (Fig. 3.13). However, it is critical that drainage trenches are backfilled with gravel. If this strategy is chosen, the parent subgrade soil must be compacted and/or a 6-mil polyethylene layer used to separate the rootzone medium from the subsoil and from the collar. It is recommended that courses with adequate financial resources not eliminate the gravel layer since it increases the probability of success.

Evaluating Gravel

Four size criteria are currently used for selecting the appropriate gravel for a rootzone composition. These include: (1) bridging determination to maintain layer integrity; (2) permeability as the gravel needs to have suitably higher permeability than the rootzone; (3) uniform particle size distribution; and, (4) certain gravel diameter size limitations.

Fig. 3.14 Two gravel sizes used in construction. The gravel on the left is used to backfill tile drainage lines. The gravel on the right is used for the 4 *in* (10 *cm*) gravel bed overlying the drainage lines

Gravel Size

The physical properties of the rootzone mix will determine the size of gravel to be used. In theory, the diameter of the gravel should be five to seven times the diameter of the sand used to construct the rootzone. This will permit stable "bridging" between the sand and gravel and prevent migration of smaller particles from the rootzone into the gravel. For example, if the sand used to construct the rootzone is approximately 1 *mm* (1/24 *in*) in diameter, then 6 *mm* (¼ *in*) pea gravel is used (Fig. 3.14). If the gravel is too coarse, the sand may migrate into the gravel layer (Fig. 3.15). Likewise, if the gravel or sand is too fine, a zone of saturation may form at the interface of the rootzone sand and gravel layer causing a deeper saturated rootzone layer (Fig. 3.16) that interferes with normal drainage and plant rooting.

In the laboratory, the correct gravel size in relation to the average size of sand particles that the gravel will contact can be determined. The properties used in specifying materials used in construction include the D_x value and percentage of particles within a desirable size range. The D_x value is the sieve opening size through which x% of particles in a sample pass. Two D_x values are commonly used, with the larger value (D_{85}) indicating the general coarseness of the sample while the lower D_x value (D_{15}) often reflects the largest makeup of particles in a sand-based rootzone mix. Using the D_x values, several factors can be calculated to determine the acceptability of gravel for use below a particular rootzone.

Fig. 3.15 If the gravel is too coarse, sand may migrate into the gravel layer. In the laboratory, the correct gravel size in relation to the average size of sand particles the gravel will contact is determined

Fig. 3.16 If the gravel's or sand's diameter is too fine in rootzone construction, the zone of saturation may extend into the rootzone, causing drainage and rooting issues

Bridging Factor

One factor to consider in sand-over-gravel profiles is the bridging factor. The bridging factor is calculated by dividing the D_{15} value of the gravel by the D_{85} value of the rootzone sand. The largest 15 % of the sand particles form a "bridge" with the smallest 15 % of the gravel particles. This bridging is caused by the irregular shape of particles, friction between particles, and the weight of the material above, all working to "lock" the smaller particles into voids in the gravel. Once the particles have locked together and "bridged" above the voids, this prevents further significant particle movement, while maintaining adequate permeability. For bridging to occur, the USGA recommends the gravel D_{15} value be *less than or equal to* 8 times the sand rootzone D_{85} value (bridging factor less than or equal to 8). In other words, the smallest 15 % of the gravel particles must have diameters no more than 8 times greater than the largest 15 % of the rootzone particles.

Sand meeting the USGA specs will have a D_{85} between 0.4 and 0.7 *mm* (0.016 and 0.03 *in*); in other words, 15 % of the sand particles will be larger than this size. Using the bridging factor, gravel D_{15} (smallest 15 % diameter of the gravel) should not be larger than 3.2 to 5.6 *mm* (8 times 0.4 to 0.7 *mm*). If the gravel is too coarse, the bridging factor will be too high, and if the sand is very dry, it may migrate into the gravel. When sand and gravel are matched to bridge, sand will not migrate into the gravel voids, even though many of these gravel voids are larger than 0.4 to 0.7 *mm*.

Permeability Factor

The permeability factor is determined by dividing the D_{15} value of the gravel by the D_{15} of the rootzone. The permeability factor indicates if the gravel can transmit the needed amount of water to the drain lines. The USGA recommends that the gravel D_{15} value be greater *than or equal to* 5 times the sand rootzone D_{15} value (permeability factor greater than or equal to 5).

Uniformity Factor

The uniformity factor is determined by dividing the D_{90} value of the gravel by the D_{15} value of the gravel. As discussed previously, this is one gradation index (or coefficient of uniformity) for a gravel, used to express a ratio of larger to smaller particles. As with sands, larger coefficient of uniformity values indicate a wider range of particles present, increasing the probability of undesirable particle interpacking (Fig. 3.17). The USGA recommends that the gravel D_{90} value be less *than or equal to 3* times the gravel D_{15} value (uniformity factor less than or equal to 3).This indicates the gravel has a desirable narrow spread in particle sizes.

Lastly, the USGA recommends that the gravel should totally pass through a 0.5 *in* (12 *mm*) sieve, with less than 10 % passing a 0.08 *in* (2 *mm*) sieve (No. 10) and *greater than* 5 % passing a 0.04 *in* (1 *mm*) sieve (No. 18) (Table 3.5).

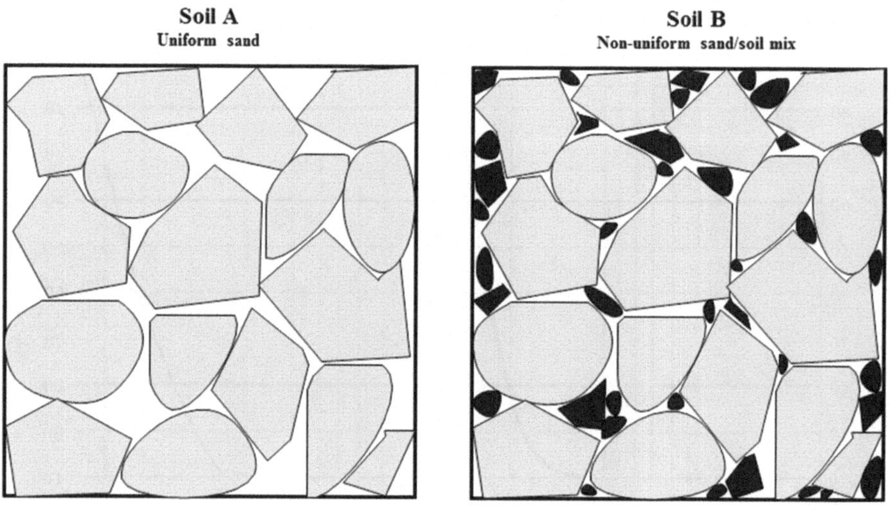

Fig. 3.17 Uniformity of coefficient (or gradation index) describes the uniformity of the particle sizes, thus, predicts the potential of particle interpacking. The larger the uniformity of coefficient value, a wider range of particles are present, increasing the probability of undesirable particle interpacking. Interpacking sands with a large gradation index reduces total pore space, reducing hydraulic conductivity

Table 3.5 Size recommendations by the USGA for gravel when an intermediate sand ("choker") layer is or is not used in golf green construction

Performance factors	Recommendation[a]
Gravel size when an intermediate ("choker") layer is not used	
Bridging factor	• D_{15} of gravel $\leq$ 8 times the D_{85} of the rootzone
Permeability factor	• D_{15} of the gravel $\geq$ 5 times the D_{15} of the rootzone
Uniformity factor	• D_{90}/D_{15} ratio of gravel $\leq$ 3.0
Additional factors	• No particles of gravel >12 *mm*
	• ≤10 % of gravel <2 *mm*
	• Not more than 5 % of gravel less than 1 *mm*
Gravel size when an intermediate ("choker") layer is used	
Gravel size	• ≤10 % of particles >12 *mm* (½ *in*)
	• ≥65 % of particles between 6 *mm* (1/4 *in*) and 9 *mm* (3/8 *in*)
	• ≤10 % of particles <2 *mm*
Intermediate (Choker) layer material size	• ≥90 % of particles between 1 and 4 *mm*

[a]D_{15}, gravel = the particle diameter below which 15 % of gravel particles (by weight) are smaller. D_{85}, rootzone = the particle diameter below which 85 % of rootzone particles (by weight) are smaller

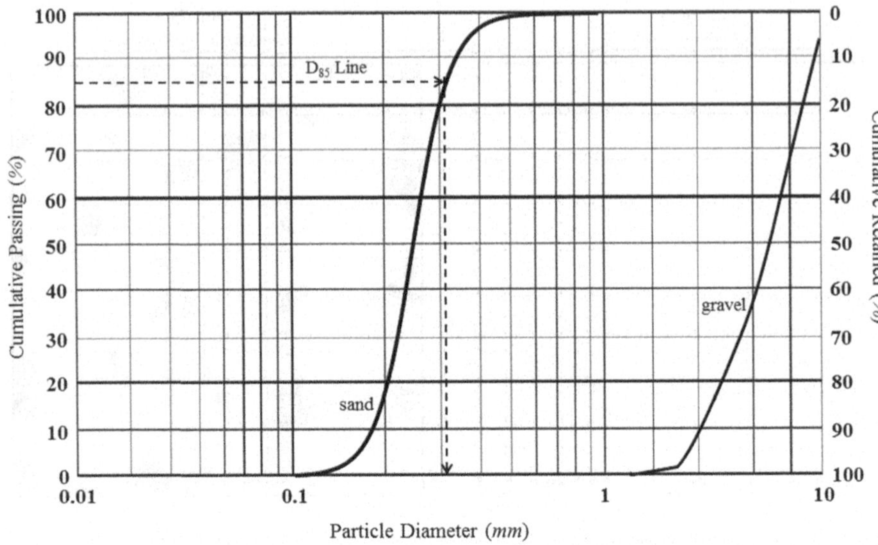

Fig. 3.18 Sand meeting the USGA specs will have a D_{85} between 0.4 and 0.7 *mm* (0.016 and 0.03 *in*); in other words, 15 % of the sand particles will be larger than this size. Using the bridging factor, gravel D_{15} (smallest 15 % diameter of the gravel) should not be larger than five times sand D_{85} (0.4 to 0.7 *mm*), or 2 to 3.5 *mm* (five times 0.4 to 0.7 *mm*). If the gravel is too coarse, the bridging factor will be too high, and if the sand is very dry, it may migrate into the gravel. When sand and gravel are matched to bridge, sand will not migrate into the gravel voids, even though many of these gravel voids are larger than 0.4 to 0.7 *mm*. Gravel D_{15} (smallest 15 % diameter of the gravel), $\leq$8 times sand D_{85} (0.4 to 0.7 *mm*) or 3.2 to 5.6 *mm* (8 × 0.4 and 0.7). Shown is sand particle and gravel size versus the cumulative percentage that has passed for a particular size and not percentage retained. For sands, this curve is often sigmoid-shaped while it is linear or quadratic in nature for gravel

Interpolation

Instead of graphing, an alternative method to determine D_{15} and D_{85} would be by interpolation, which determines a linear approximation within the range of a discrete set of known data points. In the following example, to determine D_{15} by interpolation between $D_{18.8}$ of 3.35 and $D_{4.0}$ of 2.0 *mm* for the gravel, the difference in D values is assumed proportional to the difference in %-passed (<D) for those particle diameters. It is known that D_{15} for the gravel falls between the sizes 3.35 and 2.0 *mm*, as the 15 % passed values would fall between 18.8 and 4.0 %. Interpolation would then use the ratio of the difference in particle diameters to the difference in % passed to estimate

(continued)

the difference in particle diameter (D_x) for the 'x' value nearest 15 % which in this case is 18.8 %. Therefore the change (indicated as Δ) in particle size between 18.8 and 15 % passed is estimated from the known data set of $D_{18.8} = 3.35\,mm$ and $D_{4.0} = 2.0\,mm$. This estimated difference is then subtracted from the nearest known D_x value to the desired D_x value, which in this example is 3.35 *mm* ($D_{18.8}$).

$$\frac{\Delta \text{ particle size}}{18.8\,\% - 15.0\,\%} = \frac{3.35\,mm - 2.0\,mm}{18.8\,\% - 4.0\,\%}$$

$$\Delta \text{ particle size} = 0.35\,mm$$

$$\text{Therefore } D_{15} = 3.35\,mm - 0.35\,mm = 3.00\,mm$$

This is close to the 2.95 *mm* D_{15} value determined using linear regression.

Example The following information was determined from a sieve analysis of a potential rootzone mix and gravel sample. Can this rootzone mix meet the bridging criteria to prevent migration into the gravel?

	Particle Diameter (%)								
					Sand (mm)				
	Gravel (mm)	Fine Gravel (mm)			V. coarse	Coarse	Medium	Fine	V. fine
Sample	6.3	4.75	3.35	2.0	1.0	0.50	0.25	0.15	0.05
Gravel									
% Retained (>D)	33.2	21.9	26.1	14.8	3.2	0.8	0.0	0.0	0.0
% Passing (<D)	66.8	44.9	18.8	4.0	0.8	0.0	0.0	0.0	0.0
Rootzone mix									
% Retained (>D)	0.0	0.0	0.0	0.3	5.5	36.2	38.9	13.7	3.3
% Passing (<D)	0.0	0.0	0.0	99.7	94.2	58.0	19.1	6.4	3.1

For bridging to occur: D_{15} (gravel) $\leq 8 \times D_{85}$ (rootzone).

From the data in the table, a regression equation is developed to fit the line or curve to determine D_{15} and D_{85}. When graphing these values, remember to graph size versus the cumulative percentage that has passed for a particular size and not percentage retained (Fig. 3.18). For sands, this curve is often sigmoid-shaped while it is linear or quadratic in nature for gravel. For the gravel in this example, D_{15} is 2.95 *mm* and the D_{85} value for this rootzone is 0.84 *mm* (see Fig. 3.19). Therefore,

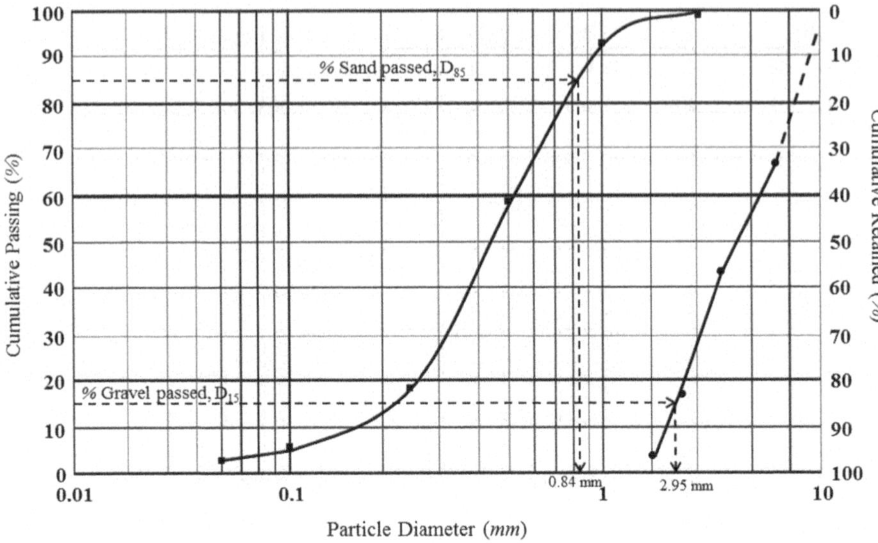

Fig. 3.19 For bridging to occur: D_{15} (gravel) $\leq 8 \times D_{85}$ (rootzone). From the data in the table, a regression equation is developed to fit the line or curve to determine D_{15} and D_{85}. For sands, this curve is often sigmoid-shaped while it is linear or quadratic in nature for gravel. For D_{15} (gravel) this is 2.95 *mm* while for the D_{85} (rootzone), it is 0.84 *mm*

$8 \times 0.84 = 6.72\,mm$ and $2.95\,mm \leq 6.72\,mm$. Bridging of the coarsest 15 % rootzone particles should occur with the finest 15 % gravel particles and this gravel is considered compatible for bridging with the sand.

For adequate permeability: D_{15} (gravel) $\geq 5 \times D_{15}$ (rootzone).

Again, a regression equation or interpolation is needed to determine D_{15} for the rootzone which is 0.22 *mm*. From this, the D_{15} (gravel) is $\geq 5 \times D_{15}$ (rootzone), which is $5 \times 0.22 = 1.10\,mm$. Therefore, adequate permeability should occur with this gravel and rootzone sand.

For Uniformity Coefficient of Gravel: D_{90} (gravel)/D_{15} (gravel) ≤ 3.0.

If Fig. 3.19 is expanded to include D_{90} (gravel), its value is then 7.92 *mm* and D_{15} (gravel) is 2.95 *mm*. Therefore, D_{90} (gravel)/D_{15} (gravel) $= 7.92/2.95 = 2.68$ which is less than 3.0. Thus, the Uniformity of Coefficient of this gravel passes the criteria.

If the proper-sized "pea" gravel is not available, then a 2 to 4 *in* (5 to 10 *cm*) layer of coarse sand (1 to 4 *mm* particle diameter) is placed on top of the gravel layer. This coarse sand layer is commonly referred to as the "choker" layer, which acts as a barrier to prevent rootzone soil particles from migrating downward into oversized gravel (Fig. 3.2). It is best to install the coarse sand layer manually to prevent mixing with or into the gravel bed. Sand should be delivered and dumped on the outside perimeter of the green and moved into place in wheelbarrows on a plywood board path. If 1/4 to 3/8 *in* (6.4 to 9.5 *mm*) "pea" gravel is available and the rootzone particle size conforms to those limits previously discussed, then the choker layer may not be necessary. Normally, it is cheaper and easier to use properly sized pea gravel alone compared to using coarser-sized gravel plus a choker layer since the choker layer must be evenly spread by hand instead of using a machine.

Gravel Composition

Superintendents should carefully choose their gravel. Several types of stone are sold for use in drainage; including crushed limestone, crushed granite, and river rock or gravel (mainly quartz). Other local sources of stone may be available. Granite and quartz gravels are best since they are strong and less likely to be crushed. Softer gravels, such as limestone (calcium carbonate), may break down over time due to the weight of the overlying soil and to chemical reactions with acidic water. Gravel suspected of being soft should be tested by a soils laboratory using the LA Abrasion Test (ASTM procedure C-131) and values should exceed 40. Gravel sources other than granite and quartz should also be analyzed to determine weathering stability using the Sulfate Soundness Test (ASTM procedure C-88). Weight loss should be less than 13 %.

Soil Modification to Improve Permeability

Soil modification to enhance internal soil moisture percolation is a common practice in the turfgrass industry. However, several misconceptions exist regarding soil modification to improve permeability. One such misconception is manifested in the practice of applying a 2 to 6 *in* (5 to 15 *cm*) layer of sand over a native soil with little or no surface slope provided and no subsurface drain lines installed. This is often referred to as the "bathtub" effect where the finer-textured native soil will not adequately drain and the area holds water like a bathtub (Fig. 3.20). Heavy rainfall then causes saturation of the added sand layer and surface water accumulates, causing poor playing conditions. This is why most heavy use turf areas need 10 to 12 *in* (25 to 31 *cm*) of modified topsoil and properly spaced drain lines to lower this excess surface moisture further down in the soil profile. The drains act similar to a drain in a bathtub, providing a means of water removal.

Another misconception is that an inch or so of a coarse sand, such as a river bottom sand, can be tilled into the top 3 to 6 *in* (7.6 to 15 *cm*) of native soil to enhance internal percolation. Unfortunately, this practice is rarely successful. First, a uniform, medium to medium-course sand should be used that has consistent particle size. River bottom sand often has a wide range of particle sizes. This variety in particle size allows smaller silt and clay particles to become dispersed among the larger sand particles, effectively reducing the pore space for water to percolate. Similarly, adding sand to native soil, which often has a high degree of silt and/or clay, often "clogs" these larger internal sand pores, again reducing internal percolation. Lastly, trying to uniformly "mix" the surface applied sand with the underlying soil is virtually impossible with a tractor-mounted roto-tiller. These

Fig. 3.20 Water accumulating in a "bath tub" effect associate with insufficient drainage when a sand-based rootzone overlies a slow-draining subsoil

machines will not provide the blended soil mix desired. Proper mixing requires "off-site" machine blending.

Table 2.1 demonstrates the results of blending high-quality (USGA specified) sand into a native Cecil clay soil. The sand:clay blend was performed "off-site" in a laboratory, providing a very uniform distribution of sand and soil in the various ratios. As shown in Table 2.1, adding just 10 % clay soil to this sand reduced its hydraulic conductivity by almost 85 % (from 58 to $9\,in\ h^{-1}$, 148 to $23\,cm\ h^{-1}$). Conductivity values quickly dropped as the clay soil content increased; for example, with a 50:50 blend, the hydraulic conductivity was less than $0.2\,in\ h^{-1}$ ($0.5\,cm\ h^{-1}$), totally unacceptable by today's standards. Furthermore, adding 20 % sand to the soil reduced drainage more than 50 % compared to straight (100 %) soil. This again represents small soil particles "clogging" the larger pores between sand particles. Refer to Chap. 2 for several equations used when mixing various sized sands and resulting effects on K_{sat}, particle-size distribution percentages, and bulk density.

Changes in Soil Characteristics Over Time

Over time, even the best built facility is subjected to changes in the soil physical and chemical characteristics. Typically, due to traffic (especially on saturated soil) and layer development, many of the initial desirable soil physical characteristics (i.e., K_{sat}, porosity) are reduced while undesirable ones (i.e., bulk density, organic

Table 3.6 Target ranges for various soil physical and organic properties of established golf greens depending on their description[a]

	Green description			
			Well-drained	
Characteristic	Native soil	Modified soil	0–4 *in*	4–12 *in*
Infiltration rate (*in* h^{-1}, *cm* h^{-1})	≥2 (5)	≥4 (10)	≥6 (15)	≥10 (25)
Noncapillary porosity (%)	≥12	≥14	~20	≥20
Capillary porosity (%)	15 to 30	<30	15 to 25	<20
Bulk Density (*g* cm^{-3})	1.35 to 1.45	1.35 to 1.45	1.35 to 1.45	1.40 to 1.50
Water retention (%)	10 to 25 %	<25	10 to 20	10 to 20
Organic content at various depths (%):				
• 0.25 to 1 *in* (0.6–2.5 *cm*)	1.5 to 2.5	1.5 to 3.0	1.5 to 2.5	0.1 to 1.0
• 1 to 2 *in* (2.5–5 *cm*)	1.0 to 2.0	1.0 to 2.0	1.0 to 2.0	0.1 to 1.0
• 2 to 3 *in* (5–7.6 *cm*)	0.5 to 2.0	0.5 to 2.0	0.5 to 2.0	0.1 to 1.0
• 3 to 4 *in* (7.6–10 *cm*)	0.5 to 2.0	0.5 to 2.0	0.5 to 1.5	0.1 to 1.0

[a]Target ranges from ISTRC (Oppold 1997)

matter, water retention) increase. For example, the %-moisture retained increased at one facility from 20 % after 1 *year*, to 23 % after 6 *years* to 31 % after 19 *years* while K_{sat} decreased from 11, to 4 to 3 *in* h^{-1} (30, 10, and 7.6 *cm* h^{-1}) over the same period (Habeck and Christians 2000). Similarly, aeration porosity decreased from 20, to 14, to 8 % while capillary porosity increased from 29, to 33, to 41 % over the same period. Besides traffic, layer development in terms of organic matter, using excessively fine topdressing material, and possibly silt and clay deposits in irrigation water all can contribute to these soil physical characteristics changes. Table 3.6 provides suggested soil physical and organic matter ranges for established greens, depending on their description.

Iron Oxides

A relatively recent discovery in some sand rootzones is the buildup of various iron oxides. These are presumed to originate from the recent agronomic shift away from high annual nitrogen rates. Iron used to provide desirable green color, therefore, has increased dramatically. The iron is often in its ferrous state (Fe^{+2}) which completely dissolves in water. Ferrous iron is also found in irrigation water in certain areas. Although the water may appear clear, if ferrous iron is present, it turns a yellow or rusty brown color after being exposed to air. Because ferrous iron dissolves in water it can easily pass through standard filtration systems. It also causes reddish-brown staining in toilets, showers, and sinks—especially where water faucets drip. Ferric iron (Fe^{+3}), unlike ferrous iron, does not dissolve in water and is precipitated out creating an orange-brown coating which can clog filters, pipes, showerheads, and soils over time (Fig. 3.21).

Ferrous iron forms under anaerobic (or saturated) conditions. It typically forms near the soil surface and once dissolved in water, becomes soluble and can easily

Fig. 3.21 A recent phenomena has been noticed in sand-based rootzone where a layer of orange to rust-colored ferric iron (Fe^{+3}) forms at the interface of the rootzone and underlying gravel drainage bed. This typically originates as a dark gray-colored ferrous iron (Fe^{+2}) at the top of the rootzone and becomes mobile when it encounters anaerobic conditions. Once the ferrous iron contacts the oxygenated gravel layer, it turns into ferric oxide (Fe_2O_3) which becomes immobile. The hard, impenetrable ferric iron layer then disrupts normal moisture drainage, causing a higher zone of saturating to develop which may eventually reach the soil surface

move (leach) downward. At this stage, the soil ferrous iron typically has a black or dark grey color, often seen as "streaks" in the profile. When the ferrous iron encounters aerobic conditions, such as the underlying drainage gravel, it transforms into ferric iron. Ferric iron is insoluble, often forming a distinct layer that becomes dense and hard, taking on an orange or rust color, and reducing or preventing drainage out of the bottom of the rootzone. Normal soil water drainage then is disrupted, often causing the rootzone to become increasingly saturated.

3.2 Bunker Sands

Several criteria are used to determine the long-term suitability of particular sand for golf course bunkers. These include:

- Particle Size Distribution
- Particle Shape
- Crusting Potential
- pH and Hardness

- Infiltration Rate
- Color
- Overall Playing Quality

Particle Size Distribution

Ideal bunker sand particle size range is between 0.25 and 1 *mm*. USGA recommends >75 % of the sand should fall between 0.25 and 0.50 *mm*. However, a sand meeting this narrow range is difficult to find and expensive to make. Therefore, a slightly wider particle size range is typically satisfactorily used for bunker sand specifications (Table 3.7).

Particle Shape

The shape of the sand particles has a strong influence on playing quality and maintenance of a bunker sand. A sand particle's shape is classified by examining both the relative sharpness of the particle's edges (termed **angularity**) and the overall shape (termed **sphericity**) of the particle itself (Table 3.8). The surface of particles can range from very angular (i.e., many sharp, well-defined edges) to well rounded (smooth surfaces). The shape of the particle can range from low sphericity (an elongated particle) to high sphericity (a particle that is nearly round).

The angularity and sphericity of the particles have a strong influence on the playing quality of the sand. For example, a low-sphericity, very angular sand generally has high resistance to "fried-egg" (or impact pressure) lies. Such sand also tends to stay in place better on the bunker faces. However, this same

Table 3.7 Sand particle size for bunkers

Size (*mm*)	Percent (%)
Fine gravel (>2.0)	<2
Very course sand (1 to 2)	<15
Coarse + medium + fine sand (0.15 to 1.0)	78 to 100
Very fine sand (0.05 to 0.15)	<5

Table 3.8 Criteria for selecting bunker sands

Test criteria	Good	Fair	Poor
Silt and clay content (<0.05 *mm*)	<3 %	3 %	>3 %
Ball penetration ($kg\ cm^{-2}$)	>2.4	1.8 to 2.4	<1.8
Crusting	None	Light	Moderate to severe
Set up (crust formation)	None	Light	Moderate to severe
Shape	Angular	Sub-angular	Round

sand would produce very firm bunkers that some players may find objectionable. Well-rounded, high-sphericity sands can produce fried-egg lies and are more likely to move off the bunker face during maintenance and irrigation or rainfall.

Sands usually consist of a mixture of particle shapes and sizes. This is important to the stability and playing quality of the sand. Again, as a general rule, sands highly uniform in size range and shape (particularly if rounded with high sphericity) tend to be less stable than sand with a wider range of particle dimensions. A soil scientist would view sand particles under magnification and classify its shape based on its angularity and sphericity.

Penetrometer

A sand penetrometer is a hand held; spring-loaded device that directly measures the resistance a particular sand has to applied pressure (or fried-egg lie). This somewhat indirectly measures the angularity and sphericity of sand particles and their stability under pressure (Fig. 3.22). Values greater than 2.4 $kg\ cm^{-2}$ are desirable (i.e., are able to better resist fried egg lies), those between 1.2 and 2.4 $kg\ cm^{-2}$ fair, and values $<1.8\ kg\ cm^{-2}$ are considered undesirable (least likely to resist fried egg lies).

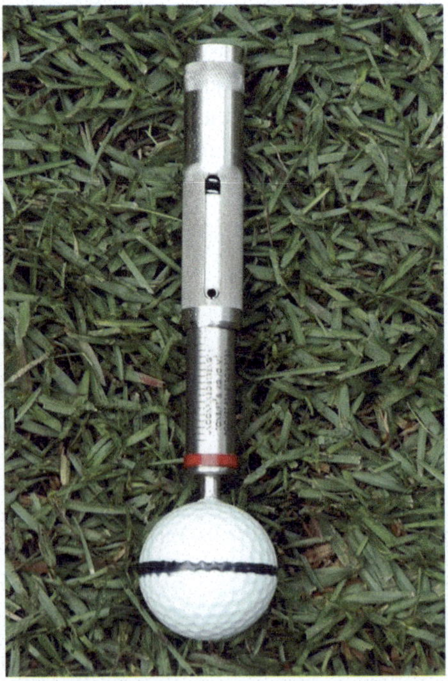

Fig. 3.22 A sand penetrometer measures the resistance a particular sand has to applied pressure

Crusting Potential

Crusting is the formation of a layer of dried, stiff sand on the surface of the bunker. Such layers typically are 1/8 to 1/4 *in* (3.2–6.4 *mm*) in thickness, and they severely decrease the playing quality of the bunker. Sands prone to crusting require more frequent raking to maintain good playing quality. If the crusting potential is high, the bunkers will require raking following each irrigation and rainfall event. This greatly increases the labor required to keep the bunkers in good conditions.

Crusting is directly related to the percentage of silt and clay in the sand. As silt and clay increase, the severity of crusting increases as well. To directly test for crusting potential, simply wet a thin layer of sand and allow it to dry overnight. Then attempt to lift the layer on the edges using a spatula. Crusting usually reported as:

N	None
L	Light
M	Moderate
S	Severe

A more quantitative method to test for crusting potential is to measure the silt and clay content of a proposed sand using the decantation procedure based on Stokes Equation (Chap. 1) (Table 3.8).

pH

Sands with an extremely high pH (>8.0) are likely to be strongly calcareous and therefore subject to physical and chemical weathering over time. However, the pH value is much less important than other test results and, on its own, should not be used to disqualify a sand being considered for use.

Hardness

The hardness value will have an influence on the mechanical weathering of a sand. Sand particles that are very soft can be crushed into smaller particles during raking; thus, a soft sand may drain perfectly at first and slowly degrade in quality as the particles are broken down.

Infiltration Rate

For bunkers the infiltration rate is usually 50+ $in\ h^{-1}$ ($127 + cm\ h^{-1}$) due to the absence of organic matter and less compaction. This can be a good test for analyzing bunker sands as readings of less than 20 *in* h^{-1} (51 *cm* h^{-1}) indicate severe contamination with silt, clay and organic debris.

Color

Generally, white-colored sands are preferred by golfers due to their aesthetic values. However, these are not always available locally; thus, another colored (and cheaper) sand can be used if it meets the above soil physical criteria.

Overall Playing Quality

This is a very subjective, but often important bunker selection criteria. One means to evaluate several potential bunker sands is to divide a large bunker into sections. Sands which meet the above physical and chemical properties are placed in separate sections in the bunker and the golf professional and interested members hit out of these. A vote is then taken by the participants to determine which sand is considered "best" for a particular club.

3.3 Sports Fields and Sand Capping

Proper construction produces a field with good surface and subsurface drainage that is easier to maintain and quicker to resume activity following heavy rainfall (Fig. 3.23). With increasing interest in playing athletic events on time to satisfy television contracts, proper drainage is a must for higher profile fields. The two main types of rootzones are native soils and sand based soils, and both have their advantages and disadvantages.

Soil Profiles

Native Soil Rootzones

Native soils can provide acceptable playing surfaces if they are properly designed and maintained. This type of field construction is best for lower profile fields with limited budgets, frequency and intensity of use. Native soils generally have higher water and nutrient holding capacities which provides a better growing medium for grass plants. Native soil fields, however, are more likely to become compacted quicker depending on the intensity and frequency of use and soil moisture content

Fig. 3.23 Proper construction produces a field with good surface and subsurface drainage that is easier to maintain and quicker to resume activity following heavy rainfall. With increasing interest in playing athletic events on time to satisfy television contracts, proper drainage is a must for higher profile fields. Unamended, native soil fields (shown) often drain poorly, eventually rutting and puddling excessively

when used. If compaction is addressed with routine maintenance and renovation, these fields have excellent traction and playability. Unfortunately these soils are influenced the most by weather conditions. If the field is used in periods of high soil moisture, the soil structure can be destroyed, causing compaction and surface rutting. If this occurs, usually renovation is the only way to restore the field.

In an effort to increase drainage and decrease compaction, sand is often added to the surface of a native soil and roto-tilled in, or added in a long-term topdressing program or following aeration. Caution should be exercised when doing this. In many cases, instead of correcting a problem, additional ones are created. In order for this strategy to succeed by increasing the permeability of a native rootzone, fields typically require 60 % or more sand on a volume basis throughout the rootzone. In heavier clay or high silt-containing soils, significant improvement in drainage and aeration properties do not typically occur until sand volume proportions exceed 80 % or more depending on particle size distribution of the sand and soil components. Proper and thorough mixing of the sand and soil also can only be achieved when the soil is mixed offsite; otherwise, a marbling effect will occur.

Sand-Based Rootzones

To improve soil water drainage and resistance to compaction, a sand based rootzone should be considered. This allows for frequent use and the ability to withstand variable weather conditions. Some disadvantages associated with sand rootzones include poor surface stability, poor water and nutrient holding capacity, and high costs for installation and maintenance.

Selecting the Right Rootzone Mix

To be successful, any sand based rootzones and their components should be analyzed for particle size range distribution, surface firmness, rootzone depth, moisture retention, and sand particle stability. As previously discussed, sand is divided into five main classes based on particle size: very coarse, coarse, medium, fine, and very fine (Table 3.9). Particle size analysis, however, is not enough information to predict how a rootzone mix will perform. Porosity (total and capillary), organic matter, bulk density, coefficient of uniformity, water retention, and water infiltration and percolation rate are other important performance criteria used to test a rootzone mix (Table 3.10). The ideal sand for such fields has particle size between 0.1 and 0.5 *mm*, the coarse to fine texture range. However, sands with a minimum component of 60 % (95 % preferably) between 0.25 and 1 *mm* diameter are acceptable (Table 3.11). **The importance of maintaining these tight soil specifications cannot be overemphasized.** Sands outside this range either become droughty and unstable or result in hard surfaces with reduced drainage potential. Sand particles larger than 2 *mm* in diameter should constitute less than 5 % of the sand by weight (Table 3.9). In most instances, concrete or coarse building sands are

Table 3.9 Suggested soil size distribution percentage ranges for sports fields as well as particle range for faster and slower draining fields

<table>
<tr><th>USDA size class (mm)</th><th>Normal drainage (%)</th><th>Faster drainage (%)</th><th>Slower drainage (%)</th></tr>
<tr><td>Gravel (>2)</td><td>—</td><td rowspan="2"><10</td><td rowspan="2"><10</td></tr>
<tr><td>Very coarse sand (1–2)</td><td><10</td></tr>
<tr><td>Coarse sand (0.5–1)</td><td>>35</td><td rowspan="3">>80</td><td rowspan="3">~60–65</td></tr>
<tr><td>Medium sand (0.25-0.5)</td><td>>50 (75[a])</td></tr>
<tr><td>Fine sand (0.1–0.25)</td><td>≤25</td></tr>
<tr><td>Very fine sand (0.05–0.1)</td><td><10</td><td rowspan="3"><10</td><td>≤15</td></tr>
<tr><td>Silt (0.002–0.05)</td><td rowspan="2"><10</td><td rowspan="2"><10</td></tr>
<tr><td>Clay (<0.002)</td></tr>
</table>

[a]At least 75 % medium sand is ideal

Table 3.10 Suggested soil physical parameters for sports fields

Parameter	Suggested
Soil mix	– 80 to 85 % sand + 10 to 15 % peat + 5 to 10 % native soil
Sand fractions	–>75 % all sand between 0.1 and 0.6 *mm*
	–<10 % in very coarse sand range (1.0 to 2.0 *mm*)
	– 70 to 90 % coarse + medium + fine range (1.0 to 0.1 *mm*)
	–<25 % very fine sand + silt + clay (<0.1 *mm*)
Gradation index (D_{90}/D_{10})	– 3.5 ± 1.0
K_{sat} ($in\ h^{-1}$)	– 6 to 16 (15–41 $cm\ h^{-1}$)
Porosity: total (%)	– 35 to 55
• Capillary (small pores)	• 15 to 25
• Non capillary (large pores)	• 18 to 25
Water retention (%)	– 14 to 20 (18 % is ideal)
Bulk density ($g\ cm^{-3}$)	– 1.2 to 1.6 $g\ cm^{-3}$ (1.4 is ideal)

Table 3.11 Two sand samples and their physical properties

		Sand sample (% by weight)		
Particle size (*mm*)	Description	**1**	**3**	USGA recommendation
>2.0	Gravel	33.6	<1	<3 %
1.0 to 2.0	Very coarse sand	5.2	1.0	<7 %
0.5 to 1.0	Coarse sand	10.7	1.0	<45 %
0.25 to 0.5	Medium sand	23.7	13	≥35 % (75 ideal)
0.15 to 0.25	Fine sand	24.4	68	≤20 %
0.05 to 0.15	Very fine sand	2.2	15	<5 %
≤0.05 *mm*	Silt + clay	0.2	1.0	≤8 %
Additional parameters				
K_{sat} ($in\ h^{-1}$)	—	13.3	4.1	>6 (>15 $cm\ h^{-1}$)
Bulk Density ($g\ cm^{-3}$)	—	1.6	1.42	1.2 to 1.6 (1.4 ideal)
Pore Space (%)	—	41	46	35 to 55

too coarse for soil mixtures as the field becomes very hard, droughty, compacted and/or the soil surface becomes unstable. Likewise, river bottom sand often has excessive coarse particles or excessive silt and/or clay, thus drains poorly and compacts easily. Washing and screening may improve these sands, but their specs should be checked by a qualified laboratory before use. Mason sand is usually concrete sand screened to remove gravel. It has more fine particles but is generally preferred as most of its particles are in the very coarse, coarse, and medium sand range. Most importantly, sands for sports fields cannot have their permeability destroyed by compaction. Also, a balance must be struck between hydraulic conductivity and moisture retention.

Sand used for field construction is often mixed with a small percentage (5–20 % by volume) of a suitable native loamy soil and/or organic amendment such as sphagnum peat, to increase water and nutrient retention (discussed later). Other organic sources such as sawdust, seed hulls, sewage sludge, and animal manure should be examined closely before use. Some of these materials are relatively short-lived, contaminated with weed seeds, or could reduce or restrict drainage. All components being considered should be sent to a soil physical laboratory for analysis and approval before use.

A final laboratory value for saturated hydraulic conductivity between 6 and 16 $in\ h^{-1}$ (15 and 41 $cm\ h^{-1}$) is desirable for most sports fields (Table 3.10). The water retention rate (capillary porosity) of the mix should also be about 18 % by weight. Higher drainage values and lower water retention rates than these require increased inputs in terms of fertilizer and irrigation while lower values may provide insufficient drainage in a given period of time. Pure sand fields are not recommended due to excessive drainage, thus the need for heavy amounts of water and fertilizer, and since most 100 % sand-based fields lack the surface stability and firmness necessary for good footing without excessive tearing of the turf (Fig. 3.24). Pure sand fields also accumulate thatch quicker than fields modified with a small amount of organic matter or native soil. Experience strongly suggests that adding 5 to 10 % of a laboratory approved native loamy soil provides footing stability without significantly sacrificing drainage. A sand-based field with 6 $in\ h^{-1}$ (15 $cm\ h^{-1}$) saturated hydraulic conductivity and 10 % silt and clay will grow in faster and be more stable than one with 20 $in\ h^{-1}$ (51 $cm\ h^{-1}$) saturated hydraulic conductivity and 3 % silt and clay.

Fig. 3.24 Pure sand fields are not always recommended due to excessive drainage, thus the need for heavy amounts of water and fertilizer, and since most 100 % sand-based fields lack the surface stability and firmness necessary for good footing without excessive tearing of the turf. Pure sand fields also accumulate thatch quicker and more so than fields modified with a small amount of organic matter or native soil

Example In Table 3.11, two sand samples are being considered for a sports field rootzone. Based on their physical properties, select one of them. First, neither sand sample falls within the recommendation guidelines suggested by the USGA. Sample 1 has almost 34 % gravel when the recommended amount is less than 3 %. It also is insufficient in medium sand (~24 % vs. minimum of 35 % recommended) and excessive fine sand (24 % compared to 20 % maximum allowed). However, its K_{sat} is sufficient to high (13 $in\ h^{-1}$, 33 $cm\ h^{-1}$). Its bulk density is at the maximum level of 1.6 $g\ cm^{-3}$. Conversely, sand sample 2 has the majority of its sand in the particle size description of fine sand (68 %) and very fine sand (15 %). The USGA recommendation for each is less than 20 % for fine sand and less than 5 % for very fine sand. Its hydraulic conductivity (K_{sat}, 4.1 $in\ h^{-1}$, 10.4 $cm\ h^{-1}$) is also below the minimum recommended level (6 $in\ h^{-1}$, 15 $cm\ h^{-1}$). Therefore, though neither sand is close to USGA recommended guidelines, sand sample 1 would likely be the better option of the two. A more detailed analysis of the particles >2.0 mm is needed. Extreme surface firmness should be expected with this sample.

Rootzone Depths

Determining depths needed for a specific rootzone mixture has been suggested by several mathematical methods (Table 3.12). The suggested depths fluctuate depending on budget constraints, field expectations, anticipated annual rainfall, rootzone mixture used, if a gravel bed is used, if drain lines are installed, and if so, their spacing. Typically, these depths range from 6 to 12 *in* (15 to 30 *cm*). The following are various means of determining rootzone depths and a discussion of each with some strengths and weaknesses.

Using Soil Moisture Retention Curves (SMRCs) for Determining Rootzone Depths

Several means to determine rootzone depths utilize SMRCs. Soil moisture retention curves are created for all rootzone blends being considered as outlined in Chap. 1. SMRCs quantify soil moisture amounts at various matric tensions. This allows for estimation of water content at various depths, based on distance above impermeable soil layers or perched water tables, and effects of soil compaction on soil moisture retention and porosity.

Sand Depth When Drain Lines Are Installed

Turf areas with insufficient natural drainage are often "sand capped" by placing sand to a particular depth above native soil in an attempt to provide a dry surface so play can resume. Sufficient depth of sand is needed to provide a desirable dry

Table 3.12 Comparing several proposed means of determining rootzone depths for sports fields and golf course fairway sand capping

Method	Brief description	Additional information	Reference
Depths based on soil moisture retention curves (SMRC)	SMRCs determine depth needed to reach 10 to 25 % aeration (air-filled) porosity. 10 % aeration porosity = minimum; 15 % = better; 35 % aeration porosity = best	When drain lines are not used, minimum sand depth is where capillary and aeration porosity lines meet on the SMRC graph. Typically, a 10 *in* (35 *cm*) minimum depth is needed	Chapters 1, 2 and 3
Depths based on adjusting air entry point (or top of the perched water table) values from SMRCs	From SMRCs, rootzone depth is based on air entry point (or top of the perched water table, also called critical tension) and these values are then adjusted. For golf greens, 4 *in* (10 *cm*) is added to the air entry point value while for fairways and sports fields, 6 *in* (15 *cm*) to the air entry point value	Hooghoudt's formula is also used in this method to determine drain line spacings. Adams and Gibbs (1994) also propose rootzone depth is based on critical tension but note water storage values are also important	McIntyre and Jakobsen (2000); Adams and Gibbs (1994)
Depths based on SMRCs over 100 to 600 *mm* range with drainage (air entry point) between 150 and 200 *mm* (6 to 8 *in*)	The suction at which air-filled porosity is ~25 % and water-filled (or capillary) porosity is ~30 % is the optimum rootzone depth	If a gravel layer is used, 5 *cm* (2 *in*) is deducted from this 150 to 200 *mm* value	Handreck and Black (2007)
Depths based on: $K_{sat} \geq 100\,mm\,h^{-1}$ for soccer to $150\,mm\,h^{-1}$ for golf. From SMRCs, capillary porosity of 15 %, air-filled porosity of ≥10 % (preferably 15 %) and a gravimetric moisture content between 10 and 18 %	Again, SMRCs are needed to determine the proposed capillary porosity and air-filled porosity depths	SMRCs are used to convert gravimetric moisture content to volumetric water content	Baker and Richards (1993, 1997); Bingaman and Kohnke (1970)
Rootzone depth is based on the critical tension formula: $\frac{71.4}{D\,(mm)}$ D = particle diameter of dominant pore	Rootzone depth is by estimating capillary rise of water in uniform diameter capillary tubes. Critical tension is inversely proportional to soil particle diameter (D in this equation)	The construction depth for a sand soil perched over a gravel drainage bed should neither be more than the critical tension nor less than half	Stewart (1994)

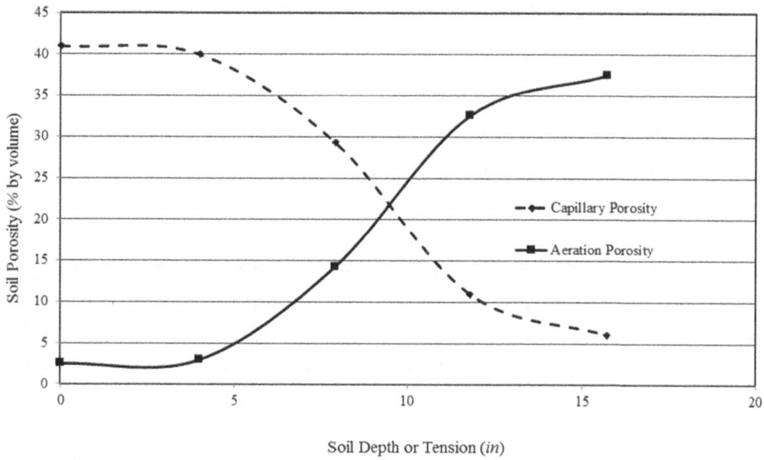

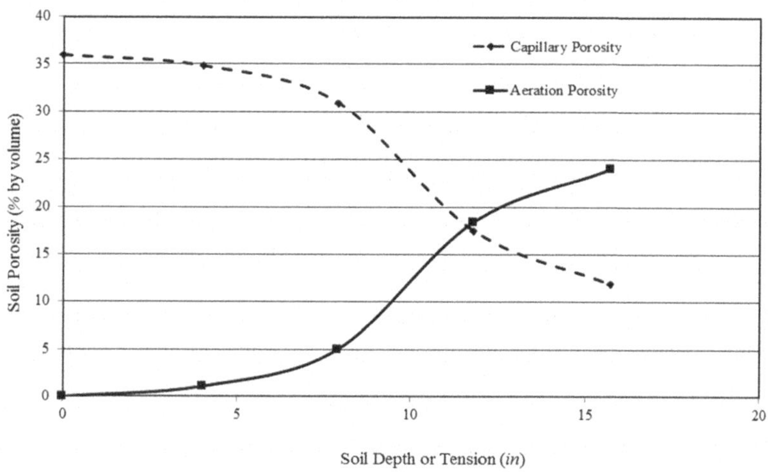

Fig. 3.25 Soil moisture retention curves for two samples being considered for sand-capping a fairway. The first sample (designated as *top*) has a saturated hydraulic conductivity (or K_{sat}) rate of 96 $in\ h^{-1}$ (244 $cm\ h^{-1}$) while sample #2 (*bottom*) has a K_{sat} value of 33 $in\ h^{-1}$ (84 $cm\ h^{-1}$)

surface or water will fill the soil pores and the surface will become saturated. To determine adequate depths, several variables need to be known: (1) porosity and K_{sat} of sand being used; (2) amount of anticipated rainfall; and, (3) drainage characteristics of the underlying native soil.

By plotting soil moisture content against soil depth (tension), one can approximate the depth needed for a sand-capping to adequately work for an anticipated rainfall event. For example, the soil samples represented in Fig. 3.25 were saturated and then at various soil depths (or tensions) the amount of remaining water held in the soil (designated as capillary porosity, % by volume) was plotted. Since total porosity remains essentially constant, a mirror-image of capillary porosity is aeration porosity, which indicates the percentage of air a soil retains as the water drains.

The USGA recommends golf green sands contain 15 to 35 % capillary (water-filled) porosity and 15 to 30 % aeration (air-filled) porosity. However, for fairway or sports soils, 10 % aeration porosity has been suggested as minimum. From Fig. 3.25 (top), to obtain 15 % aeration porosity at the soil surface for the top sand sample (#1), about 8 *in* (20 *cm*) of sand is necessary. To obtain 10 % aeration porosity, about 7 *in* (18 *cm*) of sand depth is needed. For the bottom sand sample (#2, Fig. 3.25, bottom), 10 % aeration porosity at the soil surface requires about 9 *in* (23 *cm*) of sand while 15 % aeration porosity requires about 11 *in* (28 *cm*).

This procedure makes several assumptions. The most critical one is that matric suction (tension) is zero at the bottom of the sand cap layer. For this condition to occur during drainage, the underlying soil must have an equal or greater drainage capacity than the sand being placed on top of it. Of course, this rarely occurs since the poorly draining turf soil is the reason sand-capping is being considered in the first place. If drain lines are not provided, a saturated zone will form at the intersection of the cap sand and original soil surface. Nil suction (zero matric suction) exists at the surface of the saturated zone. If insufficient depths of sand cap is provided, this saturated zone may reach the soil surface following rainfall or flooding, causing unwanted surface moisture.

To design for proper drainage line spacing, two additional values are needed. The first is the saturated hydraulic conductivity (K_{sat}) of the cap sand. The second is anticipated rainfall amount the soil will be exposed to or is being engineered for. These values, along with the depth of the sand cap, are then inserted into a simplified form of Hooghoudt's equation to determine the necessary drain line spacing for the particular situation (Chap. 2). Hooghoudt's equation estimates drainage rates in the soil furthest from the drain line which, of course, is the halfway point between two drain lines. This then provides drain line spacing for an "ideal" sand depth necessary to keep the top several inches below saturation, thus, open for play.

For example, sand sample #1 has a soil moisture release curve as depicted by Fig. 3.25, top and a K_{sat} value of 96 $in\ h^{-1}$ (344 $cm\ h^{-1}$). Sand sample #2 (Fig. 3.25, bottom) has a K_{sat} value of 33 $in\ h^{-1}$ (84 $cm\ h^{-1}$). To determine the drain line spacing to provide adequate drainage following a 1 $in\ h^{-1}$ (2.5 $cm\ h^{-1}$) rain event, Hooghoudt's equation calculates 137 *in* (11 *ft*, 3.4 *m*) drain lines spacing is needed for sand sample #1, while for sand sample #2, 103 *in* spacing (or about 9 *ft*, 3.7 *m*) is needed for adequate drainage. The greater K_{sat} value for sand sample #1 (96 $in\ h^{-1}$, 344 $cm\ h^{-1}$) allows for the wider drain spacing compared to sample #2 (33 $in\ h^{-1}$, 84 $cm\ h^{-1}$). If 15 % aeration porosity is used instead of 10 % as in the previous example, for sample #1, drain line spacing becomes 157 *in* (13 *ft*, 4 *m*) while for sample #2, this becomes 126 *in* (or about 11 *ft*, 3.4 *m*) (Table 3.13). If lower rainfall design rates are used, then the required minimum drain line spacing correspondingly increases. For example, with sand sample #1 and using 10 % aeration porosity, calculated drain line spacing is 11 *ft* (3.4 *m*) for a 1 $in\ h^{-1}$ (2.5 $cm\ h^{-1}$) rainfall event, 13 *ft* (4 *m*) for a 0.75 $in\ h^{-1}$ (1.9 $cm\ h^{-1}$) rainfall event and 16 *ft* (4.9 *m*) for an anticipated 0.5 $in\ h^{-1}$ (1.3 $cm\ h^{-1}$) rainfall event.

Table 3.13 Sand cap depths with and without drainage and drain line spacing of two sand samples based on soil moisture retention curves (Fig. 3.23) and Hooghoudt's equation using a 1 *in* h^{-1} (2.5 *cm* h^{-1}) anticipated rainfall event

Sample (K_{sat} values)	Aeration porosity 10 % Sand depth	Aeration porosity 10 % Drain line spacing	Aeration porosity 15 % Sand depth	Aeration porosity 15 % Drain line spacing	Sand depth without drain tiles: Minimum depth (using capillary & aeration porosity)	Sand depth without drain tiles: Ideal depth (using 35 % aeration porosity)
Sand #1 (96 *in* h^{-1})	7 *in*	11 *ft*	8 *in*	13 *ft*	9 *in*	10 *in*
Sand #2 (33 *in* h^{-1})	9 *in*	9 *ft*	11 *in*	11 *ft*	12 *in*	15+ *in*

Sand Depth When Drain Lines Are Not Installed

If drain lines are not used, then the depth of sand capping has to be increased so the saturated zone will not reach the soil surface following heavy rainfall. A *minimum* sand depth when drain lines are not used is where capillary and aeration porosity lines meet on the SWRC graphs. However, *ideally*, the soil depth necessary to obtain 25 % aeration porosity is used. For sample #1 in Fig. 3.25, this depth then increases to about 9 *in* (23 *cm*) as a minimum depth and 10 *in* (25 *cm*) for an ideal depth. For sample #2, minimum depth is now 12 *in* (30 *cm*) with 15+ *in* (38+ *cm*) needed for ideal depth. As a rule-of-thumb, when drain lines are not installed, a minimum of 10 *in* (25 *cm*) sand cap depth is needed for desired results. Again, these depths are necessary to compensate for the perched water table (saturated zone) that develops at the interface of the sand-cap and underlying soil and to allow adequate aeration porosity in the upper rootzone. The height of the perched water table (saturated zone) depends mainly on the saturated hydraulic conductivity of the underlying soil. Again, predicted capillary and aeration porosity values should actually be expected at soil depths above this saturated zone surface.

Obviously, there are enormous costs in properly sand-capping a fairway or sports field (Fig. 3.26). Shallower sand depths are often used due to costs and to avoid drying of the surface layer of sand during drought. This generally works except when heavy rainfall forces soil water to raise to the surface, causing saturation. Thus, facilities have to make a financial decision, balancing the number of days it is closed due to wet soils versus the costs of sand-capping and drain line installation. Overall, properly spaced drain lines are generally much cheaper (and more effective) in draining a fairway or sports field than adding sufficient sand to prevent saturated conditions from reaching the soil surface. Sand capping is best when insufficient surface contouring is provided and drain-tile daylight sites are unavailable due to naturally occurring high water tables.

Fig. 3.26 Sand capping a golf course fairway can provide desired results if appropriate soil science is used in determining needed sand depth and drain tile spacing

McIntyre and Jakobsen (2000)

These authors advocate running compacted hydraulic conductivity and water holding capacity tests to assess their performance and then using Hooghoudt's Formula to determine either drain line spacing, or rearranging the equation to determine appropriate soil depth to avoid roots from growing in the depth of free water zone (Table 3.12). Furthermore, they advocate using a soil moisture retention curve to determine various characteristics of a sand being considered. They identify critical points on the retention curve of interest: **Bottom of Capillary Fringe**, **Perched Water Table**, **Air Entry Point (AEP)**, **Aeration Porosity Depth**, **Gravitational Water**, **Field Capacity**, **Available Water**, and **Wilting Point**. From this, the optimum rootzone depth combines maximum perched water table depth plus an adequate depth of drained sand for root growth. This is performed by determining the air entry point of the sand and adding appropriate soil to account for rooting depth: for golf greens, 4 *in* (10 *cm*) is added, while for fairways/sports fields, 6 *in* (15 *cm*) is added to AEP values.

This method is similar to the first where both depend on constructing soil moisture retention curves and then using either porosity values or arbitrary soil depth values to keep potential rootzone saturation from occurring. Testing to determine saturated hydraulic conductivity and using the results in Hooghoudt's equation is the best design method for preventing the rootzone in single-tier profiles from becoming saturated during rainfall or irrigation. This method takes into account the true predicted maximum depth of the changing saturated zone during wetting and drainage. This would indicate the height of the water table between drains, and the authors suggest this could be used along with rooting depths of the turf to provide a soil depth that would prevent roots from growing into the free standing water table.

In addition, Adams and Gibbs (1994) state **critical tension** is the depth of pore water continuity required over a water-table to cause the maximum size of pore to empty, thus, permit air entry. Furthermore, with respect to a constructed profile, it is the depth to a capillary break layer of stone or gravel or the depth to the maintained water-table. They further add that sand depth over a gravel drainage bed should neither be more than the critical tension (as the soil surface may dry excessively) nor less than half the critical tension (as the sand remains excessively wet). Although the authors indicate laboratory measurements of critical tension are useful to help determine depths of constructed rootzone over a gravel drainage raft, it is not the only measurement of importance. Water storage is equally important, but they provide little guidance on appropriate values.

Handreck and Black (2007)

These authors advocate matching depths to the actual physical properties of the rootzone mixture, specifically to its moisture release curve over the 100 to 600 *mm* (10 to 60 *cm*) water suction range to see how the curve develops (Table 3.12). The point at which the mixture starts to drain (air entry point) should be in the 150 to 200 *mm* (15 to 20 *cm* or 6 to 8 *in*) range. This would set the anticipated perched water table at this depth. The suction (*mm* water) at which air-filled porosity is around 25 % and water-filled (capillary) porosity is around 20 % is the nominal optimum depth of the rootzone. Optimum means maximized perched water table depth combined with an adequate depth of drained sand for root growth. Their research recommends subtracting 5 *cm* (2 *in*) from this depth for tension from the gravel layer, netting a predicted depth of 100 to 150 *cm* (4 to 6 *in*). Again with this design approach, the water retention curve allows a more detailed and reliable way to predict the performance of a rootzone with respect to water available to plants.

Baker and Richards (1993 and 1997) plus Bingaman and Kohnke (1970)

These authors suggest minimum requirements of hydraulic conductivity of 100 $mm\ h^{-1}$ for soccer to 150 $mm\ h^{-1}$ for golf with capillary porosity of 15 %, air-filled porosity minimum of 10 %, preferably 15 %, and a gravimetric moisture content between 10 and 18 %.

Stewart (1994)

Stewart advocates using Jurin's formula when determining the rootzone depth needed by estimating the capillary rise of water in uniform diameter capillary tubes. Water will rise up the tube until the upward lift due to surface tension acting around their internal circumference of the tube balances the downward weight of water within the tube. The height of capillary rise formula is derived from the forces acting to pull the water upwards including surface tension, length of contact between water and pore or pore circumference, and the upward component of this force (cos α) and the forces acting to pull the water downward including water density, the volume of water above the free standing water surface, and gravity. This is known as Jurin's Formula (Fig. 3.27). Further discussion on the height of capillary rise formula is in Chap. 1.

$$\text{Force up} = \text{force down}$$
$$\mathrm{T}(2\pi \mathrm{r})(\cos\alpha) = \mathrm{d}(h\pi \mathrm{r}^3)(g)$$

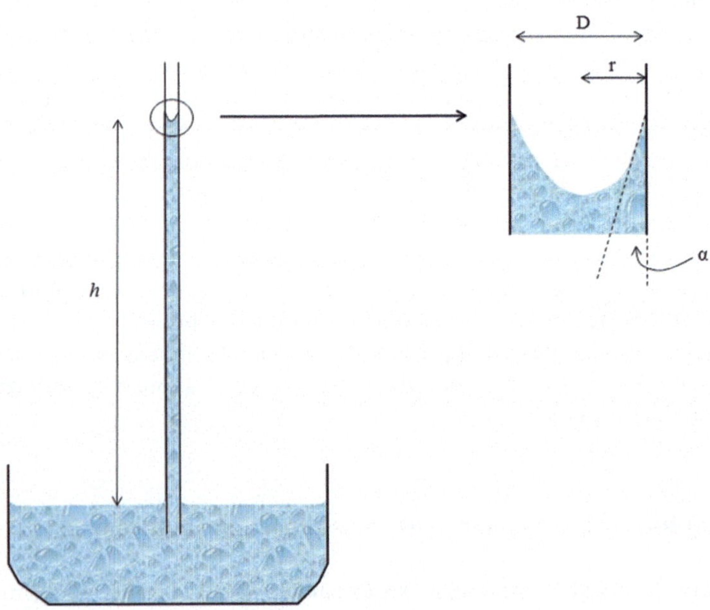

Fig. 3.27 The height (*h*) of capillary rise formula is derived from the forces acting to pull the water upwards. These include surface tension, length of contact between water and pore or pore circumference (r = radius; D = diameter), and the upward component of this force (*cos α*). Countering upward forces are those acting to pull the water downward including water density, the volume of water above the free standing water surface, and gravity

where:

h = height of the water column (cm),
T = surface tension of water (72 millinewtons or $mN\ m^{-1}$),
g = acceleration due to gravity (980 $cm\ s^{-3}$),
d = density of water (1 $g\ cm^{-3}$),
r = radius of the water-containing pore (cm).
D = diameter of the water-containing pore (cm).

The equation is then solved for h, where:

$$h = \frac{2\mathrm{T}\cos\alpha}{g\mathrm{dr}}$$

Since cos α is approximately 1, this is simplified to:

$$h = \frac{2\mathrm{T}}{g\mathrm{dr}}$$

As T, g and d can be considered constants, the equation is reduced to:

$$h = \frac{0.15}{\mathrm{r}} = \frac{0.3}{\mathrm{D}}$$

This is stating the smaller the pore, the higher water will rise in it. This, however, measures pore size and not particle size and the equation is adjusted to reflect this. Stewart further assumes that for a uniform set of spheres in closest packing, pore radius is ~0.21 times particle diameter, converting units to mm to adjust the formula to H = 15/0.21 (d). This now provides a formula that solves for the height of capillary rise based on particle diameter rather than pore size, and reduces the above equation to:

$$\text{critical tension or } h\ (mm) = \frac{71.4}{\mathrm{D}}$$

where: D = particle diameter dominating the pore system of mixed particle-size sand or gravel (mm).

This equation suggests that **critical tension** (or **air entry point**) is inversely proportional to particle diameter so that doubling the particle size will halve the critical tension. For practical evidence, the construction depth for a sand soil perched over a gravel drainage bed should neither be more than the critical tension nor less than half the critical tension.

Example According to Steward, using Jurin's Formula, a sand with an average particle diameter size of 0.4 mm has a critical tension (or perched water table depth) of what?

$$\text{critical tension or } h\ (mm) = \frac{71.4}{\mathrm{D}\,(mm)} = \frac{71.4}{0.4\,mm} = 179mm(\text{or } 7.0\,in)$$

This design technique bases sand depth on “critical tension”. This reference defines critical tension as being equal to capillary rise in pores created by particle diameter “dominating” the pore system (D).

Discussion

All of the previously discussed methods for determining optimum sand depth have scientific merit. As with most scientifically-based recommendations, the natural tendency is to come up with a “rule of thumb” for quick answers, rather than taking adequate time and effort to really evaluate a problem. For most sands, a quick answer of 25 to 30 *cm* (10 to 12 *in*) recommended depth would supply enough water storage in air-filled pores to hold water which previously made the existing conditions unsatisfactory because of excessive moisture at the soil surface.

The most useful source of valid scientific information for determining appropriate soil depth is water content related to tension, indicated by SMRCs. An adequate range of related values should be taken from a representative sand source to create a valid retention curve. The air entry point can be taken from the curve by finding the minimum tension required to cause a significant drop in water content from saturation (when the largest pores begin to empty and allow air to enter). This value is required for use in either single-tier or USGA profiles to accurately predict the depth of the perched water table. This is the depth to which rooting depth should be added, as this zone which remains saturated for significant amounts of time as the profile approaches equilibrium.

The agronomic-based design of sand depth is basically the same for profiles with or without a gravel drainage layer. The difference in recommended sand depths for profiles with no gravel layer is basically an engineering design difference. An adequate gravel layer and drain pipe system provides essentially instantaneous drainage (removal) of water in the profile below the sand layer. There is no need to add depth to the sand layer to accommodate storage of water above the amount held by the soil’s capillary capacity.

Testing to determine saturated hydraulic conductivity and using the results in Hooghoudt’s equation is the best design method for preventing the rootzone in single-tier profiles from becoming saturated during rainfall or irrigation. This method takes into account the true predicted maximum depth of the changing saturated zone during wetting and drainage. By spacing drain lines per Hooghoudt’s equation, the drains can be expected to remove rainfall or irrigation adequately to prevent rise in the saturated zone. Otherwise, additional air-filled pore space equal to the effective depth of water added to the profile must be provided by additional depth of sand.

In the case of sand capping, the physical properties of the existing site must also be considered. If drain lines are not installed below a sand cap, the infiltration rate of the existing soil will dictate actual drainage rates, and the sand's only real value is from the added storage in air-filled pores.

Example A SMRC was constructed for each sand in Table 3.13 with Fig. 3.28 (top) representing sand sample 1 in Table 3.13 and Fig. 3.28 (bottom) representing sand sample 2. Based on their SMRCs, select the minimum depth you would recommend the rootzone be constructed.

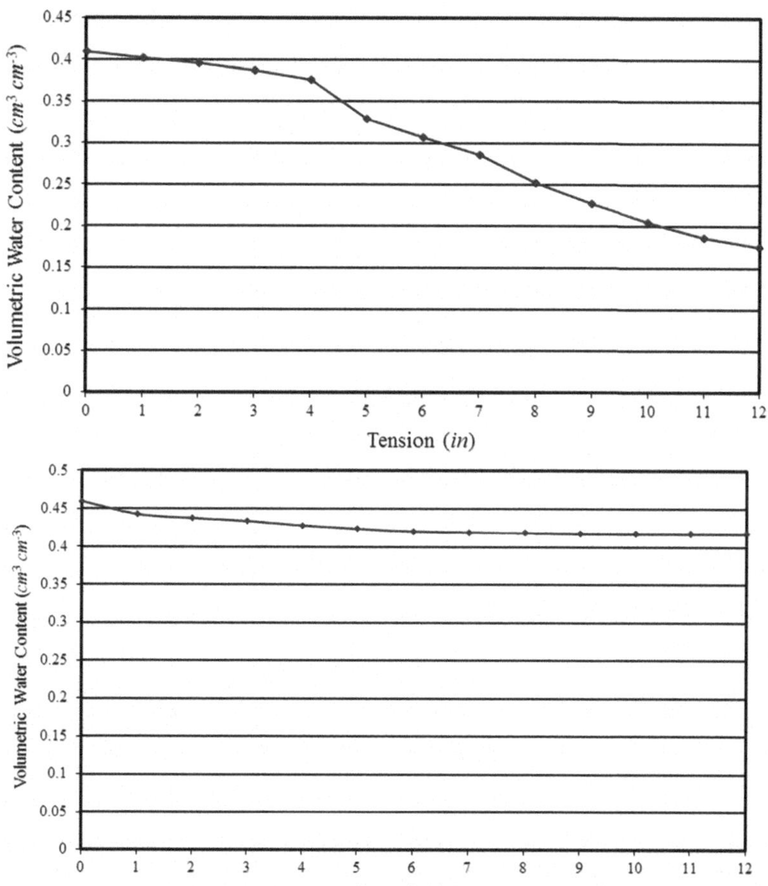

Fig. 3.28 Examples of soil moisture retention curves of two sand samples with their physical analysis listed in Table 3.12. The top sand has an air entry point starting at about 4 *in* (10 *cm*) while the bottom sand's SMRC remain essentially flat. This strongly suggests the sand will remain excessively wet even with sufficient rootzone depth as the capillary forces are too strong and successively retain soil moisture against the force of gravity

For sand sample 1, its SMRC (Fig. 3.28, top) indicates a minimum of about 4 *in* (10 *cm*) in depth would be needed before drainage would begin. Conversely, sand sample 2 (Fig. 3.28, bottom) cannot be placed deep enough to allow adequate drainage as the line is essentially flat, indicating the fine sand has sufficient capillary adhesion to attract water and not to allow it to drain, thus staying at or near saturation for extended periods. Therefore, though neither sand is ideal, sand sample 1 would be the better option of the two samples.

Summary

In summary, when determining the appropriate depth needed when sand capping or developing rootzone depths:

1. No universally accepted procedure exists for determining appropriate depth of sand capping/rootzones.
2. Optimum rootzone depth combines maximum perched water table depth (air-entry) with adequate depth of drained sand for root growth. Two of the most used methods are:

   (a) From SMRCs, determine rootzone depth (tension) needed at either:

   - 10 % aeration (air-filled) porosity (minimum);
   - 15 % aeration porosity (better);
   - 35 % aeration porosity (best).

   (b) From SMRCs, determine air entry point (top of perched water table):

   - To this value, add 4 *in* (10 *cm*) for golf greens;
   - To this value, add 6 *in* (15 *cm*) for fairways/sports fields.

3.4 Questions

1. When considering rootzone materials for golf greens, sand capping fairways, or for sports fields, what are the minimum testing requirements needed to evaluate these? (*particle size analysis, physical analysis including bulk density and soil porosity, saturated hydraulic conductivity when compacted, and soil moisture retention curves*).
2. Define and explain what a **perched water table** (zone of saturation) is in a rootzone profile and discuss if this a positive attribute? (*perched water table or zone of saturation, is a layer of water that forms at the junction of two layers of dissimilar sized particles. In turf, it most often develops at the interface of the rootzone media and underlying gravel layer. The zone of saturation will not*

move readily from the smaller pores of the finer-textured rootzone layer into the large pores of the coarser layer until the finer layer is saturated with sufficient water to break the capillary break (tension) between the two dissimilar materials. In general, a perched water table is desirable as well-draining sands will retain moisture in the rootzone, yet when sufficient rainfall occurs, drainage occurs as the rootzone now contains sufficient amount (weight) of water to break the capillary tension between the rootzone and underlying gravel layer).

3. Explain why adding several inches or centimeters of sand to an existing native topsoil and roto-tilling it in rarely produces the desired effect of improved drainage. (*the smaller sized particles of the native soil will "clog" the larger pores between the sand particles. Typically it requires 80 % or more of a topsoil mix to be sand before sufficient drainage occurs*).
4. Explain some considerations when using saturated hydraulic conductivity tests for evaluating potential rootzones. (*saturated hydraulic conductivity or* K_{sat}*measures the ratio of the flux, or infiltration and percolation rate of moisture into and through a soil profile, to the hyraulic gradient, which equals the total hydraulic head divided by the length of water movement through the soil . The soil should be compacted at field capacity to 3.03 J* cm^3*(14.3 lb-ft) using a 3 lb (1.36 kg) hammer dropped 15 times from a height of 30.5 cm (13 in. prior to running the test. Darcy's equation is then used to calculate* K_{sat}*. Typically, the* K_{sat} *of a field is reduced by about 33 % over the first several years of use due to settling of the rootzone and compaction from play and maintenance practices. Laboratory* K_{sat} *values should be between* 6 and 16 *in* h^{-1}, *15 and 41 cm* h^{-1}).
5. Soil moisture retention curves help explain water amounts at various depths of the rootzone. Discuss this and other potential benefits of SMRCs. (*To construct SMRCs, the soil is saturated and then exposed to various tension (typically 0 to 30 cm or 0 to 40 cm) to examine the pattern at which moisture is retained throughout the soil profile at these tensions. Typically a zone of saturation (perched water table) develops at the interface of the underlying gravel layer and rootzone. As one goes up the rootzone soil profile, it typically becomes drier. The critical tension (or air-entry) point is where water begins to drain from the soil's largest pores and indicates the top of the zone of saturation. The amount of moisture in the soil between air-entry point and wilting point nearer the surface is the plant available moisture. SMRCs also help predict the suitability of a soil for a rootzone by indicating moisture levels following compaction. If SMRCs indicate a significant increase in soil moisture throughout the profile following compaction, it may be too wet to be used as a desirable rootzone. Also, if the zone of saturation nears the soil surface, the soil isn't deep enough to absorb excessive rainfall without become saturated (a "mud hole") following heavy rainfall*).
6. The following is an analysis of three soils being considered for use as a golf green rootzone. Discuss the pros and cons of each, selecting the "best" one to be used and why.

Particle size distribution and soil physical properties of sand, sand/peat and sand/soil mixes.

Particle size distribution		USGA limits (%)		Sand (%)	Sand/peat (%)	Sand/soil (%)
Textural name	*mm*					
Gravel	>2	≤3	≤10	0.1	0.1	0.8
Very coarse sand	1.0–2.0	<7		7.6	7.3	12.0
Coarse sand	0.5–1.0	<45	≥60	26.0	25.4	24.6
Medium sand	0.25–0.5	>35 (75 ideal)		45.6	46.4	36.6
Fine sand	0.15–0.25	≤20		19.1	18.3	16.6
Very fine sand	0.05–0.15	<5	≤10	0.6	1.1	1.2
Silt	0.002–0.05	≤5		1.2	1.3	7.9
Clay	<0.002	≤3				
Soil physical properties						
Organic Matter (%)	—	1 to 5		1.3	2.3	2.0
Hydraulic conductivity ($cm\ h^{-1}$)		15 to 30		86.3	27.9	15.7
Bulk Density ($g\ cm^{-3}$)	—	1.2 to 1.6 (1.4 ideal)		1.75	1.57	1.74
Porosity:	—					
Total (%)		35 to 55		36.2	42.8	36.0
Capillary (%)		15 to 35		8.9	16.7	15.8
Air-filled (%)		15 to 30		27.3	26.1	20.3

(The sand/peat mix met the USGA recommendations for particle size distribution and soil physical properties. Straight sand also met the recommendations for particle size distribution, but values for capillary porosity and hydraulic conductivity were outside the range set by the USGA for root zone mixes. Particles greater than 1 mm made up more than 10 % of the sand/soil mix and bulk density was higher than recommended. Other soil physical properties were within USGA limits).

7. From the following sand particle size distribution, calculate the D_{95}/D_5, D_{60}/D_{10} and D_{90}/D_{10} and interpret the results.

Class	Particle diameter (*mm*)	% Retained
Gravel	>2.0	0.2
Very Coarse Sand	1 to 2	1.7
Coarse Sand	0.5 to 1	15.8
Medium Sand	0.25 to 0.5	67.3
Fine Sand	0.1 to 0.25	13.4
Very Fine Sand	0.05 to 0.1	1.6

First, calculate the cumulative % retained and cumulative % passing:

Class	Particle diameter (*mm*)	% Retained	Cumulative % retained	Cumulative % passing
Gravel	>2.0	0.2	*0.2*	*99.8*
Very coarse sand	1 to 2	1.7	*1.9*	*98.1*
Coarse sand	0.5 to 1	15.8	*17.7*	*82.3*
Medium sand	0.25 to 0.5	67.3	*85.0*	*15.0*
Fine sand	0.1 to 0.25	13.4	*98.4*	*1.6*
Very fine sand	0.05 to 0.1	1.6	*100*	*0*

Next, graph the sand diameter versus the cumulative % retained and passing (Fig. 3.27).

Next, determine D_5, D_{10}, D_{60}, D_{90}, and D_{95}. $D_5 = 0.15\,mm$; $D_{10} = 0.20\,mm$; $D_{60} = 0.38\,mm$; $D_{90} = 0.60\,mm$; and $D_{95} = 0.80\,mm$.

$$D_{95}/D_5 = 0.80\,mm/0.15\,mm = 5.33$$
$$D_{90}/D_{10} = 0.60\,mm/0.20\,mm = 3.00$$
$$D_{60}/D_{10} = 0.38\,mm/0.20\,mm = 1.90$$

The D_{95}/D_5 of 5.33, according to Bingaman and Kohnke (1970), this would be acceptable as it is between 2 and 6.

The D_{90}/D_{10} of 3.00, according to Adams and Gibbs (1994), this would be acceptable as it is ≤ 3.3.

The D_{60}/D_{10} of 1.90, according to Blake (1980), this would be acceptable as it is <4.

8. Indicate if the following sand and gravel samples will "bridge", preventing unwanted soil movement into the gravel layer (D_{15} Gravel/D_{85} Rootzone ≤ 8). Also indicate if these pass the Uniformity Factor (D_{90} Gravel/D_{15} Rootzone ≤ 3) and Permeability Factor tests (D_{15} Gravel/D_{15} Rootzone ≥ 5) (hint: graph % particles passing vs. diameter) (Figs. 3.29 and 3.30).

	Retained particle diameter (%)										
						Sand (*mm*)					
	Gravel (*mm*)					V. coarse	Coarse	Medium	Fine	V. Fine
Sample	13.5	9.5	6.3	4.0	3.0	1.0	0.50	0.35	0.15	0.05
Gravel % Retained (>D)	0.0	8.4	57.9	31.8	1.3	0.6	—	—	—	—
Rootzone mix % Retained (>D)	—	—	—	—	0.3	4.2	28.8	36.0	27.4	2.5

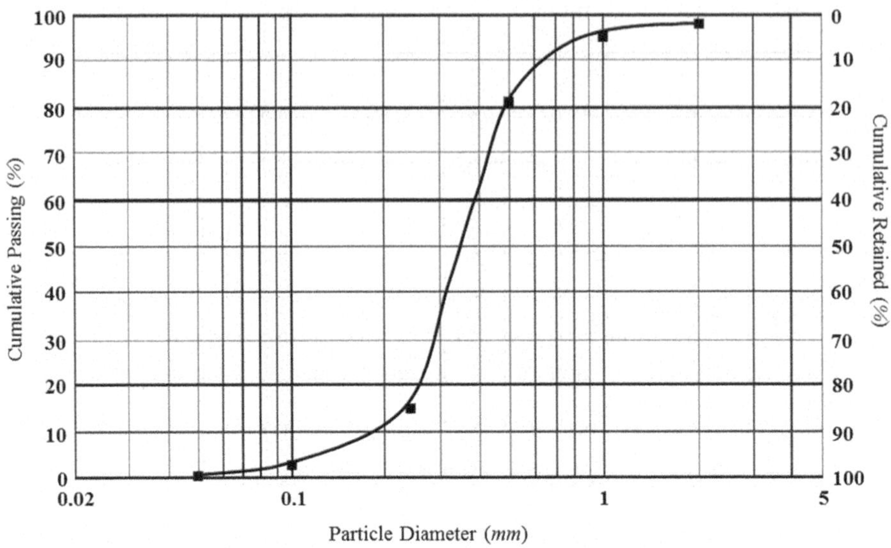

Fig. 3.29 Graph for Question #7

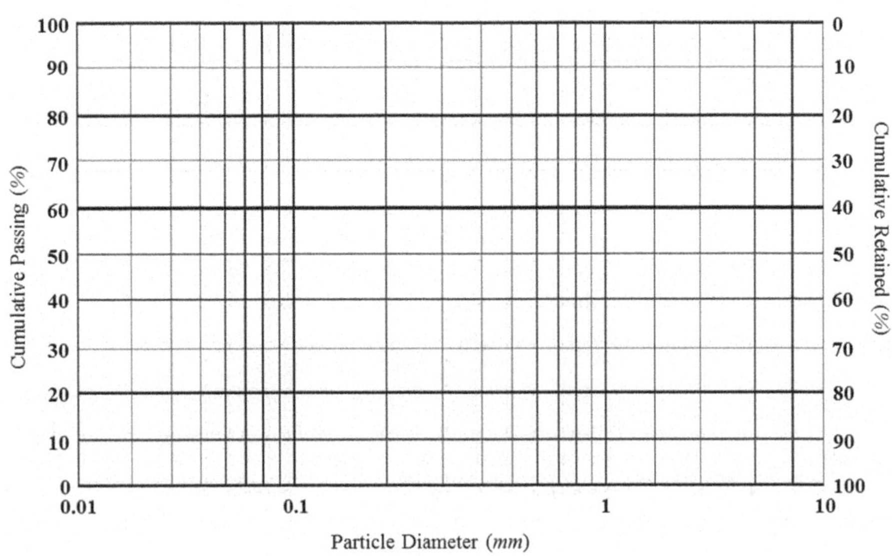

Fig. 3.30 Graph paper for Question #8

answer:

Compatibility Tests

	Passed particle diameter (%)									
						Sand (mm)				
	Gravel (mm)					V. coarse	Coarse	Medium	Fine	V. Fine
Sample	13.5	9.5	6.3	4.0	3.0	1.0	0.50	0.25	0.15	0.05
Gravel % Passing (>D)	*100*	*91.6*	*33.7*	*1.9*	*0.6*	*0*	—	—	—	—
Rootzone Mix % Passing (>D)	—	—	—	—	*99.7*	*95.5*	*66.7*	*30.7*	*3.3*	*0.8*

Data Recorded from Graph

	D_{15}	D_{85}	D_{90}
Gravel	*5.65 mm*	—	*8.5 mm*
Rootzone	*0.2 mm*	*0.78 mm*	—
	Bridging Factor (D_{15}Gravel/D_{85}Rootzone)	*Coefficient of Uniformity (D_{90}Gravel/D_{15}Gravel)*	*Permeability Factor (D_{15}Gravel/D_{15}Rootzone)*
USGA	≤ 8	≤ 3	≥ 5
Gravel	*5.65 mm/0.78 mm = 7.24*	*8.5 mm/5.65 mm = 1.50*	*5.65 mm/0.2 mm = 28.3*
Rootzone		—	

This sand is within the suggested ranges for bridging with gravel, gravel uniformity factor, and permeability factor between the rootzone and gravel.

9. From the data set in the previous question, use interpolation to generate the various values and determine if these pass Bridging Factor, Coefficient of Uniformity and Permeability Factor tests. Compare interpolated values to the graph-generated ones (Fig. 3.31).

   Interpolated Values:

Gravel D_{15}

$$\frac{\Delta \text{particle size}}{33.7 - 15\,\%} = \frac{6.3 - 4.0}{33.7 - 1.9}$$

$$\frac{\Delta \text{particle size}}{18.7\,\%} = \frac{2.3}{31.8\,\%}$$

$$\Delta \text{ particle size} = 1.3\,mm$$

$$= 6.3 - 1.3\,mm$$

$$= 5.0\,mm$$

Gravel D_{90}

$$\frac{\Delta \text{particle size}}{91.6 - 90\,\%} = \frac{9.5 - 6.3}{91.6 - 33.7}$$

$$\frac{\Delta \text{particle size}}{1.6\,\%} = \frac{3.2}{57.9\,\%}$$

$$\Delta \text{ particle size} = 0.055\,mm$$

$$= 9.5 - 0.055\,mm$$

$$= 9.45\,mm$$

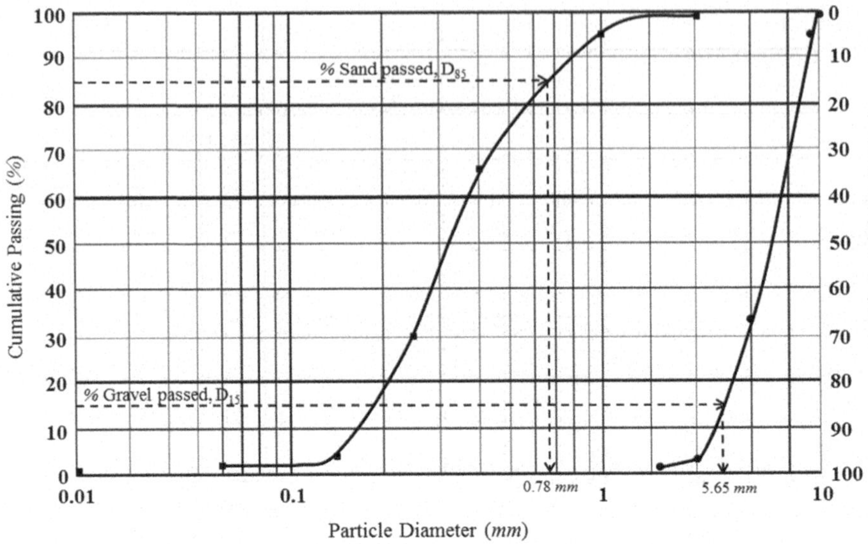

Fig. 3.31 Graph for Question #8

Rootzone D_{15}

$$\frac{\Delta \text{particle size}}{30.7 - 15\,\%} = \frac{0.25 - 0.15}{30.7 - 3.3}$$

$$\frac{\Delta \text{particle size}}{5.7\,\%} = \frac{0.1}{27.4\,\%}$$

$$\Delta \text{ particle size} = 0.06\,mm$$

$$= 0.25 - 0.06\,mm$$

$$= 0.19\,mm$$

Rootzone D_{85}

$$\frac{\Delta \text{particle size}}{95.5 - 85\,\%} = \frac{1.0 - 0.5}{95.5 - 66.7}$$

$$\frac{\Delta \text{particle size}}{10.5\,\%} = \frac{0.5}{32.8\,\%}$$

$$\Delta \text{ particle size} = 0.18\,mm$$

$$= 1.0 - 0.18\,mm$$

$$= 0.82\,mm$$

Comparing Interpolation to Graph Values

—	Interpolation	Graph
Gravel D_{15}	*5.00 mm*	*5.65 mm*
Gravel D_{90}	*9.45 mm*	*8.50 mm*
Rootzone D_{15}	*0.19 mm*	*0.20 mm*
Rootzone D_{85}	*0.82 mm*	*0.78 mm*
Bridging Factor (≤ 8) (D_{15}Gravel/D_{85}Rootzone)	*5.00 mm/0.82 mm = 6.10*	*5.65 mm/0.78 mm = 7.24*
Coefficient of Uniformity (≤ 3) (D_{90}Gravel/D_{15}Gravel)	*9.45 mm/5.00 mm = 1.69*	*8.50 mm/5.65 mm = 1.50*
Permeability Factor (≥ 5) (D_{15}Gravel/D_{15}Rootzone)	*5.00 mm/0.19 mm = 26.32*	*5.65 mm/0.2 mm = 28.3*

Data generated by interpolation mirrored closely those from the graph of %-passing vs. diameter. All data (interpolated and graph generated) also passed the compatibility tests (Bridging Factor, Uniformity Factor, and Permeability Factor).

10. From the following sieve analyses for two potential sands being considered to construct a sports field, calculate the resulting various textural fractions if the sand are mixed in the following ratios.

Sample	Gravel >2.0 *mm*	V.C. Sand 1.0–2.0 *mm*	Coarse Sand 0.5–1.0 *mm*	Med Sand 0.25–0.5 *mm*	Fine Sand 0.1–0.25 *mm*	V.F. Sand 0.05–0.10 *mm*
PB Fine (F) Sand	0	0.8	16.0	71.0	10.9	0.8
PB Coarse (C) Sand	11.1	10.5	29.1	43.1	5.0	0.4
3:1 C:F	—	—	—	—	—	—
2:1 C:F	—	—	—	—	—	—
1:1 C:F	—	—	—	—	—	—
1:2 C:F	—	—	—	—	—	—
1:3 C:F	—	—	—	—	—	—
USGA Rec.	≤3 %	≤7 %	≥60 % combined		≤20 %	≤10 %
	≤10 % combined		—	≥80 % combined		

To determine the outcome of mixing 2 sands on a volume to volume (v/v) ratio basis, the following calculations can be performed:

$$\text{New\%particle size} = \frac{[\text{sand 1 fraction\%} \times \text{ratio sand 1}] + [\text{sand 2 fraction\%} \times \text{ratio sand 2}]}{\text{ratio sand 1} + \text{ratio sand 2}}$$

Sample	Gravel >2.0 *mm*	V.C. Sand 1.0–2.0 *mm*	Coarse Sand 0.5–1.0 *mm*	Med Sand 0.25–0.5 *mm*	Fine Sand 0.1–0.25 *mm*	V.F. Sand 0.05–0.10 *mm*
3:1 C:F	*8.3*	*8.1*	*25.8*	*50.1*	*6.5*	*0.5*
2:1 C:F	*7.4*	*7.3*	*24.7*	*52.4*	*7.0*	*0.5*
1:1 C:F	*5.6*	*5.7*	*22.6*	*57.1*	*8.0*	*0.6*
1:2 C:F	*3.7*	*4.0*	*20.4*	*61.7*	*8.9*	*0.7*
1:3 C:F	*2.8*	*3.2*	*19.3*	*64.0*	*9.4*	*0.7*
USGA Rec.	≤3 %	≤7 %	≥60 % combined		≤20 %	≤10 %
	≤10 % combined		—	≥80 % combined		

11. Construct a soil moisture retention curve using Fig. 3.32 with the following data. Identify and discuss air-entry point, wilting point, total available water and calculate the equivalent depth of water in the soil (Fig. 3.33). Soil cores were 5.2 *cm* in diameter and 1.3 *cm* deep. All core materials weigh the same. Saturated hydraulic conductivity (K_{sat}) = 64.1 $cm\ h^{-1}$.

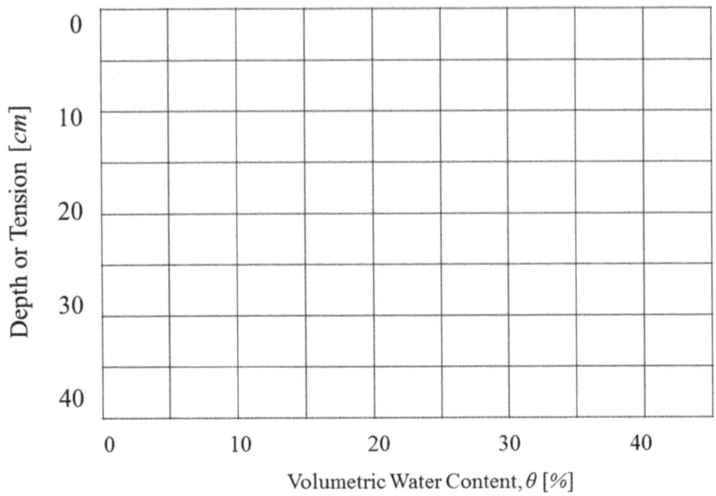

Fig. 3.32 Graph paper for Question #11

Tension (*cm*)	Weight (*g*)
0	53.013
5	53.701
10	53.434
15	51.057
20	49.946
25	47.769
30	46.934
35	45.991
40	45.814
45	45.547
Oven dry	44.436

First, determine the bulk density of the cores (equal at varying tensions). To do this, the volume of the soil is needed:

$$\begin{aligned} \textit{soil volume} &= h\pi r^2 \\ &= 1.3\ cm\,(3.14)(2.6\ cm)^2 \\ &= 27.6\ cm^3 \end{aligned}$$

$$\textit{bulk density} = \frac{\textit{dry weight}\ (g)}{\textit{soil volume}\ (cm^3)} = \frac{44.436\ g}{27.6\ cm^3} = 1.61\ g\ cm^{-3}$$

Then, determine total soil porosity (equal at varying tensions):

$$\text{total porosity (\%)} = \left(1 - \frac{\text{bulk density (g)}}{\text{particle density (cm}^3)}\right) \times 100$$

To determine θ_m and θ_v:

$$\text{gravimetric water content } (\theta_m) = \frac{\text{mass (wet soil)} - \text{mass (dry soil)}}{\text{mass (dry soil)}}$$

$$\text{volumetric water content } (\theta_v) = \text{gravimetric water content } (\theta_m) \times \text{bulk density}$$

Tension (*cm*)	Weight (*g*)	Bulk Density (*g* cm^{-3})	Total Porosity ($cm^3\ cm^{-3}$)	θ_m ($g\ g^{-1}$)	θ_v ($cm^3\ cm^{-3}$)
0	53.013	*1.61*	*0.393*	*0.193*	*0.311*
5	53.701	*1.61*	*0.393*	*0.209*	*0.336*
10	53.434	*1.61*	*0.393*	*0.202*	*0.325*
15	51.057	*1.61*	*0.393*	*0.149*	*0.240*
20	49.946	*1.61*	*0.393*	*0.124*	*0.200*
25	47.769	*1.61*	*0.393*	*0.075*	*0.121*
30	46.934	*1.61*	*0.393*	*0.056*	*0.090*
35	45.991	*1.61*	*0.393*	*0.035*	*0.056*
40	45.814	*1.61*	*0.393*	*0.031*	*0.050*
45	45.547	*1.61*	*0.393*	*0.035*	*0.056*
Oven dry	44.436	*1.61*	*0.393*	–	–

To determine air-filled (aeration) porosity:

$$\text{air} - \text{filled porosity (\%)} = \text{total porosity} - \theta_v$$

Tension (*cm*)	Weight (*g*)	Bulk density (*g* cm^{-3})	Total porosity ($cm^3\ cm^{-3}$)	θ_m ($g\ g^{-1}$)	θ_v ($cm^3\ cm^{-3}$)	Air-filled porosity ($cm^3\ cm^{-3}$)
0	53.013	*1.61*	*0.393*	*0.193*	*0.31*	*0.082*
5	53.701	*1.61*	*0.393*	*0.186*	*0.30*	*0.057*
10	53.434	*1.61*	*0.393*	*0.180*	*0.29*	*0.068*
15	51.057	*1.61*	*0.393*	*0.149*	*0.24*	*0.153*
20	49.946	*1.61*	*0.393*	*0.134*	*0.22*	*0.193*
25	47.769	*1.61*	*0.393*	*0.075*	*0.12*	*0.272*
30	46.934	*1.61*	*0.393*	*0.056*	*0.09*	*0.303*
35	45.991	*1.61*	*0.393*	*0.035*	*0.06*	*0.337*
40	45.814	*1.61*	*0.393*	*0.031*	*0.05*	*0.343*
45	45.547	*1.61*	*0.393*	*0.025*	*0.04*	*0.337*
Oven dry	44.436	*1.61*	*0.393*	–	–	–

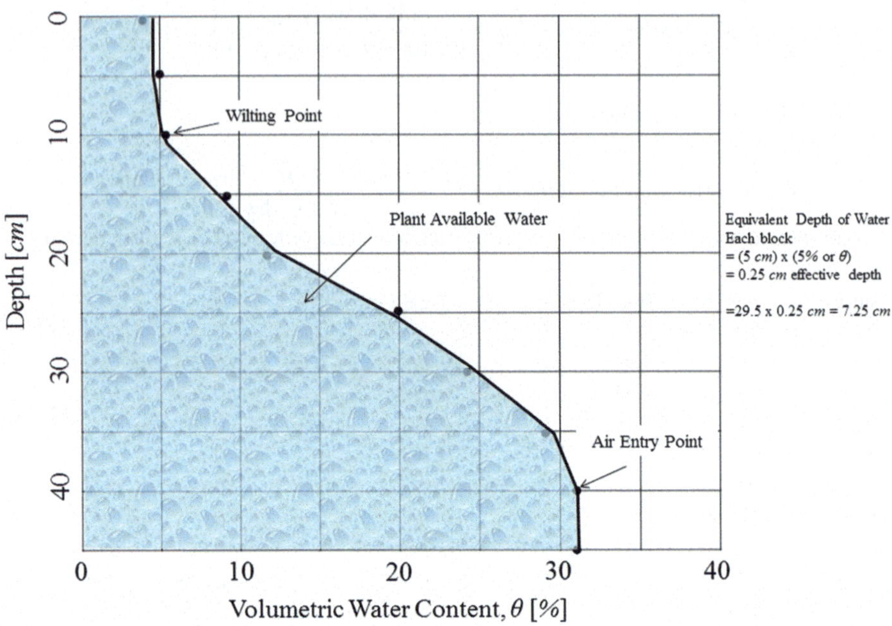

Fig. 3.33 Graph for Question #11

From Fig. 3.33, approximately 29.5 blocks are filled with water and each block has an equivalent depth of water of 0.25 cm (5 cm × 5 % or θ). Therefore the total amount of water in this soil is approximately 7.25 cm (29.5 blocks × 0.25 cm equivalent depth of water per block) or ~2.9 in.

12. Five different methods were presented determining the necessary depth for a rootzone or sand capping a poorly drained soil. With the data from Fig. 3.34 plus the following soil analysis data set, compare these five methods.

	Particle size (% by weight) (*m*)						
Sample	>2.0	1.0	0.50	0.25	0.15	0.05	<0.05
rootzone	0.0	4.9	37.1	43.8	13.7	0.4	0.3

(a) *Using Soil Moisture Retention Curves of Aeration Porosity vs. soil depth.*

Aeration (air-filled) porosity	Corresponding soil depth (cm)
10 %	*10 cm*
15 %	*15 cm*
25 %	*25 cm*

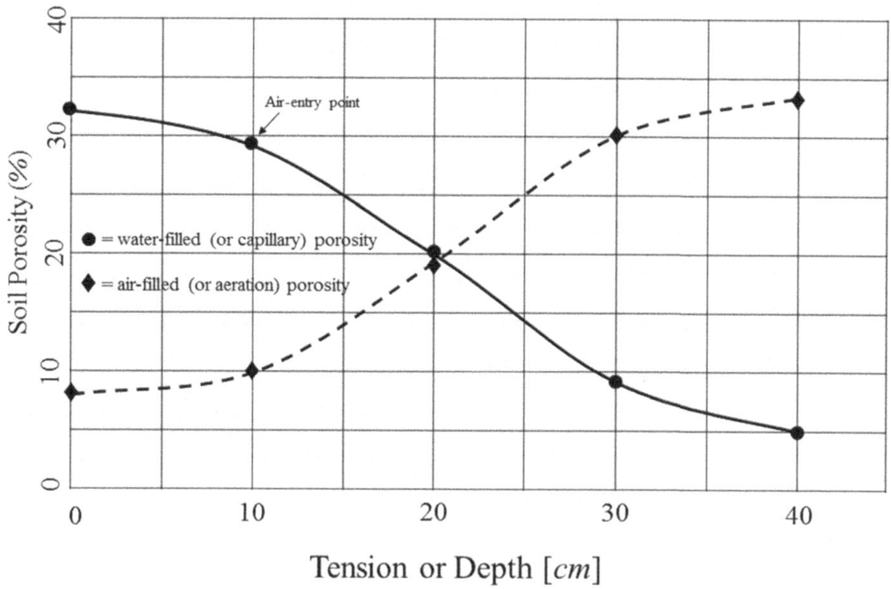

Fig. 3.34 Graph for Question #12

According to the SMRC (Fig. 3.34), a minimum of 10 cm (4 in. would be needed, a better depth of 15 cm (6 in), with an ideal depth of 25 cm (10 in).

(b) *Depths depend on adjusting air entry point (top of the perched water table) on the SMRC. For golf greens, 10 cm (4 in. are added to air-entry value while for fairways and sports fields, 15 cm (6 in. are added. Therefore, for this particular sand with an air entry value of 10 cm, for golf greens the recommended depth would be 20 cm (or ~8 in. and for the fairway/sports fields, 25 cm (or 10 in).*

(c) *Optimum rootzone depth is where air-filled porosity on a SMRC is ~25 % and capillary (water-filled) porosity is ~20 %. With the rootzone in the example, this intersection is between 20 and 25 cm (~8–10 in).*

(d) *From SMRCs, depth is based on capillary porosity of 15 %, air-filled porosity of ≥10 % (preferably 15 %) and a gravimetric moisture content between 10 and 18 %. This method is similar to the first one presented. From the SMRC, this depth would be between 16 and 25 cm (~6 and 10 in).*

(e) *Rootzone depth is based on the critical tension formula: 71.4/D (mm) where D is the dominant soil particle diameter. The rootzone depth is by estimating capillary rise of water in uniform diameter capillary tubes. Critical tension is inversely proportional to soil particle diameter with shallower rootzones needed for soils dominated by larger sized diameter particles. The dominant particle size in this sample is 0.25 mm (43.8 %).*

$$71.4/D = 71.4/0.25 = 286\ mm\ (or\tilde{\ }11\ in)$$

Therefore, in conclusion, the following suggested rootzone depths were calculated from the various means:

Method	Calculated rootzone soil depth (cm)
Soil Moisture Retention Curves	*10 cm at 10 % aeration porosity*
	15 cm at 15 % aeration porosity
	25 cm at 25 % aeration porosity
Air Entry Point plus add appropriate soil depth	*50 cm (greens); 55 cm (fairways/ sports fields)*
Junction of Aeration & Capillary Porosity on SMRCs	*20 cm*
Capillary porosity of 15 % & air-filled porosity of ≥10 % on SMRCs	*15 cm*
Based on the critical tension formula: 71.4/D (mm)	*28.6 cm*

Discussion

A range of suggested rootzone depths were produced by the various five methods. These ranged from 10 (question a) to 29 cm (question e) deep. It appears between 20 and 25 cm (~8–10 in.) is the overall best depth range for this rootzone sample. For areas with heavier rainfall, 25 cm (10 in. should be considered while areas with lower rainfall amounts should consider 20 cm (~8 in).

Chapter 4
Water Management and Conservation

Plants typically contain between 75 and 85 % water by weight, and begin to die if their water content drops to 60–65 % within a short period of time. Water acts as a buffer in plants against extreme temperature fluctuations, ensuring damage from high or low temperatures occur slowly.

Unfortunately, rainfall does not occur frequently enough in most cases to provide adequate water to sustain turfgrasses, especially with the limited root systems associated with most closely-mowed turf areas and soils used with low water-holding capacity. This situation is further intensified by warm weather and the high aesthetic demands by clientele. To ensure efficient watering, turf facilities require well-designed irrigation systems based on soil infiltration rates, soil water-holding capacity, anticipated annual rainfall, plant water-use requirements, depth of rootzone, conveyance losses from the surrounding area, and desired level of turfgrass appearance and performance.

Water loss from a turf area occurs through evaporation, transpiration, run-off, leaching, and conveyance losses. Turf managers have a degree of control over these water-loss mechanisms; therefore, they should have a good understanding of each mechanism in order to maximize water conservation. In addition to water quantity, water conservation also encompasses irrigation water quality which will also be covered in this chapter.

4.1 Water Use

Determining When to Irrigate

There are a number of methods used to determine how much water turf requires at any given time, under any given environmental conditions. Several are indirect and base their estimates on measuring soil moisture. Others simulate evapotranspiration from the canopy, while yet others make direct soil measurements.

L.B. McCarty et al., *Applied Soil Physical Properties, Drainage, and Irrigation Strategies*, DOI 10.1007/978-3-319-24226-2_4

Fig. 4.1 Turfgrass discoloration such as a *blue-green* to *grayish* color is a key indicator of drought occurrence

Visual Symptoms

A simple method used to determine when to irrigate is to observe visual symptoms of moisture stress. Moisture-stressed grass appears blue-green or grayish-green in color (Fig. 4.1), recuperates slowly (>1 min) after walking or driving across it ("foot-printing") (Fig. 4.2), or wilts continuously. These symptoms occur when plant moisture is insufficient to maintain turgor pressure, due to more water being lost than taken up. As a result, the plant rolls its leaves to minimize exposed leaf surface and wilts to conserve moisture. Golf course managers should avoid prolonged moisture stress, especially on greens. This method is best used for low-maintenance turf such as golf course roughs, out-of-play sports fields, or home lawns.

While visual observation for stress symptoms may be the simplest method, it does have some drawbacks. Waiting for wilting symptoms is a good method of determining when the turf needs water, but not necessarily how much water is needed. Turf managers also cannot afford to wait until drought symptoms appear on putting greens since this causes unacceptable turf quality. Certain areas or patches of turf will tend to wilt prior to others due to poor irrigation distribution, localized dry spots, poorly developed root systems, or variation in soil texture. Watering the whole turf area to eliminate these "hot spots" will waste water; thus, extensive hand watering is often needed.

Fig. 4.2 Foot printing or traffic patterns as an indicator of drought occurrence

Evaporatory Pans

Another method of irrigation scheduling is the use of evaporatory pans. A U.S. Weather Service Class A Evaporatory Pan is 122 *cm* in diameter, 25 *cm* deep, and is supported 15 *cm* above the ground (Fig. 4.3). Evaporatory pans are filled with water and placed in a representative location, where water loss is measured over time. The amount of water evaporating from the pan correlates to that lost by evapotranspiration (ET). This correlation is generally accurate except during windy conditions which tends to exaggerate the amount of water lost by the evaporatory pan compared to actual ET rates.

The water quantity lost through evaporation correlates with turfgrass ET, but is not exactly the same; turfgrasses use less water than the quantity evaporated from the pan. A crop coefficient (K_c) value is needed to adjust this correlation (Table 4.1). Warm-season grasses use 55 to 65 %, and cool-season grasses use 65 to 90 %, of pan evaporation. Thus, if the evaporative pan shows a 1 *in* (2.5 *cm*) water loss, a bermudagrass turf would actually have lost approximately 0.60 *in* (1.5 *cm*) while bentgrass would have lost approximately 0.85 *in* (2.2 *cm*).

Soil Moisture Measuring Devices

Soil moisture measuring devices have been developed with the goal of indicating how much moisture is available to plants. Soil moisture is measured in two

Fig. 4.3 Evaporatory pan used to measure daily evapotranspiration water losses

Table 4.1 Crop coefficient (K_c) values for a class A evaporative pan or the Penman-Monteith equation

Grass	Class A-evaporative pan K_c values	Penman-Monteith K_c values
Bermudagrass	0.55 to 0.65	0.70 to 0.80
Tall Fescue	0.65 to 0.75	0.75 to 0.95
Perennial Ryegrass	0.65 to 0.75	0.80 to 1.0
Kentucky Bluegrass	0.70 to 0.80	0.85 to 1.0
Creeping Bentgrass	0.75 to 0.90	0.95 to 1.0

distinctly different methods—*quantitatively* (or *volumetric*), the actual amount of moisture in the soil, and *qualitatively* (or *tensiometric*), how tightly water is held by soil. Though numerous means of measuring these exist, the more common ones including gravimetric water content, TDR, tensiometers, and FDR (or hand-push) probes.

The water content of different soils varies due to large differences between soils in their total particle surface areas. For example, moisture levels at field capacity for sands may be as low as 7 % whereas clays may have as much as 40 % moisture content at field capacity. In another example, the permanent wilting point volumetric water content may range from 1 to 2 % for sandy soils to 25 or 30 % for clay (finer-textured) soils. This variation demonstrates that a measure of soil water (volumetric) content does not necessarily indicate the amount of water available to plants. A better indicator of a plant's soil-water availability is the energy status of water (called tensiometric or water potential) which measures the relative amount of work (or energy) needed to remove a unit of water from a particular soil.

Quantitative Methods

Quantitative methods for measuring soil moisture include gravimetric sampling, neutron probe (or scatter), and dielectric constant (Time Domain Reflectometry and Frequency Domain Reflectometry, TDR and FDR) probes. The most accurate is the gravimetric water content method where a volume of soil is weighed, dried, and then reweighed (Table 4.2). The impracticality of this method and expense (>$5000) for neutron probes have led to the development of other techniques.

Dielectric constant methods measure the soil's ability to transmit electricity (electro-magnetic waves or pulses) with the value increasing as the water content of the soil increases (Fig. 4.4). The permittivity constant for air is approximately 1; dry soil between 3 and 5; and about 80 for water. Values are related through calibration to known soil moisture content determined using either a neutron probe or the gravimetric sampling technique. The equipment consists of an electronic meter connected to 2 to 4 rods placed into the ground. The instrument sends an electrical signal through the soil and the rods serve as the transmitter and receivers. TDR and FDR probes are currently the most commonly used dielectric devices. Although these devices are able to detect the amount of moisture in the soil, they do not determine how much of it is available to plants.

Advantages of using dielectric devices to quantify soil moisture include:

- ability to leave soil moisture sensors in place to continuously monitor soil moisture content,
- repeatability of measurements,
- sensitivity to small changes in soil moisture content,
- precise resolutions with depth due to the narrow vertical zone of influence.

Disadvantages include:

- need for soil specific calibration for best accuracy,
- relatively small zone of measurement,
- possibility of soil salinity influencing probe reading,
- sensitivity to air gaps,
- probe length should equal rooting depths.

Time Domain Reflectometry (or TDR)

These systems measure the travel time of an electromagnetic wave between sending the pulse and receiving it, and is the preferred tool for researchers. With TDR, a pair of parallel metal rods connected to a signal receiver is inserted into the soil. The rods serve as conductors while the soil is the dielectric (a nonconductor of

Table 4.2 Comparison of common techniques of soil moisture content measurement

Technique	Measurable range	Advantages	Disadvantages
Gravimetric water content. Measures soil moisture by weighing-drying-reweighing.	Full range of water content (%).	– simple equipment needs, – highly accurate, – easy interpretation.	– destructive sampling, – labor intensive, – collection, transport, and time restraints.
Time Domain Reflectometry (TDR). Measures time for an electromagnetic wave to travel using soil medium as a dielectric. Moisture slows this down.	Up to 50 % volumetric water content (0.50 kg water kg^{-1} soil)	– accurate, – minimal soil disturbance, – soil specific-calibration is optional, relatively insensitive to temperature, – also estimates, with limited accuracy, soil EC.	– expensive, – accuracy decreases in high saline (>25 $dS\ m^{-1}$) conditions or heavy clay soils, – relatively small sensing volume (~ 1 *in*, 2.5-*cm*, radius around probe).
Frequency Domain Reflectometry (FDR) or **Hand-push probes**. Measures the change in frequency of a capacitor using soil medium as a dielectric.	Up to 70 % volumetric water content (0.70 kg water kg^{-1} soil) or to −7.0 MPa	– relatively inexpensive, – can be automated with irrigation, – stable in different soil types and over a large range of moisture contents.	– needs soil-specific calibration for accuracy, – samples small volume of soil (~ 4 *in*, 10 *cm*, radius around probe), – sensitive to soil air gaps, saline soils and temperature.
Tensiometers. Measure how tightly (the "tension") water is held by soil.	0 to −0.08 MPa or 0 to −80 kPa	– direct readout of soil water potential (or tension), – inexpensive, – can be automated with irrigation, – relatively reliable, – good accuracy, – unaffected by soil salinity.	– soil moisture retention curve needed to relate to soil water content, – samples a small area near cup, thus, multiple samples are needed in larger areas, – doesn't measure soil salinity content, – exposed gauges, sensitive to disturbance and soil air gaps.
Unit conversions: 1 $bar \approx 1\ atm = 14.7\ psi$; 1 $kPa = 0.001\ MPa = 0.01\ bar = 1\ cb = 10\ cm$ H_2O			

Fig. 4.4 Soil moisture dielectric probe with two sets of probes used to measure quantitative levels of soil moisture near the soil surface and several inches (turfgrass rootzone) below this

electricity). The presence of water (higher dielectric constant) proportionally slows the speed of the electromagnetic wave. Traditionally, TDR instruments were more expensive due to the advanced electronics needed to provide this series of precisely-timed electrical pulses and ability to read these. However, recent technology has allowed TDR moisture sensors to be priced closer to the less accurate FDR based alternatives.

Frequency Domain Reflectometry (or FDR)

These are also known as *hand-push probes* and as *dielectrical capacitance probes*. Like TDR systems, FDR are also dielectric sensors as their electrodes are separated by the dielectric (soil). One or two pairs of electrodes (either an array or parallel spikes or circular metal rings) form a capacitor, with the soil acting as the dielectric in between. This capacitor works with the oscillator to form a tuned circuit and changes in soil water content are detected by changes in the reflected frequency. Most of these sensors operate at low frequencies (100 MHz or less) compared to higher (~250 MHz) operating frequency for TDR probes. The high frequency used for TDR probes allows less dependency on soil specific properties like texture, salinity or temperature. The greater the soil moisture content, the smaller (or greater change in) the frequency. The dielectric reading is then converted to volumetric water content (m^3 water m^{-3} soil or θ_v) with readout in percentage (% volume).

In general, FDR probes perform best in coarser-textured, non-saline soils and often require specific soil calibration, limiting their use or comparison between different soils or locations. Less precise electronics are needed vs. TDR, thus FDR probes are cheaper. All electronic resistance probes are influenced by temperature, soil composition and bulk density, and the solute concentration (EC) of the soil solution and since moisture content is a non-linear curve, calibration equations are required for specific soils. Probes should also be at least 6 *in* (15 *cm*) in length to reduce wavelength reflection which produces erroneous readings as do EC levels at and above 25 $dS\ m^{-1}$.

Electrical Conductivity Probes

These are a commonly available low-cost means of measuring soil moisture in the soil based on the soil's ability to pass a current of electricity between two probes. In many ways the concept is similar to resistance blocks but the probes (electrodes) have direct contact with the soil and are not buffered as in resistance blocks (discussed below). The more moisture in soil the better the conductivity or the lower the electrical resistance. This method is very sensitive to probe spacing as well as being influenced by soil type and salts, primarily in the form of fertilizers. Because of this strong correlation, these probes are more commonly used to measure salt content in soils.

Qualitative Methods

These methods measure how tightly soil moisture is held by soil particles but do not directly measure the quantity of water contained in it. As the tension increases, water extraction becomes more difficult for the plant. Tensiometers and porous blocks (i.e., gypsum, ceramic, nylon, and fiberglass) are qualitative methods.

Tensiometers

These are sealed, water-filled tubes with a vacuum gauge on the upper end and a porous ceramic tip on the lower end (Fig. 4.5). Water in the tensiometer comes to equilibrium with water in the soil and provides an indication of how difficult it is (or tension required) for the plant to obtain water from the soil, but does not directly provide information on soil water content. To obtain this, a soil moisture release curve is needed. A lower reading indicates more available water. Though very accurate when scheduling irrigation, tensiometers are often not practical in turfgrass applications as their presence disrupts play and/or maintenance practices.

Electrical Resistance Blocks

Electrical resistance blocks measure soil moisture tension with two electrodes imbedded in a porous material such as gypsum, nylon, fiberglass, or a sand-ceramic

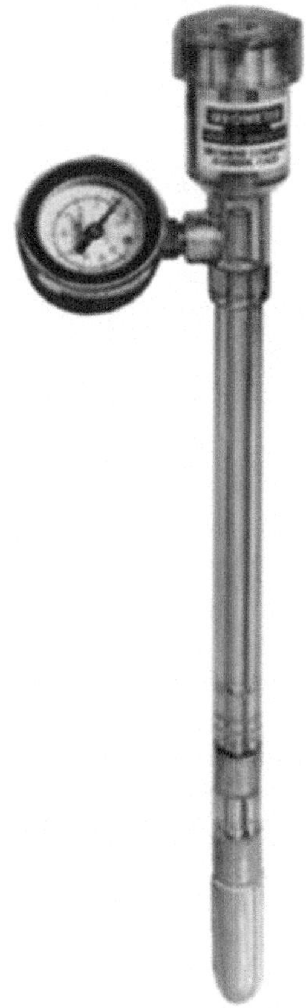

Fig. 4.5 Tensiometer used to determine how much tension is required for plants to extract soil moisture

mixture. Gypsum or similar material is used to buffer against salts that would affect resistance readings. Moisture is allowed to move in and out of the blocks as the soil dries or becomes moist. The electrodes measure resistance to electric current when electrical energy is applied. The more moisture in the block, the lower the resistance reading indicating more available moisture. These are accurate when measuring low soil moisture content and can be left in place for extended periods. They are, however, sensitive to saline conditions, and like tensiometers, measure soil moisture only at the area immediately surrounding them. They also are not as accurate in predominately sandy soil.

Calibrating Soil Moisture Devices

When using any soil moisture measuring device for irrigation purposes, three critical soil moisture levels need to be quantified: (1) gravitational water; (2) field capacity; and, (3) wilting point. A SMRC will indicate these (Fig. 1.25) but they

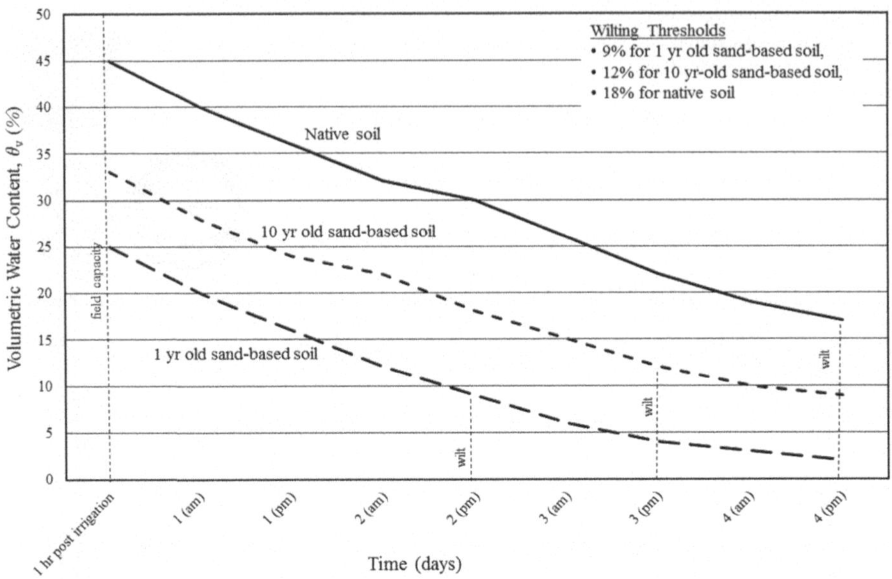

Fig. 4.6 Moisture levels often vary with different soils. Shown are differences in volumetric moisture content for three different soils at field capacity and wilting points. Field capacity was highest for the native soil at approximately 45 % and lowest at 25 % for a 1 year old sand-based rootzone. Wilting occurred earliest for the 1 year old sand-based rootzone (2 days after irrigation) while it was latest for the native soil (4 days after irrigation). Redrawn and modified from Karcher (2013)

also should be confirmed with field readings. Moisture levels typically vary with soil type and uniformity and readings vary between soil measuring devices (Fig. 4.6). Therefore, it is highly recommended soil moisture measuring devices be calibrated for the particular soil in consideration.

The amount of moisture between field capacity and wilting is that available for plant use. Gravitational water is typically that in the soil following heavy rainfall and puddle disappearance. About 24 *h* later, a reading should approximate field capacity. Field capacity is also the amount of soil moisture present about 1 *h* following heavy irrigation. Multiple readings should be taken over the whole area and the probe length should extend to just below the average rooting depth. Readings should be taken at least twice daily until significant plant wilting is observed.

Example From the use of a hand-held TDR probe, determine how much water is needed to return the total moisture levels to field capacity once it reaches wilting point for two sands with volumetric soil water content at field capacity (θ_{fc}) of 0.35 and 0.25 $cm^3\ cm^{-3}$ and volumetric water content at wilting point (θ_{wp}) of 0.025 and 0.015 $cm^3\ cm^{-3}$ for sands 1 and 2, respectively. The TDR probe measures moisture in the top 10 *cm* (4 *in*) of the soil profile.

Determine available water for each sand using the equation:

$$D_e = soil\ depth\ (\theta_{fc} - \theta_{wp})$$

where:

D_e = equivalent depth of available water in the top 10 *cm* (4 *in*),
θ_{fc} = volumetric water content at field capacity,
θ_{wp} = volumetric water content at wilting point

$$\text{For sand 1}: \ D_e = \text{soil depth}\ (\theta_{fc} - \theta_{wp}) = 10\ (0.35 - 0.025) = 3.25\ cm\ (1.28\ in)$$
$$\text{For sand 2}: \ D_e = \text{soil depth}\ (\theta_{fc} - \theta_{wp}) = 10\ (0.25 - 0.015) = 2.35\ cm\ (0.93\ in)$$

Therefore, for sand 1, 3.25 *cm* would be needed to bring the soil moisture level back to field capacity once it reach wilting point while 2.35 *cm* would be required for sand 2.

Predictive Models or Evapotranspiration Feedback

Predictive models, such as the modified Penman-Monteith ET (also known as FAO 56) model, based on weather station data and soil types also are available. These are often referred to as Irrigation or ET Controllers in the industry. They estimate or predict ET of the turf. These are relatively accurate and applicable, especially as long-term predictors of yearly turf water requirements. Models, however, are only as effective as the amount of data collected and the number of assumptions made. Weather data such as rainfall, air and soil temperature, relative humidity, and wind speed are incorporated into certain model formulae, and estimated soil moisture content is made. Accessible weather data, as well as specialized computer equipment and programs, must be available (Fig. 4.7).

Evapotranspiration feedback strategies are also used to schedule irrigation. Weather station or evaporative pan data can be used to calculate water use. This value is referred to as potential ET (ET_p) and is used as a reference point. Actual turf water use usually is not quite as high as ET_p, so a factor called the **crop coefficient** (K_c) is used to convert ET_p to actual turf ET (as discussed in the previous section on evaporative pans). Crop coefficients are fairly constant for a given species, but vary considerably between species (Table 4.1). For example, the K_c of bermudagrass is about 0.75. This means bermudagrass will use about 75 % as much water as is predicted from using environmental data to calculate ET_p. If environmental data indicates the theoretical reference crop used 2.2 *in* (5.6 *cm*) of water for a given week in the summer, multiply 2.2 by 0.75 indicates 1.65 *in* (4.2 *cm*) of water is actually used by bermudagrass. Most cool-season grasses have a K_c of approximately 0.85, indicating cool-season grasses actually require 1.87 *in* (4.75 *cm*) of water in the previous situation. These calculated water use rates

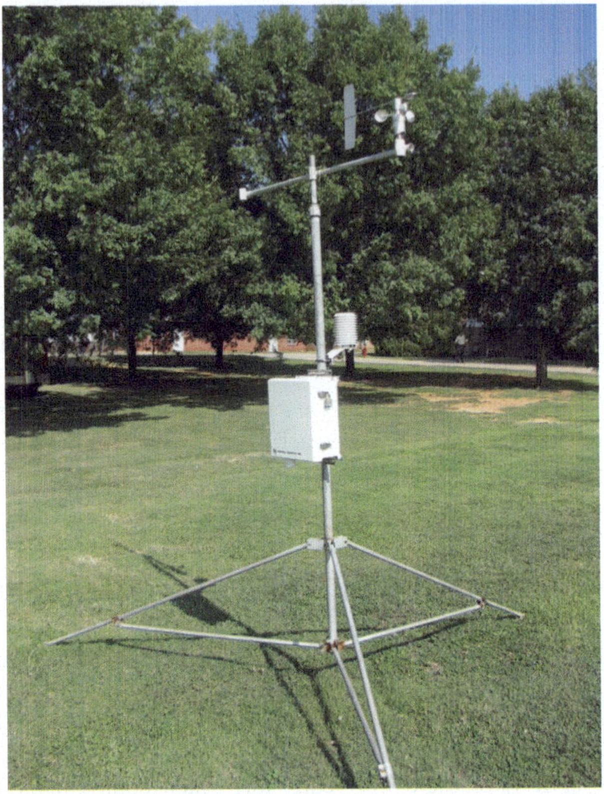

Fig. 4.7 Automatic weather station used to construct predictive models on how much soil moisture is present

are the "feedback" used to determine irrigation rates. Using the site information and weather data, ET feedback controllers run a "water balance" that keeps track of how much water is in the soil. Controllers then adjust the run timers (or amount) of water applied to the turf.

Atmometer

The atmometer (also referred to as the "ET gauge" or Bellani plate) also can be used to estimate evaporative demand (Fig. 4.8). This relatively inexpensive device consists of a water reservoir connected to a porous plate covered by green fabric designed to simulate a leaf surface. Water from the reservoir is wicked through the plate to the fabric, where it evaporates. The drop in the reservoir is then easily measured on a daily basis, much like checking a rain gauge. Rates of water loss are directly related to weather conditions, especially temperature, wind, and humidity, and have been found to correlate very well with turfgrass water demand. Atmometers may be an attractive alternative to the more costly weather station-based system while still supplying similar information.

Fig. 4.8 Atmometer which is a water-filled container connected to a porous plate covered by green fabric to estimate evapotranspiration rates of turfgrasses

The atmometer should be located in a sunny turf-covered area representative of the majority of the golf course. Additional units may be necessary for varying microclimates such as shady, windy, or stagnant areas, and irrigation rates should be adjusted accordingly. Atmometers require calibration and provide only an estimation of watering needs.

4.2 Evapotranspiration Rates

Plants absorb water from the soil and lose water to the atmosphere. Only about 5 % of all water consumed by turf is used in photosynthesis, carbohydrate synthesis, and other metabolic reactions. About 95 % of this water is lost as vapor from the leaves to the atmosphere, by the process of **transpiration**. Water is also lost by **evaporation** from soil and leaf surfaces. Evaporation is typically much lower than transpirational losses in a mature turf. The combined total of water lost through transpiration and evaporation is termed **evapotranspiration**, abbreviated ET. Evapotranspiration is usually expressed in inches or millimeters per day, week, or month. Since ET is the total water lost from the turf system, it represents the water demand, or the total amount that must be replaced to maintain a healthy turf. Environmental parameters largely controlling ET are light intensity and duration, relative humidity, wind velocity, and temperature. Increasing solar radiation, temperature, and wind increases ET, while increasing relative humidity decreases ET. Other parameters affecting ET to a lesser extent include soil-water content, turf-root system development, inherent turf water needs and dehydration avoidance mechanisms, and turf cultural practices.

Transpiration occurs through tiny pores in the leaf, called the **stomata**. Stomata are usually open, allowing water vapor and oxygen to move out of the leaf and carbon dioxide to move in for photosynthesis. To conserve water, stomata often close during periods of peak water demand (hot, windy afternoons), but will usually reopen after environmental conditions moderate. Under prolonged stress, however, stomata may close for extended periods, which in turn affects other plant functions.

Although it might seem like transpiration is just a waste of water, it is in fact critically important as it cools the leaf. If not for transpirational cooling, a leaf could reach 120 °F (49 °C) or higher during midsummer, a lethal temperature for most plants. Fortunately, transpiration keeps leaves much cooler, usually below 90 °F (32 °C), due to the **latent heat of vaporization** for water, or the large amount of energy needed to convert liquid water to water vapor via evaporation. For example, for every *calorie* of solar energy absorbed by the plant, 1 *g* of turfgrass tissue (mainly water) will increase in temperature by nearly 2 °F. Ten *calories* of solar energy could warm 1 *g* of turf tissue by about 18 °F. However, it takes a lot of energy, 539 *cal*, to evaporate 1 *g* of water. By transpiring only 1 *g* of water, a turf plant loses enough energy to cool 539 *g* of plant tissue by roughly 2 °F. Multiply this by the millions of grams of water a turfgrass area loses daily and the incredible cooling capacity of transpiration becomes evident. Humans use a similar process when perspiration evaporates, cooling their bodies.

Transpiration is also directly involved in mineral nutrition, both by causing soluble nutrients to be drawn to the roots along with soil water and by moving nutrients and certain hormones from roots to shoots. It is the diffusion of water through the stomata that creates a lower pressure potential, allowing water to overcome gravitational potential and move upward from the soil, into the plant roots, and upward in the plant. When the transpiration stream is lacking, as when plants are grown in a saturated atmosphere (100 % *RH*, $\Psi_t \sim 0\,MPa$), nutrient deficiency symptoms, especially for nitrogen and iron, often develop or a term called **wet wilt** occurs where plants collapse in the presence of water when evaporation exceeds root water uptake.

Environmental Influence on Evapotranspiration

Environmental parameters that control plant ET include relative humidity, temperature, solar radiation, and wind. Of these, solar radiation is the driving force for evaporative demand by stimulating stomata opening. Cloudiness can decrease ET by blocking incoming radiation.

Atmospheric relative humidity and wind velocity also influence ET rates. As air becomes more saturated at higher humidities, the vapor pressure gradient between leaves and air is reduced, resulting in less ET. Under calm air conditions, the existing vapor pressure tends to form an external layer of still air adjacent to the leaf called the **boundary layer**. The boundary layer, if not disturbed, acts as an insulator by protecting the leaf from sudden vapor pressure changes, and thus reduces ET. The boundary layer thickness is determined by wind speed. With increasing wind, the boundary layer decreases and ET increases. As a result, ET rates tend to increase with higher temperatures, light, and wind, but decrease with higher atmospheric relative humidity and cloud cover. Minimal ET rates occur when dark, cloudy days with high relative humidity, low temperatures, and little wind occur. Conversely, the highest ET rates occur on bright sunny days with low relative humidity, high temperatures, and moderate-to-high winds.

Turfgrass Water-Use Rates

Water-use rates are usually expressed in inches or centimeters of water lost per day or per week. In general, warm-season grasses use less water due to their greater resistance to water stress compared to cool-season grasses (Table 4.3). This ranges between 35 and 50 % less water required to maintain desirable warm-season grass color compared to cool-season grasses. Bermudagrass ET is between 0.3 and 0.9 cm water day^{-1}, while tall fescue water use ranges from 0.4 to 1.3 $cm\ day^{-1}$. Lower values are associated with cooler or more humid regions of the United States, while higher values are typical of warm arid regions. Tall fescue has the highest potential

Table 4.3 General mean summer turfgrass evapotranspiration (ET) rates

	Summer ET rates			
Turfgrass	$in\ day^{-1}$	$mm\ day^{-1}$	$in\ week^{-1}$	$cm\ week^{-1}$
Bahiagrass	0.25	6.2	1.75	4.4
Bermudagrass	0.12–0.30	3.1–8.7	0.84–2.10	2.1–5.3
Buffalograss	0.20–0.30	5.3–7.3	1.40–2.10	3.6–5.3
Centipedegrass	0.15–0.33	3.8–8.5	1.05–2.31	2.7–5.9
Creeping bentgrass	0.19–0.39	5.0–9.7	1.33–2.73	3.4–6.9
Kentucky bluegrass	0.15–0.26	3.7–6.6	1.05–1.82	2.7–4.6
Perennial ryegrass	0.15–0.44	3.7–11.2	1.05–3.08	2.7–7.8
Seashore paspalum	0.25–0.31	6.2–8.1	1.75–2.17	4.4–5.5
St. Augustinegrass	0.13–0.37	3.3–9.6	0.91–2.59	2.3–6.6
Tall fescue	0.15–0.50	3.6–12.6	1.05–3.50	2.7–8.9
Zoysiagrass	0.14–0.30	3.5–7.6	0.98–2.10	2.5–5.3

Low values within a range represent humid conditions; high values are for arid conditions (compiled from Beard 1985; Carrow 1995; McCarty 2011). ET rates during non-summer months generally are much lower

ET rates, but avoids drought stress due to its deep and extensive root system and ability to go dormant for short periods without lethal consequences.

Potential Evapotranspiration (ET_p) Rates

As previously discussed, another method to schedule irrigation is the development of ET feedback systems based on an estimate of the potential ET (indicated as ET_p) developed from climatic data or weather pan evaporation. The ET is then adjusted to actual plant ET use with an appropriate crop coefficient (K_c) that more accurately reflects actual ET for the particular turfgrass under irrigation:

$$ET_p = K_c \times pan \quad evaporation$$

Currently, K_c for warm-season grasses ranges from 0.60 (moderate stress) to 0.90 (nonstressed) and from 0.80 to 0.85 for cool-season grasses. General estimates of ET_p may be calculated using the following values for K_c:

$$\text{Warm-season grasses : } ET_p = 0.75 \times \text{pan evaporation rate}$$

$$\text{Cool-season grasses : } ET_p = 0.85 \times \text{pan evaporation rate}$$

Scheduling Irrigation Based on ET Rates

Potential ET rates can be calculated from a variety of equations. In general, by using historical climatological data as a reference and incorporating this in the modified Penman or McCloud equation to determine specific ET rates, potential ET rates have been calculated at various locations throughout the country. From this, normal net irrigation requirements to maintain low-to-medium maintenance grass are estimated.

When using any predictive equation to determine ET rates or net irrigation requirements to maintain grass, a series of assumptions must be made. These assumptions influence actual amounts of net irrigation requirements since each location and golf operation is designed and built differently. Allowances are needed to account for these and to adjust for any differences.

1. The net irrigation requirement is affected by irrigation system efficiency or **distribution uniformity** (designated DU). To determine the actual irrigation quantity needed to provide the minimum intended amount uniformly across the turf, the following equation is used:

$$\text{actual irrigation needed} = \frac{ET_p}{\text{Distribution Uniformity}}$$

For example, if 1.0 *in* (2.54 *cm*) of water is needed as determined by multiplying pan evaporation rate by K_c to achieve ET_p with a 75 % efficient (or DU) system, then 1.33 ($1.0 \div 0.75$) in of total "applied" water is required to uniformly apply this minimum 1.0 *in* (2.5 *cm*) over the whole turf area.

2. Environmental parameters at the time of application also influence the amount of water delivered to plants. Applications made during hot temperatures, windy conditions, and when relative humidity is low, as well as with fine mist irrigation nozzles, can result in extensive evaporation (up to 30 to 50 %) of irrigation prior to reaching the turfgrass. Irrigation should not be scheduled during such periods. However, special practices such as establishing new turf areas, and watering-in fertilizer or pesticide applications, often necessitate irrigation during adverse conditions.
3. Net irrigation requirements listed are for taller-mowed grass. Closely maintained grass, such as golf greens and tees, have significantly less rooting depth compared to taller-mowed plants; thus, they require more frequent, shallow irrigations and have less room for error if not properly and adequately watered during periods of heat and drought stress.
4. Rainfall amounts used in these calculations are averages based on historical climatological data. Deviations from these averages usually occur, and net irrigation amounts during exceptionally dry years will have to be increased to compensate for this. Values listed also assume even rainfall distribution over the entire period. If uniform rainfall distribution does not occur, irrigation amounts higher than those listed in Table 4.3 are required.
5. "On-site" computer-assisted ET-predicted models calculate water needs based on local conditions. Generally, a range of ET models are used that estimate between 0.8 and 1.2 of actual ET.

4.3 Irrigation Strategies

With potential shortages of irrigation water, it is in the best interest of a turf facility to conserve water whenever possible and to design irrigation programs that provide quality turf with minimum water use. Irrigating too heavily not only wastes valuable water, but it invites the potential for increased disease incidence, turf thinning, shallow rooting, reduced stress tolerance, and increased soil compaction and turf wear. Inefficient use of electricity and excessive wear and tear on the irrigation pumps and other components of the system also are reasons to maximize water use.

Playing conditions are also influenced by watering practices. Overwatered golf courses tend to play much longer and have slower putting greens. Conversely, drier turf results in quicker putting surfaces and more bounce and roll; in effect, shortening the course. However, if allowed to dry excessively, this increases the risk of losing turf from moisture stress and causing a reduction in aesthetic quality. Many courses also are restricted in the amount of water they can use and may be mandated to irrigate based on ET data, soil moisture levels, or other water need indicators.

Steps in formulating an irrigation strategy include:

1. Calibrate an irrigation system's output and distribution uniformity (or DU).
2. Determine daily ET rates or soil moisture status by one of the methods discussed. A reasonable estimate of daily summer mean ET rates for various grasses are provided in Table 4.3.
3. Accurately track daily rainfall and ET rates so a water budget can be set-up and followed.
4. When irrigation is needed, use the appropriate crop coefficient to find daily ET rate and incorporate distribution uniformity (DU) of the irrigation system as shown earlier and below.
5. Make adjustments for rainfall, varying microclimates, and forecasted weather.

Irrigation System Calibration

The first step in irrigation scheduling is to determine how much water the irrigation system applies, typically expressed as inches per hour ($in\ h^{-1}$). This information is central to water management. The easiest and most common way to determine application rate is by "canning" the turf area. For small areas, a dozen or so empty tin cans are placed in a grid system across the turf with the location of each catch-can recorded (Fig. 4.9). It is important the cans are the

Fig. 4.9 Conducting a distribution uniformity test when calibrating an irrigation system to determine how uniform water is being applied

same size, have a consistent cross-section, and are fairly tall; soup or vegetable cans work well. The irrigation system is then activated for a timed period, usually 15 to 30 min, to let the cans collect a ¼ to ½ *in* (6.4 to 13 *mm*) of water. The average amount of water in each can is then measured with a ruler and adjusted to the amount of water caught per hour. These cans are all emptied into a single can and the water depth is measured with a ruler. The depth is then divided by the total number of cans to get the average depth per can. This value must be divided by the time period to calculate the application rate. For example, assume 12 cans were used to collect irrigation for a 30-min period. The total depth of all cans was 4.4 *in* (11 *cm*). Dividing 4.4 *in* (11 *cm*) by 12 gives 0.37 *in* (0.94 *cm*) per can. Now multiply the average depth, 0.37 *in* (0.94 *cm*) per one-half hour, by 2 to calculate the application rate of 0.74 $in\ h^{-1}$ (1.9 $cm\ h^{-1}$).

The canning method also helps indicate the distribution uniformity (DU) of the irrigation system which is the ratio of under-watered area to the average applied within the sprinkler coverage area. The most common measure of distribution uniformity is the "low-quarter" method. With this method, distribution uniformity is determined by identifying the depth of irrigation applied to the driest 25 % of the test area and dividing it by the mean depth of water in all cans. The equation of DU involves:

$$\text{distribution uniformity (DU)} = \frac{\begin{array}{c}\text{average } \textbf{least} \text{ amount of water depth}\\ \text{collected in 25\% of all cans}\end{array}}{\text{average depth of water collected for all cans}}$$

Typical DU values range from 55 to 80 %; even rainfall is not 100 % uniform. The lower the value, the less uniformity with which an irrigation system applies water; thus, the more water and energy requirements are needed to uniformly meet plant needs. Obtaining 80 % DU is considered excellent (achievable), 70 % as good (minimum) and 55 % or less as poor. Means of improving existing DU values include: (1) changing sprinklers and sprinkler nozzles; (2) pressure changes (increases); and, (3) changing sprinkler spacing.

Example

1. Determine the distribution uniformity (DU) of the following conditions. A can test was performed with 20 cans evenly spaced 5 *ft* (1.5 *m*) apart in a grid system. After a 15-min run cycle, the average depth in the five **least**-filled cans was 0.2 *in* (0.5 *cm*). The average depth measure in all cans was 0.33 *in* (0.84 *cm*). The irrigation rate is then adjusted from inches per 15-min to inches per hour by multiplying the 0.33-*in* (0.84-*cm*) and 0.2-in (0.5-*cm*) by 4 to achieve 1.32-$in\ hr^{-1}$ (3.36-$cm\ hr^{-1}$) and 0.8-$in\ hr^{-1}$ (2.0-$cm\ hr^{-1}$) respectively. The DU value is then determined:

$$\text{distribution uniformity (DU)} = \frac{\text{average } \textbf{least} \text{ amount of water depth collected in 25\,\% of all cans}}{\text{mean depth collected for all cans}}$$
$$= \frac{0.8\,in}{1.32\,in}$$
$$= 0.61\ (\text{or } 61\,\%)$$

2. How much water would be needed to apply 0.5 *in* (1.3 *cm*) over the entire area? $\frac{0.5\ in}{0.61\ \text{DU}} = 0.82\ in\ (2.1\ cm)$ of irrigation needed to apply at least 0.5 in (1.3 cm) over the area.
3. How long would the irrigation system need to run to apply 0.88 *in* (2.2 *cm*)?
   From the above information, it was determined the irrigation system delivered 1.32 $in\ hr^{-1}$ (3.4 $cm\ hr^{-1}$), therefore,

$$0.88\ in \times \frac{1\ hr}{1.32\ in} \times \frac{60\ min}{hr} = 40\ min$$

Irrigation system calibration, but not DU, can also be determined by knowing the amount (gal) of water applied per irrigation head, the sprinkler spacing (*ft*), and by using one of the formulas listed in Table 4.4. Different formulas are needed depending on whether the sprinkler head design is on square spacing, triangular spacing, or single row design. For example, to determine inches of water applied per hour for an irrigation system designed with triangular spaced heads 50 *ft* apart that apply 30 $gal\ min^{-1}$ of water per head, use the following equation from Table 4.4.

$$\text{in water applied } h^{-1} = \frac{96.3 \times gal\,min^{-1} \text{ per head}}{(\text{sprinkler spacing, } ft)^2 \times 0.866}$$
$$= \frac{96.3 \times 30\,gal\,min^{-1} \text{ applied per head}}{(50ft)^2 \times 0.866}$$
$$= 1.33\,in\,h^{-1}\,(3.4\,cm\,h^{-1})$$

Table 4.4 Irrigation application rates per head based on the head spacing pattern

Square spacing head design
$\frac{96.3 \times gal\ min^{-1} \text{applied per full circle head}}{(\text{sprinkler spacing, } ft)^2} = in\,h^{-1}$
Triangular spacing head design
$\frac{96.3 \times gal\ min^{-1} \text{applied per full circle head}}{(\text{sprinkler spacing, } ft)^2 \times 0.866} = in\,h^{-1}$
Single row spacing head design
$\frac{96.3 \times gal\ min^{-1} \text{applied per full circle head}}{\text{sprinkler throw diameter } (ft) \times 0.80 \times \text{sprinkler spacing } (ft)} = in\ h^{-1}$

Table 4.5 Conversions and calculations for determining turfgrass irrigation needs

1 *ac-in* (water needed to cover 1 *ac* to 1 *in* depth)	$= 27,154\,gal$ $= 43,560\,in^3$ $= 3,630\,ft^3$	1 *ac-ft* (water needed to) cover 1 *ac* to 1 *ft* depth)	$= 325,851\,gal$ $= 43,560\,ft^3$
$1\ in\ 1,000\,ft^{-2}$	$= 623.33\,gal$ $= 83.33\,ft^3$	7.48 *gal*	$= 1\,ft^3$ $= 1728\,in^3$
1 *gal*	$= 0.134\,ft^3$ $= 231\,in^3$ $= 8.34\,lb$ water	1 *ft3* 1 *psi* 1 *ft* of head	$= 7.4805\,gal$ $= 2.31\,ft$ of head $= 0.433\,psi$
1 *lb* of water	$= 0.1199\,gal$ $= 27.7\,in^3$	1 million *gal*	$= 3.07\,ac\text{-}ft$

Example

1. If 46 *ac* (18.6 *ha*) of turf were to receive 1 *in* (2.5 *cm*) of water, what is the total amount of water, in gallons, needed? From Table 4.5, 1 *ac-in* of water equals 27,154 *gal*; thus, 27,154 *gal* × 46 *ac* = 1,249,084 total *gal* water needed (4.7 million *L*).
2. If water costs are $\$0.03\,ft^{-3}$ of water, what is the total cost of this volume? From Table 4.5, 1 ft^3 equals 7.48 *gal* of water; thus,

$$\frac{1\,ft^3}{7.48} \times 1,249,084\,gal \quad total \times \frac{\$0.03}{ft^3} = \$5,010$$

Determining Irrigation Rates and Frequency

In addition to the application rate and uniformity, the turf manager should know how much water the turf is using. This can be determined using reference ET from a weather station/computer system plus a crop coefficient specific for the turf species from data in Table 4.3, or with data from an atmometer or other devices as previously discussed. Historical weather information may also provide reasonable estimates of average water use. Managers also need to know where the roots are in the soil profile and approximately how much available water is held by the soil.

The amount of water needed to moisten the soil to a given depth depends on soil type, water infiltration and percolation rates, and surface slope. Figure 4.10 presents the amount of water needed to wet different soils to various depths. Soils severely sloped, compacted, or clayey in nature may have low infiltration rates. As a result, the soil may not be able to absorb the required amount of irrigation at one time. Managers may have to irrigate using multiple cycles until the desired amount is applied. After an irrigation event, managers should double-check the depth of moisture penetration using a soil probe or screwdriver so they can fine-tune their timing.

As previously noted, evaporation during hot, windy, and dry periods can reduce irrigation efficiency. Superintendents can avoid this by irrigating early in the morning before the temperature rises and humidity drops. Early morning irrigation also removes dew from the leaves, and helps prevent diseases favored by irrigating in the evening.

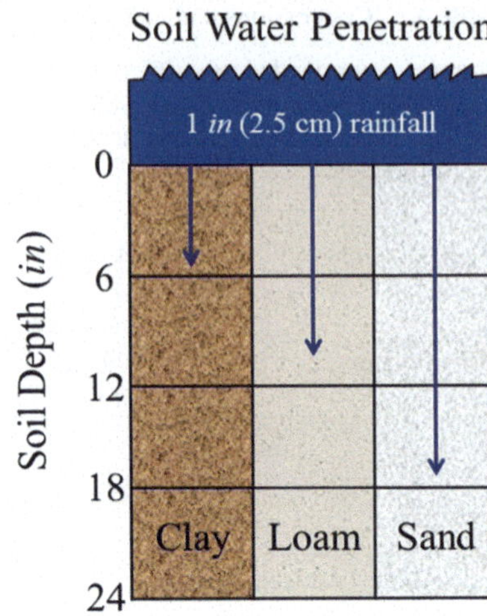

Fig. 4.10 Approximate penetration depth of water applied to three types of soil

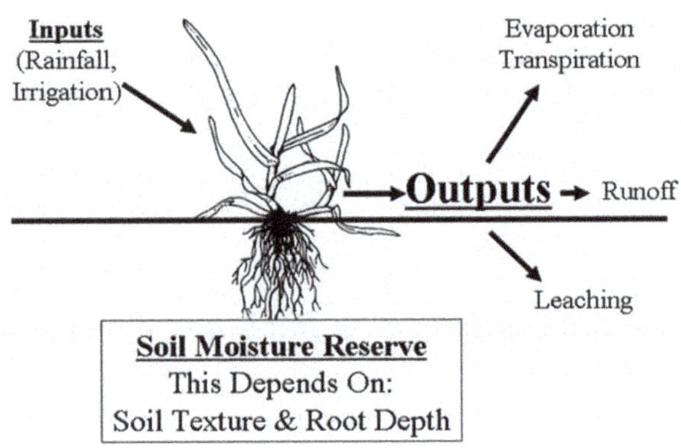

Fig. 4.11 Inputs and outputs when developing a water budget to determine irrigation needs

Water Budgeting

Budgeting water is analogous to handling money in a checking account (Fig. 4.11). There are inputs (deposits), outputs (withdrawals), and a certain amount of water in the soil (standing balance). The flow of water (money) into and out of the "checking account" (the rootzone) is simply followed over time. If the roots penetrate 12 *in* (30 *cm*), the checking account is the water held in 12 *in* (30 *cm*) of soil. If the roots penetrate only 2 *in* (5 *cm*), the checking account is considerably smaller. Irrigation is applied to wet the rootzone, no more, no less. Generally, most of the roots on putting greens and tees are in the top 6 *in* (15 *cm*) of soil, whereas roots on fairways and roughs often penetrate 12 *in* (30 *cm*) or more.

Consider a silt-loam soil at field capacity, which is roughly 2.0 *in* (5 *cm*) of water per foot of soil (see Chap. 1). A 12 *in* (30 *cm*) deep bermudagrass root system growing in this soil will have access to 2.0 *in* (5 *cm*) of available water.

Weather station data and a predictive model estimate over a 6-day period that 1.8 *in* (4.6 *cm*) of water was used by the theoretical reference crop. Correcting this reference value using a K_c of 0.7 for bermudagrass, estimates the turf actually uses about 1.3 *in* of water ($1.8 \times 0.7 = 1.26$ *in*, 3.2 *cm*). Subtracting this from the original 2.0 *in* (5 *cm*) of available water gives about 0.7 *in* (1.8 *cm*) of water left in the soil. Should the turf go another day before irrigating? No, it's time to water, since it is never a good idea to deplete most of the available water. Approximately 1.5 *in* (3.8 *cm*) of irrigation should be applied to replace the 1.3 *in* (3.3 *cm*) lost from the system. The soil is returned to field capacity without irrigating excessively and wasting water.

For most turfgrass examples, the amount of water at wilting point is negligible. Turfgrass rooting depth should be used instead of soil rootzone depth since most moisture obtained by plants in a reasonable time frame will be in the rooting depth and not below it.

Determining Approximate Intervals (in Days) Between Irrigation Cycles

$$\text{irrigation interval (days)} = \frac{\text{soil water content at field capacity} \times \text{rooting depth } (in)}{\text{daily ET rate } (in\ day^{-1})}$$

Example Determine the time between irrigation cycles for a sand soil with a volumetric water content of 15 % at field capacity, a rooting depth of 4 *in* (10 *cm*), and a summer daily ET rate of 0.20 *in day*$^{-1}$ (5 *mm day*$^{-1}$):

$$\frac{0.15 \times 4\,in}{0.20\,in\,day^{-1}} = 3 \text{ days between irrigation cycles, which brings the soil back to field capacity}$$

If rainfall occurs and it is more than the amount of water depleted during the period (1.3 *in*, 3.3 *cm*), the rootzone is returned to field capacity and any excess is ignored since it will drain and not be stored in the rootzone. If it rains less than actual ET, the running deficit is calculated over several days, and irrigation is scheduled when ET has depleted the soil moisture to a bit more than 50 % of the 0.6 *in* (0.15×4 *in*) of available water. A good rain gauge is needed to keep track of precipitation, and it is a good idea to use automatic pump shutdown switches to prevent irrigation after a significant precipitation. Conversion factors in Table 4.5 indicate gallonage required to apply certain amounts.

Example From the use of a hand-held TDR probe, determine a soil moisture management program including when to irrigate and how much water is needed to return the total moisture levels to field capacity for two sands with $\theta_{fc} = 0.32$ and 0.22 $cm^3\ cm^{-3}$ and $\theta_{wp} = 0.02$ and 0.01 $cm^3\ cm^{-3}$ for sands 1 and 2, respectively. The TDR probe measures moisture in the top 10 *cm* (4 *in*) of the soil profile.

step 1: Determine available water for each sand using the equation:

$$\mathrm{D_e} = soil\ depth\ \left(\theta_{\mathrm{fc}} - \theta_{wp}\right)$$

where:

$\mathrm{D_e}$ = equivalent depth of available water in the top 10 *cm* (4 *in*),
θ_{fc} = volumetric water content at field capacity,
θ_{wp} = volumetric water content at wilting point

$$\text{For sand 1}: \quad \mathrm{D_e} = \text{soil depth}\left(\theta_{\mathrm{fc}} - \theta_{\mathrm{wp}}\right) = 10\,(0.32 - 0.02) = 3.0\,cm(1.18\,in)$$

$$\text{For sand 2}: \quad \mathrm{D_e} = \text{soil depth}\left(\theta_{\mathrm{fc}} - \theta_{\mathrm{wp}}\right) = 10\,(0.22 - 0.01) = 2.1\,cm(0.83\,in)$$

step 2: If the effective rootzone is 10 *cm* (4 *in*) deep and the turfgrass being used has an average ET rate of 0.2 *in* day^{-1} (0.5 *cm* day^{-1}), the days between watering for each sand would be:

$$\text{sand 1}: \quad 3.0\,cm\,\text{rootzone moisture} \times \frac{1\,\text{day}}{0.5\,cm\,\text{moisture used}} = 6\,\text{days}$$

$$\text{sand 2}: \quad 2.1\,cm\,\text{rootzone moisture} \times \frac{1\,\text{day}}{0.5\,cm\,\text{moisture used}} = 4.2\,\text{days}$$

Therefore, for sand 1, 3.0 *cm* of water would be needed every 6 days while for sand 2, 2.1 *cm* would be needed every 4.2 days to return each to field capacity.

With information on ET rates and sprinkler calibration available, each sprinkler's run time can be calculated. The daily ET rate is divided by the sprinkler output. For example, if the day's ET rate is 0.3 *in* (7.6 *mm*) and the sprinkler output is 0.01 *in min*$^{-1}$ (0.25 *mm min*$^{-1}$), the irrigation time needed would be 30 *min*. However, this is adjusted according to the appropriate crop coefficient (e.g., 0.85 for bentgrass); therefore, 30 *min* is multiplied by 0.85 to give 25 *min* of run-time needed. Distribution uniformity considerations should then be incorporated to ensure enough water is being applied uniformly across the turf area.

Example Water use engineers employed at a municipality require a golf course to justify their water use permit in terms of total amount of water requested and how they determined this value (patterned after Green 2005).

A: Determine average yearly ET rate from one of the methods listed previously. In this example, 56.37 *in* (4.7 *ft*, 1.4 *m*) is used.
B: Determine normal yearly precipitation rate. In this example, 10.67 *in* (27 *cm*) is used.
C: Area of irrigated turfgrass. In this example, 110 *ac* is used (3.1 *ac* for greens, 3.7 for tees, 43.7 for fairways, and 59.5 for roughs).
D: Determine the irrigation efficiency (DU). In this example, 70 % is used.

	Greens	Tees	Fairways	Roughs
E. turf area (*ac*)	3.1	3.7	43.7	59.5
F. Turfgrass	Bentgrass	Bermuda overseeded Oct–May	Bermuda overseeded Oct–May	Bermuda
G. K_c (crop coefficient)	0.8	0.75	0.75	0.65
H. Turf Water Use [A × G] (which is ET × K_c)	45.1	42.3	42.3	36.6
I. 25 % precipitation (*in*): [B × 0.25][a]	2.7	2.7	2.7	2.7
J. Water use adjusted for 25 % precipitation (in): [*H-I*]	42.4	39.6	39.6	33.9
K. Irrigation water use (*in*): [J/D]	60.6	56.6	56.6	48.4
L. K converted to feet: [K/12] (12 *in* = 1 *ft*)	5.1	4.7	4.7	4.0
M. Annual irrigation use (*ac-ft*): [E × L]	15.8	17.4	205.4	238.0
N. Annual irrigation water use: [sum of M for all turf areas]	477 *ac-ft* (or 155,430,927 *gal*)			

[a]Water use regulators often use a precipitation efficiency adjustment value to reflect the amount (percentage) of usable precipitation by plants. Rainfall is often at inefficient amounts (too high or low) or at the wrong agronomic time.

In the above example, to compare calculated annual irrigation use to the overall formula, ET × area, the following was determined:

O. ET × 110 *ac*: *A* (*ft*) × *C* (total turfgrass area) or 4.7 *ft* × 110 *ac* = 517 *ac-ft* predicted by the simple formula,

P. Calculation efficiency for water budget: *N*/*O* × *100* or 477 *ac-ft* ÷ 517 *ac-ft* × 100 = 92 %. This value indicates the simple formula of ET × area overestimated water needs by 8 % compared to the Water Budgeting process above.

4.4 Managing Irrigation Water Quality Problems

Turf facilities are increasingly using poorer quality irrigation sources. Wells, ponds, retention ponds, canals, streams, rivers, lakes, and waste treatment plants are common water sources for irrigation. Water from waste treatment plants may contain elevated nutrient and trace element concentrations. Successful irrigation management requires regular monitoring of both soil and water chemistry, especially salt content. The following tests provide information concerning soil and water quality:

- **Water soluble salts** (or **Salinity drought hazard**)—Total salt content as measured by the electrical conductivity (EC_w) or total dissolved salts (TDS) of water. Excessive salts produce plant physiological drought.

- **Sodium status**—Soil sodium level proportionally to Ca and Mg ions as measured by sodium adsorption ratio (SAR), exchangeable sodium percentage (ESP), or adjusted SAR (adj. SAR). SAR also is used to assess the sodium levels of water. Excessive sodium causes soil structure deterioration.
- **Specific ions toxicity**—Toxic ion levels, especially boron, chloride, fluoride, sulfate and nitrate-nitrogen.
- **Alkalinity**—Bicarbonates and carbonates as measured by residual sodium carbonate (RSC).
- **pH and lime requirement.**
- **Suspended solids**, as measured by total suspended solids (TSS).
- **Soil nutrient imbalance based on**:
  - Sufficiency levels of available nutrients and cation ratio.
  - Soil cation exchange capacity (CEC).
  - Percent base saturation.
  - Percent organic matter.

Salts

A salt is a combination of positively charged ions (cations) and negatively charged ions (anions). Cations include calcium, magnesium, sodium, ammonium, and potassium; while anions include carbonates, bicarbonate, nitrate, sulfate, chloride, and boron. Table salt (sodium chloride) is found in some soils. Insoluble salts (i.e., gypsum and lime) occur, but excessive soluble salts are the primary ones that may impede plant growth rather than the insoluble ones. High soluble salts in the soil solution reduce water availability, causing the turfgrass to be prone to drought stress. This is the most important or most common salt problem involved with turfgrasses.

The amount of salt in water determines the degree of salinity and, to a large extent, the overall water quality. The following equation determines the amount of salt applied when irrigating with saline water:

$$\begin{aligned} lb\,\text{salt applied}\,ac^{-1} &= \text{irrigation water salinity level}\ \left(ppm\,\text{or}\,mg\ L^{-1}\right) \\ &\quad \times 2.72\ \text{million}\,lb(\text{weight of water per}\,ac\text{-}ft) \\ &\quad \times ac\text{-}ft\,\text{water applied} \end{aligned}$$

Example How much salt is applied, if 1 *in* (2.5 *cm*) of water with salinity levels of 640 *ppm* is used? 1 *in* = 0.083 *ft*.

$$\frac{640\,parts}{1{,}000{,}000} \times \frac{2{,}720{,}000\,lb}{ac\text{-}ft\,\text{water}} \times 1\,ac\text{-}in\,\text{water} \times \frac{1\,ft}{12\,in} = 144\,lb\,\text{salt applied per acre}$$

To determine the amount of salt applied per 1000 ft^2, divide 144 *lb* salt ac^{-1} by 43.56 = 3.3 *lb* salt applied per 1000 ft^2, when 1 *in* (2.5 *cm*) of irrigation water with a salinity level of 640 *ppm* is used.

Two types of salt problems exist: (1) those associated with the total salinity, and (2) those associated with sodium. Water with high salinity becomes toxic to plants and poses a **salinity hazard**. As mentioned, soil salt accumulation is the most common cause of plant injury from poorer quality water but normally must occur over an extended period of time before this is seen. Combinations of saline irrigation use, low precipitation, poor soil drainage, and the use of cool-season turfgrasses increase the likelihood of salinity problems. Salt soils may cause direct injury to turfgrass growth or indirect injury due to soil physical properties. Direct stresses include moisture stress as roots are unable to absorb tightly held soil moisture, ion toxicity, or nutrient (ion) imbalances. In saline soils, water moves from an area of lower salt concentration (plant roots) to an area of higher salt concentration (the soil). This causes plant water stress and wilt even though the soil may be wet. Indirect stress occurs from high soil sodium by destroying soil structure, thus, reducing water infiltration, drainage, and soil oxygen levels. Salinity problems are less likely to develop with high rainfall and cooler climates, use of salt-tolerant warm-season grasses, and soils that are well-drained.

Drought stress symptoms from salinity stress include turf developing bluish-green color, wilting, leaf rolling or folding, and eventual leaf firing (yellowing and death) (Fig. 4.12). Direct ion toxicity to plants can occur from excessive soil levels of sodium (Na^+), chloride (Cl^-), boron (B^-), bicarbonate (HCO_3^{-2}), and high pH

Fig. 4.12 Direct salinity damage to turf resembles drought symptoms or fertilizer burn. Shown is salt-damaged turf from an ocean storm surge across a golf course fairway

Fig. 4.13 Salt and bicarbonate build up on a soil surface from inadequate flushing and soil drainage

[hydroxyl (OH^-)] ions. Nutrient imbalance from high soil levels of sodium, chloride and other ions can also occur. High levels of these can cause deficiencies of calcium (Ca^{+2}), potassium (K^+), nitrates (NO_3^-), magnesium (Mg^{+2}), manganese (Mn^{+2}), and phosphorus (P).

Salts also can move upward from groundwater. Water is drawn to the surface when evaporation exceeds the amount of water being applied and is deposited on the soil and plant surface through the process of capillary rise. Formation of a white crust on the soil surface indicates salt accumulation, as does shoot browning (Fig. 4.13). Many arid and semiarid soils, especially when annual rainfall is <15 *in* (38 *cm*), are salt affected due to insufficient leaching to remove salts that accumulate from the weathering of minerals, groundwater, and rain. In arid and semiarid regions, sodium and sulfate salts (Na_2SO_4, K_2SO_4, $CaSO_4$, and $MgSO_4$) usually dominate, reflecting the composition of the soils parent material.

Measuring and Classifying Irrigation Salinity

Salinity hazard is determined by measuring the ability of water to conduct an electrical current. Salty water is a good conductor of electrical current, whereas pure water is a relatively poor conductor. Salinity is expressed in two different ways, either as **electrical conductivity** (EC_w) or **total dissolved salts** (TDS) (also reported as **total soluble salts**, TSS). There are several units commonly used to express EC_w: deciSiemens per meter ($dS\ m^{-1}$), millimhos per centimeter ($mmhos\ cm^{-1}$), or micromhos per centimeter ($\mu mhos\ cm^{-1}$). The relationship between these units is:

$$1 dS\ m^{-1} = 1\ mS\ cm^{-1} = 0.1\ S\ m^{-1} = 1\ mmhos\ cm^{-1} = 1000\ mmhos\ cm^{-1} \\ = 640\, ppm\, \text{TDS}$$

Total dissolved salts are expressed in parts per million (*ppm*) or milligrams per liter ($mg\ L^{-1}$) and are generally not measured directly, but calculated from an EC_w measurement.

$$\text{TDS}\ \left(mg\ L^{-1}\ or\ ppm\right) = \text{EC}_\text{w}\left(mmhos\ cm^{-1}\ or\ dS\ m^{-1}\right) \times 640$$

Individual components of salinity (such as sodium) may also be reported in milliequivalents per liter ($meq\ L^{-1}$). To convert ppm to $meq\ L^{-1}$, divide the *ppm* of the ion by its equivalent weight. The ratio of total dissolved salt to EC_w of various salt solutions ranges from 550 to 740 *ppm* per $dS\ m^{-1}$. The most common salt in saline water, sodium chloride, has a TDS of 640 *ppm* at an EC_w of 1 $dS\ m^{-1}$. Most laboratories use this relationship to calculate TDS from EC_w, but some multiply the amount by 700.

Example

1. An irrigation source has an EC_w of 0.53 $mmhos\ cm^{-1}$. What would the EC_w be in $dS\ m^{-1}$, $\mu mhos\ cm^{-1}$, and *ppm* TDS?

   (a) Since 1 $dS\ m^{-1} = 1\ mmhos\ cm^{-1}$, then 0.53 $mmhos\ cm^{-1} = 0.53\ dS\ m^{-1}$
   (b) Since 1 $mmhos\ cm^{-1} = 1000\ \mu mhos\ cm^{-1}$, then

$$0.53\ mmhos\ cm^{-1} \times \frac{1,000\ \mu mhos\ cm^{-1}}{1\ mmhos\ cm^{-1}} = 530\ \mu mhos\ cm^{-1}$$

   (c) To convert $mmhos\ cm^{-1}$ to *ppm*, multiply by 640:

$$0.53\ mmhos\ cm^{-1} \times 640 = 339\ \ ppm\ \ \text{TDS}$$

2. The salt content of a water sample is 1121 $mg\ L^{-1}$ TDS. What is the salt content in $dS\ m^{-1}$ and $\mu mhos\ cm^{-1}$?

   (a) To convert TDS ($mg\ L^{-1}$ or *ppm*) to $dS\ m^{-1}$, divide by 640:

$$1,121\ mg\ L^{-1} \div 640 = 1.75\ \ dS\ m^{-1}$$

   (b) To convert $dS\ m^{-1}$ (or $mmhos\ cm^{-1}$) to $\mu mhos\ cm^{-1}$, multiply by 1000:

$$1.75\ \ dS\ m^{-1}\left(or\ mmhos\ cm^{-1}\right) \times 1,000 = 1,750\ \mu mhos\ cm^{-1}$$

Table 4.6 Salt concentration hazard levels for irrigation water

Hazard	EC_w (*dS m*$^{-1}$)	Total dissolved salts (*ppm*)
Low	0.75	500
Medium	0.75–1.5	500–1000
High	1.5–3.0	1000–2000
Very high	>3.0	>2000

3. Convert 100 *ppm* Ca to *meq* L^{-1}. The equivalent weight of Ca^{+2} is 20.

$$100\,ppm \text{ Ca} \div 20 = 5\,meq\ L^{-1} \text{ of Ca}$$

Water sample salinities are often compared to those of seawater with an average EC_w of 54 *dS m*$^{-1}$ or about 34,500 *ppm* dissolved salts.

Irrigation water is classified based on the salinity hazard, which considers the potential for damaging plants and the level of management needed for utilization as an irrigation source (Table 4.6). Water with EC_w readings of less than 0.75 *dS m*$^{-1}$ is suitable for irrigation without problems. Successful use of water with EC_w values above 0.75 *dS m*$^{-1}$ depends upon soil conditions and plant tolerance to salinity. Generally, higher salinity levels can be used on sandy soils where salts can be flushed. Similar values on poorly draining clay soils that may cause problems. Under typical summer stress, EC_w of turfgrass irrigation should ideally not exceed 1.25 *dS m*$^{-1}$soluble salts. Salinity levels above 3.0 *dS m*$^{-1}$ are unsuitable for any length as an irrigation source.

Water Sodium Hazard

The primary cause of sodic or saline-sodic soil is using high sodium (Na^+) content irrigation water. While EC_w is an assessment of all soluble salts in a sample, **sodium hazard** (termed sodic or saline-sodic soil) accounts for sodium's specific detrimental effects on soil physical properties. The potential for irrigation water to have poor infiltration properties or sodium hazards is assessed by determining the **sodium adsorption ratio** (SAR) and the electrical conductivity (EC_w) of the water. The sodium adsorption ratio relates the concentration of sodium to the concentration of calcium and magnesium. Calcium and magnesium counter the negative effects of sodium on soil structure. The higher the sodium level in relation to calcium and magnesium, the higher the SAR, the poorer the water infiltration, and the more increased problems with soil deflocculation (deterioration—swelling, dispersion, and permeability reduction). The collapse of aggregates from dispersion of clay tends to clog large pores, particularly at the soil surface. Salt concentration and exchangeable sodium percentage then become problems with the loss of permeability. Calcium will hold soil together (or flocculate), while sodium pushes (or disperses) soil particles apart. The dispersed soil readily crusts and poses water infiltration and permeability problems.

SAR is defined as:

$$\text{SAR} = \frac{Na^+}{\sqrt{\frac{Ca^{+2} + \text{Mg}^{+2}}{2}}} \quad or \quad \frac{Na^+}{\sqrt{Ca^{+2} + \text{Mg}^{+2}}}$$

Ion concentrations in the equation above left are expressed in milliequivalents per liter ($meq\ L^{-1}$) while those in the equation above right are expressed in millimoles per liter ($mmol\ L^{-1}$). Milliequivalents describe the molecular weight adjusted for the valence number (number of positive charges) of the ion. The SAR is determined by the number of milligrams per liter ($mg\ L^{-1}$ or ppm) of Na^+, Ca^{+2}, and Mg^{+2} in a water sample. To convert ppm (or $mg\ L^{-1}$) to $meq\ L^{-1}$, use the following equation and equivalent weights for Na^+, Ca^{+2}, and Mg^{+2} of 23, 20, and $12.2\,mg\ meq^{-1}$, respectively. Use of the saturated paste extract method, rather than other soil test extraction methods, is necessary for determining soil Na^+, Ca^{+2}, and Mg^{+2} levels for the SAR equation.

$$meq\,L^{-1} = \frac{\text{concentration}\left(ppm\,\text{or}\,mg\,L^{-1}\right)}{\text{equivalent weight}\left(mg\,meq^{-1}\right)}$$

Example A water sample test reports $1000\,mg\ L^{-1}$ Na^+, $200\,mg\ L^{-1}$ Ca^{+2}, and $100\,mg\ L^{-1}$ Mg^{+2}.

Find the SAR value in $meq\ L^{-1}$.

step 1: Calculate the concentration ($meq\ L^{-1}$) of each ion:

$$\begin{aligned}
\text{Na}^+ &: 1000\,mg\,L^{-1} \div 23\,mg\,meq^{-1} = 43.5\,meq\,L^{-1} \\
\text{Ca}^{+2} &: 200\,mg\,L^{-1} \div 20\,mg\,meq^{-1} = 10\,meq\,L^{-1} \\
\text{Mg}^{+2} &: 100\,mg\,L^{-1} \div 12.2\,mg\,meq^{-1} = 8.2\,meq\,L^{-1}
\end{aligned}$$

step 2: Place these values into the SAR equation:

$$SAR = \frac{Na^+}{\sqrt{\frac{Ca^{+2} + \text{Mg}^{+2}}{2}}} = \frac{43.5}{\sqrt{\frac{10 + 8.2}{2}}}$$
$$= 14.4\ meq\ L^{-1}$$

Example A water analysis indicates a Na^+ concentration of $85\,meq\ L^{-1}$, a Ca^{+2} concentration of $33.3\,meq\ L^{-1}$, and a Mg^{+2} concentration of $7.1\,meq\ L^{-1}$. What is the SAR value for this water?

Table 4.7 SAR values, categories, and precautions for irrigation sources with $EC_w \geq 1\ dS\ m^{-1}$

SAR or adj SAR ($meq\ L^{-1}$)	Category	Precaution
0–10	Low sodium water	Little danger from structure deterioration to almost all soils. For ornamentals, water SAR values should be <10.
10–18	Medium sodium water	Problems on fine-textured soils and sodium-sensitive plants, especially under low-leaching conditions. Soils should have good permeability.
18–26	High sodium water	Problems on sodium accumulation on most soils. Good salt-tolerant plants are required along with special management, such as good drainage, the use of gypsum, and leaching. Generally, high and very high EC water should not be used for irrigating turfgrasses long term.
>26	Very high sodium water	Unsatisfactory except with high salinity ($EC_w > 2.0\ dS\ m^{-1}$), high calcium levels, and the use of gypsum.

$$\mathrm{SAR} = \frac{Na^+}{\frac{\sqrt{Ca^{+2} + \mathrm{Mg}^{+2}}}{2}} = \frac{85}{\frac{\sqrt{33.3 + 7.1}}{2}} = 18.9 \quad meq\ L^{-1}$$

Since salts and sodium do not act independently, the effect of sodium on soil particle dispersion, thus permeability, is counteracted by high concentrations of soluble salts (measured as EC_w) in the irrigation water. The effects of high SAR on irrigation water infiltration are dependent on the electrical conductivity of the water (Table 4.7). For a given SAR, the lower the EC_w, the greater dispersion or poorer infiltration properties; the higher the EC_w, the better the infiltration. For example, irrigation water with a $\mathrm{SAR} = 15\ meq\ L^{-1}$ has poor infiltration properties with an $EC_w = 0.5\ dS\ m^{-1}$, but good infiltration properties with an $EC_w = 2.0\ dS\ m^{-1}$. As a rule-of-thumb, if the SAR is more than ten times greater than the EC_w, then poor water infiltration is likely to occur. When the $EC_w = 0.5\ dS\ m^{-1}$ or less, the water has very few minerals to flocculate soil particles. Thus, irrigating with this **pure water** strips minerals from cation exchange capacity (CEC) sites, causing dispersed particles to settle closely next to each other. The result is a compacted soil surface which forms a thin crust layer, impeding water flow into the soil. Problems can develop quickly when $EC_w \leq 0.2\ dS\ m^{-1}$. In the case of pure water, the problem exists regardless of the SAR value since very few minerals are present to begin with.

Clay-textured soils can have structural permeability problems if a water $\mathrm{SAR} > 9\ meq\ L^{-1}$ is used over an extended period that reduces infiltration, percolation, and drainage, often causing low soil oxygen problems. In the earlier example where the water sample had an SAR of $14.4\ meq\ L^{-1}$, problems could occur if this water source was used long term on finer-textured soils.

Example A superintendent has two water sources to choose from based on their sodium hazard.

Sample 1. SAR = 5.0 and EC = 0.5.
Sample 2. SAR = 5.0 and EC = 1.5.
Which one is more suitable?

Sample 1. Water infiltration problems may occur, especially on finer textured clay or silt-based soils.

Sample 2. This sample is less likely to cause soil water infiltration problems.

Soil Sodium Permeability Hazard

Although high sodium levels in irrigation water can be directly toxic to plants (especially ornamentals), its most deleterious effect is on soil structure. Since sodium ions (Na^{+}) are monovalent (have only one positive charge), two sodium ions are needed to displace divalent (two positive charged) ions such as calcium (Ca^{+2}) or magnesium (Mg^{+2}). This concern is greater on fine-textured soils such as clays and silt loams. Salts often accumulate in high, exposed sites such as hilltops while low areas may accumulate salts from runoff.

High soil sodium causes finer-textured soil clays and organic matter to disperse (termed **deflocculation**) to where aggregates break down into smaller units and smaller clay minerals and organic particles plug soil pores, reducing water infiltration and soil aeration. Soil then seals and becomes hard and compacted, reducing soil water and oxygen movement. The higher the clay and organic matter content of the soil, the greater the effects of sodium. Typically, for soil structure breakdown, sodium levels exceed calcium levels by more than 3 to 1. These soils are characterized by pools of standing water after irrigation. To counteract the negative effects of sodium, increasing calcium and magnesium concentrations in clay soils will cause the soil to flocculate (have good structure). A key management step is to prevent soil structure breakdown.

Soil structure can be destroyed by continued use of water containing high levels of sodium. This results in reduced water infiltration, drainage, and soil oxygen. The sodium ions replace calcium and magnesium ions on the clay CEC sites, destroying its structure plus reducing pore continuity, thus reducing infiltration, percolation, and drainage.

Assessing Soil Salinity

Saline soils are classified based on two criteria: (1) the **total soluble salt** or **salinity content** based on electrical conductivity of a saturated extract (EC_e), and (2) **exchangeable sodium percentage** (or, more recently, **sodium adsorption ratio**). Additional information is also often used, such as carbonate content and potential toxic ions.

Soluble salts are measured in soils by the same basic method as used for water samples. A conductivity instrument measures electrical conductivity in an extract

(EC_e) either from a **saturated paste** (preferred method) or from a **soil:water dilution**. As total salt concentration increases, EC_e also increase. The SAR is a calculated value from a saturated paste extract sample based on milliequivalents per liter of Ca, Mg, and Na. The saturated paste extract is the most precise method to determine soil EC, SAR, and boron levels. A soil sample is brought just to the point of saturation using the irrigation source, allowing it to equilibrate for several hours, and then is subjected to vacuuming to extract the soil solution through filter paper. Spectrophotometers and other analytical equipment are then used to quantify the soil solution. Using the saturated paste extract, soils with EC_e readings $<1.5\,dS\ m^{-1}$ are considered to have low salt levels. Soils with EC_e readings of 1.6 to $3.9\,dS\ m^{-1}$ have medium levels. When soil readings are above $4.0\,dS\ m^{-1}$, soils are considered to have high salt levels and only salt-tolerant turfgrasses normally survive.

Soil water dilution ratios are either a 1:2 dilution (one part dry soil:two parts water) or a 1:5 dilution (one part soil:five parts water). Electrical conductivity readings from these three methods are not comparable, so the method used must be known in order to interpret the EC_e reading. Soil testing laboratories frequently use a 1:2 dilution method because it is more rapid than obtaining a saturated paste extract. The EC_e of a 1:2 extract is on average 20 % of the EC_e of a saturated paste extract, on sand-based greens. To estimate the EC_e of a saturated paste from a 1:2 extract, multiply the EC_e of the 1:2 extract by 5.

Assessing Soils for Sodium Problems

Sodicity refers to high concentration of sodium (Na^+) while **salinity** refers to high concentrations of total salts including NaCl, Ca^{+2}, Mg^{+2}, and SO_4^{-2}. Salt-affected soil can be classified as **saline**, **sodic**, and **saline-sodic** soils. **Saline soils** are the most common type of salt-affected soil and the easiest to reclaim. Saline soils are plagued by high levels of soluble salts, primarily chloride (Cl^-), sulfate (SO_4^{-2}), and sometimes nitrate (NO_3^-). Salts of low solubility, such as calcium sulfate or gypsum ($CaSO_4$) and calcium carbonate ($CaCO_3$), may also be present. Because exchangeable sodium is not a problem, saline soils are usually flocculated with good water permeability. Saline problems generally occur when: (1) there is insufficient rainfall to leach salts through the soil profile, (2) drainage is impaired, or (3) irrigation water contains high levels of salts.

Sodic soils, or soil structure deteriorated soils, have high levels of exchangeable sodium and low total soluble salt content, $HCO_3 > 120\,mg\ L^{-1}$ or $CO_3^{-2} > 15\,mg\ L^{-1}$. These soils tend to disperse, reducing water infiltration. Sodic soils also have a pH between 8.5 and 10 and are often called black alkali soils because the organic matter in the soil tends to disperse creating black-colored puddles (Fig. 4.14). Calcium and magnesium ions in sodic soils tend to form insoluble calcitic lime, leaving low soluble calcium and magnesium levels to displace sodium ions, allowing the sodium problems. The high sodium concentration of a sodic soil not only injures plants directly, but also degrades the soil

Fig. 4.14 Black-colored soil resulting from dispersed soil organic matter in sodic soils. This condition is referred to as **black alkali**

structure, termed "sodium hazard." Sodic soil cannot be improved by leaching the sodium from the soil profile alone. Soil amendments are required to replace the sodium in the soil, commonly a calcium containing amendment, in conjunction with leaching with acidified water.

Saline-sodic soils contain both high soluble salts and high exchangeable sodium. Saline-sodic soils, like sodic soils, are best reclaimed by adding a calcium-containing amendment and then leaching to remove excess soil sodium ions.

Two laboratory measurements are used to assess whether soils contain excessive sodium levels and if poor drainage and aeration are likely to occur. These measures are the **exchangeable sodium percentage** (ESP) and the **sodium adsorption ratio (SAR)**. The ESP identifies the degree or portion of the soil cation exchange capacity (CEC) occupied or saturated by sodium, and is calculated as follows:

$$\text{ESP}\,(\%) = \frac{\text{exchangeable sodium}\ (meq\,100\,g^{-1})}{\text{cation exchange capacity}\ (meq\,100\,g^{-1})} \times 100$$

ESP does not consider the quantity of calcium and magnesium ions relative to sodium ions present like SAR does.

Example A soil test indicates the Na^+ content of a soil is 6.9 meq 100 g^{-1} and the CEC of the soil is 17.3 meq 100 g^{-1}. Find the exchangeable sodium percentage (ESP) of this soil.

$$\begin{aligned}\text{ESP} &= \frac{\text{exchangeable sodium } (meq\,100\,g^{-1})}{\text{cation exchange capacity } (meq\,100\,g^{-1})} \times 100 \\ &= \frac{6.9\ meq\,100\,g^{-1}}{17.3\ meq\,100\,g^{-1}} \times 100 \\ &= 40\,\%\end{aligned}$$

Soil SAR is a second, more easily measured property, analogous to the irrigation water SAR discussed earlier which considers calcium and magnesium ion content in the soil. Soil SAR is calculated from soil-test extractable levels of sodium, calcium, and magnesium (expressed in $meq\,100\,g^{-1}$ or $mmol\,L^{-1}$).

ESP indicates the probability a soil will disperse, thereby reducing the permeability of soil to water and air. In the environment, salts and sodium do not act independently. High-soluble salt concentration can negate the soil particle dispersal (thus, impermeability) effects of sodium. Usually, little or only minor problems occur when ESP values are less than 13–15 %. An ESP >15 % or a soil SAR >13 $meq\,100\,g^{-1}$ indicates a **sodic** soil, where sodium causes soil colloids to disperse and plug the soil's drainage pores, thereby reducing the permeability of the soil to water and air. Sodic soils become saturated with sodium ions compared to calcium and magnesium ions, especially if bicarbonate ions are present. Symptoms of reduced permeability include waterlogging, reduced infiltration rates, crusting, compaction, disease occurrence, weed invasion, and poor aeration. Sodic soils often have considerable amounts of clay that is sticky due to the sodium. ESP and SAR are related and can be estimated by:

$$\text{ESP} = \frac{1.475 \times \text{SAR}}{1 + (0.0147 \times \text{SAR})}$$

Managing Poor Quality Water Use Sites

Managing salinity, sodicity, and alkalinity problems requires constant attention (Table 4.8). Management practices that aid in remedying these problems include:

1. Site assessment to determine which, if any, water and soil treatments are best.
2. Utilizing salt-tolerant grasses—warm-season turfgrasses generally are less salt-sensitive compared to cool-season turfgrasses, while most ornamentals are more salt-sensitive.
3. Diluting or blending poor quality water with good quality water.
4. Leaching excess salts by applying extra water.
5. Modifying soils with various amendments to replace and leach sodium from the soil.
6. Amending irrigation water to correct sodium and bicarbonate problems.

Table 4.8 Water and soil salinity problems with potential management solutions

Soil salinity problem	Potential solutions
Total Irrigation Salt Content (EC)	– Leaching; blending water sources; increase drainage and aeration; use salt tolerant varieties.
Soil SAR/adj. SAR	– Apply calcium amendment; apply sulfur alone (in calcareous soils) plus lime (in acidic soils); blending water sources; acid or sulfur irrigation injection in severe cases.
Exchangeable Sodium Percentage (ESP)	– Apply calcium amendment such as gypsum; apply sulfur alone (in calcareous soils) plus lime (in acidic soils); or sulfur irrigation injection in severe cases.
Soil Residual Sodium Content (RSC)	– Irrigation acid injection; sulfur generator; sulfur application in calcareous soils; blending water sources.
Soil Infiltration/Permeability (EC_w plus SAR)	– Gypsum additions to either: (a) low EC_w plus low SAR water; or, (b) low to moderate EC_w plus high SAR water; blending water sources.
Specific Ion Toxicity	– Establish tolerant varieties (especially ornamentals); blending water sources.
Total Suspended Solids	– Irrigation line filtration; use of settling ponds.
Nutrient Imbalances	– Adjusting fertility programs.

7. Enhancing soil drainage by using sands and installing subsurface drain lines plus intensive cultivation to enhance infiltration, percolation, and drainage of salt-laden water (see earlier).
8. Using cytokinin and iron-containing biostimulants as salt-stressed plants often exhibit low cytokinin activity, as well as using wetting agents and appropriate fertilizers.
9. Raising the mowing height to promote more stress-tolerant plants.
10. Routine use of wetting agents to help maintain good water infiltration and percolation to flush salts and sodium below the rootzone.

Blending Water Sources for Reducing Salinity

High salinity water that is unacceptable for use can be made suitable as an irrigation source by diluting it with nonsaline water. Enough nonsaline water must be available to create a mixed water of acceptable quality (i.e., not making a less-saline water that is still unacceptable). The quality of a poor water source should improve proportionally to the mixing ratio with better quality water. For example, a water source with an $EC_w = 5\,dS\ m^{-1}$ mixed equally with a source with an $EC_w = 1\,dS\ m^{-1}$ should reduce salinity in the blend to approximately $3\,dS\ m^{-1}$. A chemical analysis of the blend should be performed to confirm this. The salinity of the mixture can be calculated with this equation:

$$EC_w(\text{blend}) = \frac{\text{volume (water A)} \times EC_w\ \text{(water A)} + \text{volume (water B)} \times EC_w\ \text{(water B)}}{\text{volume (water A)} + \text{volume (water B)}}$$

Example Two water sources are available for irrigation. One has an EC_w of $3.0\, dS\, m^{-1}$ and the other, $0.6\, dS\, m^{-1}$. The water will be blended in equal amounts. What would the resulting EC_w of the blended water be?

$$
\begin{aligned}
EC_w(\text{blend}) &= \frac{\text{volume (water A)} \times EC_w\ \text{(water A)} + \text{volume (water B)} \times EC_w\ \text{(water B)}}{\text{volume (water A)} + \text{volume (water B)}} \\
&= \frac{[1\, gal \times 3.0\, dS m^{-1}] + [1\, gal \times 0.6\, dS m^{-1}]}{1\, gal + 1\, gal} \\
&= \frac{3.6\, dS m^{-1}}{2} \\
&= 1.8\, dS m^{-1}
\end{aligned}
$$

Mixing of irrigation sources can occur in irrigation ponds or within the irrigation system itself. When mixing water sources in irrigation ponds, the nonsaline water should be added immediately prior to being used so as to reduce evaporative losses. Evaporation of surface water is not only an inefficient use of water, but it also increases the salinity of the water remaining in the pond. If blending is not an option, alternating irrigating with saline followed by fresh water helps leach salts.

Leaching or Flushing Soils to Remove Salts

Salt buildup from salt-laden irrigation water occurs when rainfall is low and evaporative demand is high (Fig. 4.13). As water evaporates from the soil surface, salt deposits are left behind. Applying water in an amount greater than ET to cause the applied water to flow (or leach) through the rootzone and wash away salts is the goal of leaching salt-laden soil. Steps involved when leaching or flushing soils to remove salts include:

1. Perform soil and water test to determine the extent of salinity levels present.
2. Aerify or vent the soil. Soils which do not drain well will not benefit greatly from flushing as the salts must be removed by leaching. Also, standing water in summer is often detrimental to certain plants, such as bentgrass. Aerifying or venting the soil by slicing are two ways of improving internal soil drainage. If drainage is still inadequate, then "pulse" irrigation may work. Pulse irrigation is a series of short-run irrigation cycles where water is to match infiltration rates or added until puddling occurs. Once the surface water drains, another pulse of irrigation is applied.
3. Apply gypsum (calcium source) to replace soil sodium ions removed and also add wetting agents to improve water infiltration and assist in soluble salt removal from the rootzone.
4. Perform leaching or flushing. Several techniques to determine the amount of water needed to accomplishment the goal(s) of leaching/flushing are presented.
5. Add leached nutrients. Leaching/flushing to remove salts also often removes other elements, especially nitrogen and potassium. Add a scheduled fertilizer

Fig. 4.15 Flushing (excessive irrigation) is the best means of overcoming irrigation salinity problems. Excess moisture must be applied and excellent soil drainage are needed for this strategy to succeed

following leaching/flushing to ensure sufficient potassium levels are maintained to help combat future added sodium ions.

Measuring the EC of the soil is the best way to determine the extent of salt accumulation. When the EC exceeds the tolerance level of the turfgrass, the soil should be leached to move the salt below the rootzone. For example, 6 *in* (15 *cm*) of water is required to leach 80 % of salt out of the top 1 *ft* (30 *cm*) of a sand loam soil and about 1.5 *ft* (45 *cm*) of water is required to leach 80 % of the salt out of the top 1 *ft* (30 *cm*) of a clay loam. Typically, a course should plan on an additional 10–20 % of water needed yearly for turf growth to provide water for adequate leaching.

Frequent flushing of the soil with good quality irrigation water or rainfall is the best method of preventing excessive salt accumulation (Fig. 4.15). Unfortunately, low salinity irrigation sources are not always available and frequently saline irrigation water must be used to manage soil salinity. However, as long as the salinity of the irrigation water is acceptable, it can be used to leach accumulated salts from the turf rootzone. The goal is to maintain a soil salinity level that is not increased through salts added by irrigation and yet can support turfgrass growth. The use of soil amendments, such as gypsum, should be considered in conjunction with leaching irrigation applications in saline-sodic soils.

If saline water is used to reduce the salt level of the soil, irrigation must be applied at rates exceeding evapotranspiration to leach (or flush) excess salts out of the rootzone. Leaching of soluble salts in the soil solution is much more rapid and easier than removing sodium on the CEC sites of sodic soils. On sodic soils, the sodium is chemically bonded and must be replaced by calcium before the sodium can be leached from the soil solution. Soluble salts are already in the soil solution, thus, are more easily leached. To determine the amount of excess water required to leach salt below the rootzone, the following **leaching requirement** equation is often used.

Leaching requirement is the amount of extra water needed to leach salts from the rootzone and is defined as:

$$\text{leaching requirement} = \frac{EC_w}{EC_{dw}} \times 100\,\%$$

EC_w equals the electrical conductivity of the irrigation water and EC_{dw} is electrical conductivity of a saturated paste extract that can be tolerated by the turfgrass being grown.

Example An irrigation water source has a salinity level of 2 $dS\ m^{-1}$. The turfgrass being grown has a tolerance of 4 $dS\ m^{-1}$. What would be the recommended amount of water needed to leach salt from the rootzone?

step 1: Determine the leaching requirement for this sample and turfgrass.

$$\begin{aligned}\text{leaching requirement} &= \frac{EC_w}{EC_{dw}} \times 100\,\% \\ &= \frac{2}{4} \times 100\,\% \\ &= 50\,\%\end{aligned}$$

step 2: Fifty percent additional water above that normally applied would be needed to leach the salt from the soil. If 2 *in* (5 *cm*) of water are normally used, adding 50 % would equal 3 *in* (7.6 *cm*). Table 4.9 lists these irrigation guidelines for leaching salts from soil with saline water.

Table 4.9 Irrigation guidelines for leaching salts from soil with saline water

Irrigation water EC_w ($dS\ m^{-1}$)	Maximum plant EC_{dw} tolerance level, measured by saturated soil paste extract ($dS\ m^{-1}$)		
	4 (low)	*8 (medium)*	*16 (high)*
	(in water to replace weekly ET losses and provide adequate leaching in rootzone[a])		
0.00	1.5	1.5	1.5
1.00	2.0	1.7	1.6
2.00	3.0	2.0	1.7
3.00	6.0	2.4	1.8

[a]Multiply inches by 2.54 to convert to *cm*

Leaching requirements depend on the salt levels of the irrigation water, ET rates, and the salt tolerance of the affected plants. As the irrigation water becomes saltier or the soil heavier, the leaching requirement becomes larger, meaning more water must be added for leaching to avoid salt accumulation. A guideline is for about 70 % of the total soluble salts to be removed by leaching, 3 *in* (7.6 *cm*) of water is needed per 12 *in* (30 *cm*) of soil depth of a sandy soil, 6 *in* (15 *cm*) of water per 12 *in* (30 *cm*) of a medium-textured loam soil, and 9 *in* (23 *cm*) of water per 12 *in* (30 *cm*) of a clay (fine-textured) soils. Leaching Na^+ also removes nutrients such as K^+, Mg^{+2} and others. These should be monitored and replaced, if necessary, following leaching. Heavy Ca^{+2} applications may also cause other cations imbalances, such K^+ or Mg^{+2}, thus they may need replacing. It generally is better to have periodic leaching events (i.e., 2 to 4 times monthly at 0.2 to 0.4 *in*, 5 to 10 *mm*, per application) compared to heavier, infrequent events (i.e., once monthly) which may cause puddling.

If saline water is the only source of water available for irrigation, it is helpful to predict how the leaching fraction of known irrigation water salinity will influence soil salinity over an extended period of time. Applying a leaching fraction of 10 % will lead to an EC_e of $EC_w \times 2.1$, 15 to 20 % will lead to an EC_e of $EC_w \times 1.5$, and 30 % will lead to an $EC_e = EC_w$.

Finally, plants tolerate higher soil salinity levels if water stress is avoided by maintaining soil moisture. Adequate surface moisture also prevents capillary rise of subsurface water and salts.

Flushing Steps

1. Aerify to break hardpans or organic zone surface tension.
2. Add green's grade gypsum at 7 to 12 *lb* 1000 ft^{-2} (3.4 to 5.9 *kg* 100 m^{-2}).
3. Start flushing, usually about an hour. If puddling occurs, stop, allow it to percolate, then resume.
4. Assess flushing length of time by measuring EC_w of discharged water with a portable meter. Once readings stabilized or fall below pre-set thresholds, stop.
5. Afterwards, N and K may need to be added as they are commonly stripped by flushing.

Good Soil Percolation and Drainage

As previously mentioned, leaching works well only with soils possessing good drainage (Fig. 4.16). If compacted zones or abrupt changes in soil texture exist, less leaching occurs as water movement through the soil is reduced. Good soil drainage through modifying rootzones, increased deep tine aerification, and use of drain lines are used for carrying away salty water. Drain lines, spaced no more than 20 *ft* apart (6.1 *m*), are used on golf greens for this purpose. Aerification also initiates deep root development prior to summer heat and salt stress by reducing soil compaction and

Fig. 4.16 Salinity problems will be magnified in areas of insufficient drainage as salts will remain at or on the soil surface

disrupting soil layering. Native (or pushup) greens with limited drainage often fail when effluent water is used unless these techniques are aggressively incorporated (Fig. 4.17).

For fairways, deep aerification has become standard on effluent-using courses to increase soil drainage and provide deep channels for incorporation of soil amendments. On tees and greens, deep aerification in spring and fall are typical along with supplemental monthly venting by spiking, slicing, quadratining, hydrojetting, or other techniques. For soils with limited infiltration properties, pulse irrigation is more effective where water is applied and allowed to infiltrate before reapplying.

Clay type also influences sodium tolerance. Nonexpanding or 1:1 kaolinite, hydrous oxide clays tolerate high soil sodium content better than expanding or 2:1 vermiculite or montmorillonite clays. The percolation in expanding 2:1 clay initially is high until the cracks seal with clay swelling. Illite is intermediate in sodium sensitivity.

Salt damage also is typically experienced in low-lying areas where water accumulates. Drain line installation helps remove this excessive water, preventing toxic accumulation of salts and sodium. Sand topdressing of fairways is also becoming more prevalent to improve playing conditions, degrade thatch, and to help remove excess surface water.

Leaching is typically performed monthly during high-stress summer months but soils should be checked periodically if problems develop. Soil salinity levels should be monitored before and after leaching to determine if salts have sufficiently been moved below the rootzone. Finally, routine leaching will also remove certain soil nutrients such as nitrogen and potassium. Although most effluent sources contain small levels of these and others, monitoring of soil nutrients should occur following a leaching cycle.

Fig. 4.17 Turf replacement in areas of insufficient drainage and use of high salinity irrigation water

4.5 Water Conservation

Daily water conservation practices integrate many of the previously mentioned practices and technology. Using computerized irrigation systems to better pinpoint irrigation needs for various soil types or turfgrass use, utilizing weather stations to determine daily ET rates, installing soil moisture sensors to monitor soil moisture levels, and using automatic pump shutdown switches when significant rainfall occurs are examples of water conservation techniques.

A holistic approach to water conservation is required. If not, turf water conservation will probably be mandated by governing bodies and may include: (a) changing the grass species, (b) allowing only native grasses and Xeriscape designs, (c) reducing the area of irrigated turf, or (d) improving (updating and expanding) current irrigation designs to become more efficient. Steps to develop best management practices (BMPs) for turfgrass water conservation include:

1. Site assessment and initial planning (i.e., documenting grasses, soils, microclimate, and existing management practices).
2. Evaluating and implementing water conservation strategies.
3. Analyzing benefits and costs of water conservation measures.

Site Assessment and Initial Planning

An extensive irrigation/water audit is needed to assess current water usage rates and efficiencies. This includes identifying currently implemented water conservation measures, estimating their costs to implement, and how they have improved water-use efficiency for the facility. This helps indicate to regulatory agencies that water conservation BMPs have been in place at considerable cost and effort and the course is committed to as efficient use of available resources as possible. Examples of current conservations measures can include:

- Irrigation scheduling based on scientific principals and experiences which measure plant water requirements.
- Providing educational and demonstration opportunities for the crew and course membership.
- Irrigating in early morning or at night to reduce wind losses and to take advantage of efficient water pressure.
- Using required irrigation backflow preventers, valves, heads, and permit requirements as per local code.
- Periodically checking valve boxes for leaks or disconnected wires and open and close valves manually to confirm proper operations. Also, inspecting for and eliminating pipe leakage.
- Use a pilot tube and gauge to check pressure at the head to ensure maximum efficiency and to regulate water use.
- Periodically check the height of heads to prevent mower and other equipment damage and to check coverage, water-discharge patterns, and to raise low heads.
- Use low-maintenance turf, landscape plants, and native grasses whenever possible.
- Use mulch (>3 *in*, 7.6 *cm* deep) around landscaping to reduce evaporation and weeds.
- Use drip irrigation or low emitter heads for landscapes.
- Use multiple irrigation cycles to allow infiltration without surface runoff.
- Have water harvesting and collection sources feeding into irrigation ponds.
- Having access to color weather radar or other devices to track and predict local showers.
- Matching the application rate to the soil infiltration rate.
- Using an irrigation company with local service support and readily available parts.
- Identifying cultivations programs (i.e., mowing heights, fertility programs, aerifications) and equipment which improve water infiltration and enhance rooting.
- Use of appropriate soil amendments and wetting agents to provide efficient water infiltration, water retention, and to minimize runoff.
- Reducing or eliminating irrigation in low priority play areas.

Following identification of existing water conservation measures, the next step is to assess the current resources and infrastructure available. This can be a time-consuming and costly assessment, especially if alternative irrigation sources are explored or when major irrigation system design changes are needed.

- Hire an irrigation design specialist.
- Use an irrigation system design that provides uniform application to minimize wet and dry areas and limits run-off or leaching.
- Identify and provide cost considerations of alternative water sources such as reclaimed water.
- Identify irrigation design changes necessary for improved efficiencies.
- Assess current soil, microclimatic, and plant conditions affecting irrigation system design including zoning and scheduling issues.
- Add sufficient wire in the irrigation system to accommodate future expansions or added heads per zones.
- Provide single-head irrigation control.
- Use a variable frequency drive pump to gradually reduce water flow after pump shut-off and gradually increase water flow when turned on to reduce strain on the pipe. These motors only expend enough energy to meet the demands of the pumps.
- Have an on-site weather station or access to regional weather information to calculate daily ET rates and possibly use soil moisture sensors to monitor irrigation efficiency.
- Safeguard against water hammer when systems are pressurized by installing check values where water drains from low heads to prevent damage.
- Consider using ductile fittings and gasketed joints instead of glue due to their longer life expectancy.
- Consider looping the irrigation system to allow watering from two directions.
- Use multirow irrigation design systems compared to single rows for better coverage and less water waste.
- Using multiple short duration irrigation cycles to reduce runoff compared to single, heavy-use cycles.
- Zone irrigation heads of similar areas together (greens, tees, bunkers, fairways, and roughs).
- Isolate as many areas of the golf course as possible with individual shut-off valves from the main line.
- Use low-volume heads when possible and low trajectory heads in windy areas.
- Use the biggest irrigation pipe that is affordable—ideally, pipes should be sized for water velocities of about $3 ft\,\mathrm{s}^{-1}$ ($0.9\,m\,s^{-1}$).
- Mainline pipe should be a minimum of 4 to $6\,in$ ($10–15\,cm$) in diameter, preferably larger. Successive branches of an irrigation line should be reduced by $2\,in$ ($5\,cm$).
- Have controller flexibility to develop the most efficient irrigation program.
- Avoid placing heads in a depressed area as seepage or bleeding may occur. Use seals if this is unavoidable.

- Use part-circle pattern heads and proper design to place water only on intended turf areas and not on unintended natural areas, mulched areas, water bodies, edges of fairways or primary rough, off-property, and other such areas.
- Incorporate efficient drainage designs that allow water harvesting or recapturing in ponds or catch basins.
- Consider a remote (radio) controller to enable quicker response time to a problem.
- Using appropriate soil amendments which are known to help retain soil moisture without negatively affecting the turf.
- Use pressure-regulating stems on spray heads to prevent water waste when operated outside the designated window of pressure.

Evaluating and Implementing Water Conservation Strategies

Once existing and potential irrigation practices have been identified, they must be sorted through and the ones practical for a specific course can be implemented. These strategies are generally site-specific, driven by water-allowances and conservation goals, member expectations, and of course, financial and other resources available or required. Key components of water conservation strategies include:

- Alternative irrigation water sources, their availability, costs, quality, reliability, use requirements, suitability for a particular site, and long-term effects. Probably the major problem with effluent water is not quality but quantity. Courses find themselves having to accept a certain amount per day, whether it is needed or not. Storage of this water is a concern and must be addressed early in the planning process of using effluent water.
- Practical extent of implementing efficient irrigation design, scheduling, operations, and monitoring devices as discuss previously.
- Considerations on selecting turfgrass and landscape plants, such as the quality they can produce, their water use requirements, and quality of water needed.
- Changes in management practices which enhance water conservation as discussed previously.
- Holistic course water conservation, including landscaped areas, club house use, pool water conservation, etc.
- Educating the crew, owners, and membership on water conservation and management plans to obtain these.
- Developing a formal written water BMP conservation plan for the course and for regulatory agencies.
- Monitoring and revising the conservation plan periodically to assess the success of the plan and to identify limiting factors to achieving water conservation goals.
- Inform members, owners, crew, and concerned citizens of water conservation efforts with proper signage and other communication avenues.

Assessment of Water Conservation Costs and Benefits

To track costs and benefits which are critical information to demonstrate the facility has developed and is implementing long-term BMP water conservation efforts, a follow-up detailed review and documentation phase is necessary. Costs include labor, facility costs as outlined in evaluating and implementing water conservation strategies section, and costs associated with more stringent water restrictions, such as revenue loss, job loss, reduced and possible hazardous turf quality, etc.

The following is an example of questions to answer when developing Best Management Practices for Water for a particular golf facility (modified from the Georgia Golf Course Superintendent's Association; 2015, http://www.ggcsa.com/-best-management-practices-for-water-conservation).

Best Management Practices for Water Use and Conservation

1. Site Assessment for greens, tees, fairways, roughs, landscapes, and club grounds, including
   a. Area size involved, (ft^2, *ac*, m^2, or *ha*).
   b. Turf (plant) species involved.
   c. General factors such as mowing height, soil type, special technology, other pertinent information.
   d. Irrigation Audit:
      - Pump station—year, type, pump size(s), gallonage, safety features, condition, maintenance schedule, other information.
      - Controls—year, system type, number of field controllers, condition, other information.
      - Irrigation system—year, type, valves, output and distribution efficiency (DU) for greens, tees, fairways, roughs, plus other information.
2. Overall Water Needs
   a. Metering—number of meters, location(s), other information.
   b. Record keeping—yearly usage, scheduling, other information (attach 1 year of records).
   c. Water testing—schedule, other information (attach most recent tests).
   d. Reservoir—size, type of water, source of water, other information.
   e. Alternative water sources (yes or no), If yes, explain.
   f. Future needs—explain in detail.
3. Best Management Practices and Current Conservation Measures
   a. Current Irrigation Control/Costs—for pump station, controllers/computer, irrigation system components (sprinklers, pipe, valves, fittings, etc.), preventative maintenance of all these, other.

b. Staffing Control/Maintenance Costs—supervisor time, irrigation technician time, other assistance time (include diagnosis, repairs, recordkeeping, inventory, scheduling, etc.).
c. Scouting Costs—daily scouting time (explain).
d. Hand Watering Costs—daily hand watering time (explain).
e. Night Watering Capability—explain how this reduces loss and reduces disease occurrence.
f. Rain, Leak Loss Costs.
g. Traffic Controls/Costs—daily traffic control time (explain).
h. Management for Water Conservation (describe each):

- Mowing heights.
- Soil cultivation (number times yearly for greens, tees, and fairways).
- Evapotranspiration utilization—List source for monitoring weather data to schedule irrigations events based on ET values.
- Landscape material selection explanation.
- Natural areas.
- Fertilization—yearly rates, slow vs. quick release, stress nutrient use.
- Pest management (explain IPM programs).
- Wetting agent use (explain products, timings, etc.).
- Soil moisture sensors—calibrating and determining thresholds (saturation, field capacity, wilting point).

i. Record Keeping:

- Scouting labor hours and costs.
- Hand watering hours and costs.
- Irrigation repair hours and costs.
- Repair parts costs.
- Water usage weekly, monthly, and yearly.
- Water quality tests.
- Pesticide and fertilizer applications (in relation to irrigation).
- Other methods.

j. Irrigation Methods—combination of plant based, soil based, atmosphere based, and budget report.
k. Goal Setting—explain.
Education: for example.

- Benefits of Golf Course and Turf—i.e., economic contributor, carbon dioxide exchange for oxygen, temperature moderation, erosion control, water filtering for improved water quality, wildlife sanctuary, recreational benefits, community outreach (i.e., First Tee Programs), others.
- Publish this Best Management Plan for use at Club—articles in the Club newsletter or web page explaining proper water use and efforts towards water conservation.

l. During drought, display water conservation plans (posters) in the pro shop and locker rooms and to patrons for use at home.

4. Water Conservation Plan

a. List reasons for Water Conservation, for example:

- Proper water management dictates that overwatering is unacceptable.
- Economic considerations that inefficient watering costs money.
- Depleted water supplies and reduced water quality.
- Other reasons.

b. List Measures Implemented by the Course to Reduce Drought Effects, for example:

- Raise mowing heights where possible.
- Stop mowing non-irrigated areas.
- Increase hand watering and wetting agent use.
- Improve uniformity by improving pressure regulations, leveling heads, etc.
- Other reasons.

c. Irrigation Upgrades Implemented by the Course for Increased Water Conservation.

- List possible options and costs.

d. List and Describe Actual Plans for Water Conservation at Various Mandated Drought Levels.

5. Attachments, for example:

- Pump station records.
- Most recent water quality test results.
- Man-hour records.
- Budgets.
- Repair records.
- Copies of publications.

4.6 Hydrophobic Soils and Their Management

Hydrophobic Soils

Hydrophobic (or "water-hating") soils such as those associated with localized dry spots, occur as organic matter decomposes and humic and fulvic acids (nonpolar) produced eventually coat individual sand grain particles. Sands are more prone to develop water repellency than finer soil textural classes due to the low or smaller

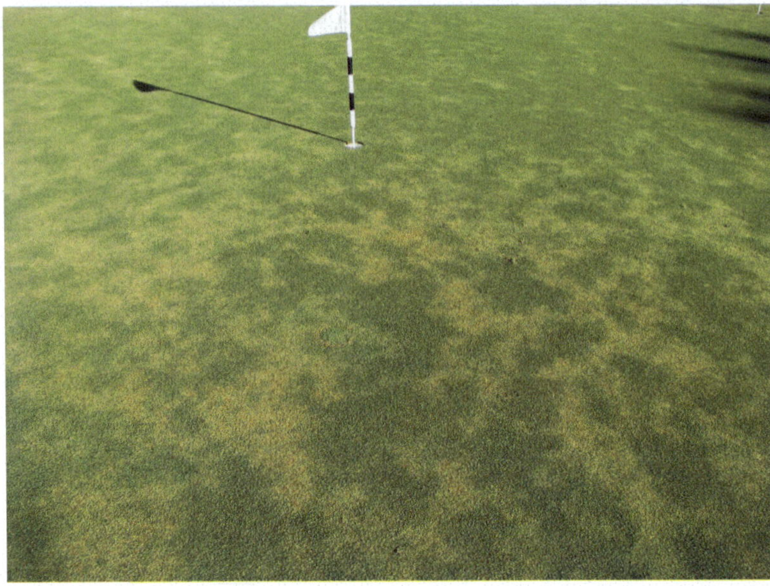

Fig. 4.18 Typical localized dry spots occurring on a golf course putting green. Powder dry soil is typically adjacent to moist soil, reflecting unhealthy and healthy plants

surface area to volume ratio of sand particles. Certain fairy ring fungi also produce a mat of below-ground hyphae which often becomes hydrophobic. These acids have extruding non-polar ends which repel water particles, much like wax, leading to the dry spots. These conditions can be so severe that normal irrigation is often ineffective in restoring adequate soil moisture. Repeated wet and dry soil cycles aggravate hydrophobic soils. Other potential sources of organic acids which coat soil particles include exudates from turfgrass roots, lipids from decomposing organic matter, surface waxes from cuticles of turfgrass plant leaves, and fungal or soil microbial by-products.

Hydrophobic soils can cause problems on golf courses (especially golf greens) and other turf areas, in nurseries and greenhouses, and in open fields (Fig. 4.18). Localized dry spots tend to be a surface phenomenon, in the top 2 *in* (5 *cm*), but can occur up to 6 *in* (15 *cm*) deep. Nursery operators sometimes encounter hard-to-wet media in pots and greenhouse beds. Farmers who work organic soils or "salt-and-pepper" soils complain that the soil wets too slowly, reducing crop productivity. Problems with hydrophobic soils are also commonly associated with citrus production areas, where mine spoils have been deposited, and with burned-over forestland and grassland.

If water cannot readily penetrate and wet the soil, the availability of moisture to plants is reduced, decreasing the germination rate of seeds, the emergence of seedlings, and the survival and productivity of plants. Lack of sufficient water in the soil also reduces the availability of essential nutrients to plants, further limiting

Fig. 4.19 Hydrophobic soils typically repel water, decreasing the efficiency of a water management program

Fig. 4.20 Hydrophobic soils are often quantified using a water droplet test. With this test, the amount of time necessary for a drop of water to penetrate a soil profile is used to determine the degree of hydrophobicity of a soil

growth and productivity. In addition, water that cannot penetrate the soil runs off the surface and increases soil erosion (Fig. 4.19).

A soil water repellency water drop penetration test is used to measure how hydrophobic a soil may be (Fig. 4.20). Droplets of water are placed every inch down a soil core and the time required for the droplet to penetrate the soil reflects the soil's degree of repellency (Table 4.10).

Table 4.10 Degree of soil hydrophobicity (or repellency) based on the water drop penetration test

Water drop penetration time (*s*)	Degree of repellency
0 to 5	None
5 to 60	Slight
60 to 600	Moderate to high
600 to 3600	Severe
>3600	Extreme

Adjuvants

An adjuvant is a spray additive that helps modify the surface properties of liquids to enhance their performance and handling. 'Adjuvant' is a broad term and includes surfactants, wetting agents, crop oils, crop oil concentrates, activators, anti-foaming agents, detergents, drift control agents, emulsifiers, fertilizers, spreaders, sticking agents, dispersing agents, penetrants, pH modifiers and compatibility agents.

Surfactants

Surfactants are adjuvants that produce physical or chemical changes at the interface of a liquid and another liquid, solid, or gas. These typically lower the surface tension of a liquid, allowing easier spreading, and lower the interfacial tension between two liquids. Since this occurs at the surface, the term "surfactant" is short for surface active agents. These facilitate emulsifying, dispersal, wetting, spreading, sticking, penetrating, or other surface-modifying properties of liquids into plants and soil (Fig. 4.21). Surfactants are widely used in everyday life in medicines, medical care, fire extinguishers, paints, inks, adhesives, waxes, laxatives, hair conditioners, and agriculture. Surfactants include emulsifiers, detergents, dispersants, penetrants, soaps, spreaders, stickers, and wetting agents.

To understand how surfactants work, it helps to understand how water works. Each water molecule is bipolar, meaning it has a negative and a positive charge, similar to a magnet. When several water molecules come into contact with each other, these positive and negative forces attract each other. This attraction of water molecules for each other is termed cohesion. The molecules on the surface of a water droplet are held together with more force than those of the interior water molecules. This causes surface tension, which causes the droplet to behave as if a thin, flexible film covered its surface, tending to keep the water molecules apart from other substances, and can prevent many things from going into solution and getting wet. This surface tension is the tendency of the water surface molecules to be attracted toward the center of the liquid, causing a water droplet with a dense, elastic membrane around it. Wetting agents help break this surface tension, thus the water droplets break down allowing dispersal. Adhesion, the attraction of water

Fig. 4.21 Wetting agents are often used to reduce the angle of beading water molecules possess. Areas outside the green rectangular areas have not been treated with a wetting agent, retaining dew as large droplets of water

molecules to other substances, is the force causing water molecules to adhere to other objects, such as soil particles.

The effects of these forces can be illustrated by placing a drop of water on a napkin and another drop on a piece of waxed paper or newly waxed vehicle. On the napkin, the force of adhesion between the water molecules and the paper molecules is greater than the force of cohesion that holds the water molecules together. As a result, the water droplet spreads out and soaks into the paper. Certain organic substances such as wax, however, do not have an adhesive force for water. On the waxed paper, therefore, the water "beads up"—that is, the droplet remains intact. The water molecules are not attracted to the wax that coats the paper's surface; instead, the water molecules cohere to each other. When the adhesive forces between water molecules and an object are weaker than the cohesive forces between water molecules, the surface repels water and is said to be hydrophobic (Fig. 4.22).

Surfactants are composed of two parts, a water-soluble end which is polar or hydrophilic, meaning it is attracted to water, and an oil soluble hydrocarbon chain which is lipophilic or nonpolar, meaning it is attracted to oil and not water. Water forms bonds with polar molecules but does not bond to non-polar molecules and is repelled by these. Chemists manipulate the ratio of the hydrophilic (polar) portion

Fig. 4.22 Severe localized dry spots on a golf course putting green. Managing consistent playing surfaces and turf health are challenges under such severe conditions

Fig. 4.23 Demonstration of unwanted run-off of untreated water (*left*) on a hydrophobic soil compared to water infiltration by a water source treated with a wetting agent

of the molecule to the lipophilic (nonpolar) to produce different surfactants, with different molecular weights, and different characteristics. Thousands of potential combinations exist, thus the reason for the hundreds of surfactants available. These components of a surfactant molecule help break water surface tension, allowing the solution to be more evenly dispersed on a surface and to reach its target (Fig. 4.23). Two major types of surfactants are emulsifiers and wetting agents.

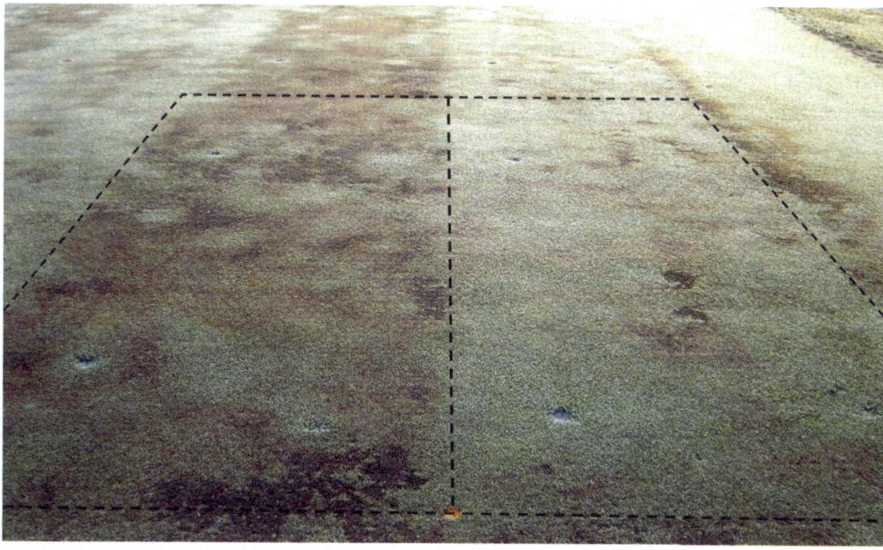

Fig. 4.24 Healthier turf (*right*) that has been treated with a wetting agent compared to severe localized dry spots which has not (*left*)

Wetting Agents

Wetting agents are a type of adjuvant that reduce interfacial tensions and cause a liquid to spread more easily over, or to penetrate, a solid surface, thus making more contact with treated surfaces (Fig. 4.24). They can reduce this surface tension by 50 to 60 % or more. In plant and soil sciences, wetting agents have a number of uses including reducing soil hydrophobicity (i.e., localized dry spots), reducing dew and frost formation, firming bunker sand, improving irrigation efficiency, reducing vehicle path dust, improving soil water infiltration, improving pesticide efficacy, and others.

Wetting agents are classified based on how they ionize or separate into charged particles in water. Four types of wetting agents are:

1. **Anionic**—negatively charged. These are often used for dispersion of clays in wettable dry granulars as well as detergents, and degreasers. They may burn plants.
2. **Nonionic**—neutrally charged. Also referred to as polyoxyethylene or alkylphenol ethoxylate. Often used to enhance water movement into soil.
3. **Cationic**—positively charged, often used as biocides (disinfectants), soaps, shampoos, and fabric softeners. Strongly adsorbed to soil particles with high plant burn potential. Rarely used.
4. **Amphoteric**—charge is pH dependent of the solution. Little use on plants.

Nonionic surfactants do not ionize, thus remain uncharged. This is the most commonly used type of surfactant and is compatible with most pesticides. They are unaffected by water containing high levels of calcium, magnesium, or ferric ions. They also can be used in strong acid solutions. Anionic wetting agents ionize with water to form a negative charge while cationic ones ionize with water to form a positive charge. Anionic wetting agents may deleteriously impact soil structure (negative soil charges repel the negatively charged anionic wetting agents) and are often phytotoxic to plants. Amphoteric surfactants can be either anionic or cationic depending on the acidity of the solution. Cationic materials are strongly adsorbed to soil particles and may become ineffective. If used in hard water, anionic and cationic surfactants can cause an insoluble precipitate or foam to form. These are only occasionally used. Soaps and detergents are types of surfactants but typically are anionic and react with salts in hard water and form a precipitate (scum), foam, or are phytotoxic to plants.

As mentioned, chemists are able to manipulate the ratio of the hydrophilic (polar) portion of the molecule to the lipophilic (nonpolar) to produce different surfactants, with different molecular weights, and different characteristics. Within the nonionic surfactant chemistry, two main groups of wetting agents are currently available: soil penetrants and water retainers.

1. *Soil Penetrants.* These are often characterized as "water-moving" chemistry, characterized by having ethylene oxide terminal functional groups. Ethylene oxide groups are hydrophilic, being able to attract or disperse water molecules. Soil penetrating wetting agents generally increase water infiltration and percolation through the rootzone, providing more uniform soil moisture distribution within the profile, leading to "fast and firm" playing conditions.
2. *Soil water retainers.* These are often characterized as "water-holding" chemistry, containing propylene oxide terminal functional groups. Propylene oxide groups are hydrophobic, thus repel water molecules. These are used where moisture retention is needed, especially sand-based rootzones with little organic matter and high infiltration and percolation rates. These are especially useful to help retain moisture during drought periods.

To take advantage of both types of wetting agents, many newer commercial products are blends of each. Extensive research has been conducted on hydrophobic soils and on the effectiveness of wetting agents. Localized dry spots in turf grown on naturally sandy soils, and on formulated materials high in sand content, become a serious turf management problem during the summer months, especially during periods of drought, windy weather, and low humidity. Despite frequent irrigation, the soil in these spots resists wetting, resulting in patches of dead or severely wilted turf. The water applied wets the turf but does not adequately penetrate the soil surface to reach the rootzone. Wetting agents or surfactants do not aid in decomposing thatch, alleviating black layer, or reducing soil compaction.

When a wetting agent is applied, its non-polar ends react (or align) with the non-polar ("water-hating") ends of the acid coated sands. The polar ("water-loving") ends of the wetting agent then are exposed outward and can attract water, restoring wettability. Wetting agents, however, do not substantially remove the hydrophobic acid coating. For most products, to minimize phytotoxicity, irrigation after wetting agent application is critical as well as not treating when temperatures are extreme. When soil organic matter content exceeds 3.5 %, this organic matter may dry down slower when treated with a wetting agent. Increasing the use rates above label recommendations generally does not increase the longevity or effectiveness of products and increases the chance of plant damage. Wetting agents can improve the efficiency of irrigation and when water repellent soil conditions occur, wetting agent use may improve root growth and survival. Efficacy differences and length of control does vary between wetting agents. Soil organisms and natural breakdown of the materials eventually occur, causing the need for repeat applications for extended results. When treating golf greens, it generally is best to treat the whole green versus just treating the hot spots. Wetting agents do not solve a subsurface drainage problem but may help leach salts from the rootzone under certain situations.

In general, studies have shown that the extent of improvement in infiltration rate is affected by the type of wetting agent used, its dilution, previous use of wetting agents on the soil, and the water content of the soil at the time water is applied. Several studies have shown that the infiltration rate of a hydrophobic soil, once it has been wetted, remains higher than it was before it was wetted, even if it is allowed to dry out again. Applying wetting agents often reduces the severity of the condition, but best use is in combination with coring—making small holes in the soil surface to allow water to pass through the hydrophobic surface layer. Also, keeping the soil moist seems to be the best defense against the development of dry spots as allowing the soil to dry out intensifies the problem. For maximum efficiency, if your goal is to rewet a dry, hydrophobic soil, the area should be aerified, followed by pre-wetting the area, applying the wetting agent or soil surfactant, and then watering it rapidly and liberally into the soil profile.

4.7 Questions

1. Soil moisture measuring devices have been developed with the goal of indicating how much moisture is available to plants. List and discuss the major means by which soil moisture is currently measured.

Quantitatively (or volumetric):

Gravimetric water content—*Measures soil moisture by weighing-drying-reweighing to provide a full range of water content (%). Simple equipment is needed, it is highly accurate, and data is easy to interpret. However, it involves destructive sampling, is labor intensive, and involves collection, transport, and time restraints.*

Time Domain Reflectometry (TDR)—*Measures time for an electromagnetic wave to travel using soil medium as a dielectric. Moisture slows this down.*

Up to 50 % volumetric water content (0.50 kg water kg^{-1}soil) can be measured. TDR is accurate, has minimal soil disturbance, soil specific-calibration is optional, relatively insensitive to temperature, and also estimates, with limited accuracy, soil EC. Limitations include being expensive, accuracy decreases in high saline (>25 dS m^{-1}) conditions or heavy clay soils, and involves relatively small sensing volume (about 1 in, 2.5 cm, radius around probe).

Frequency Domain Reflectometry (FDR) or Hand-push probes—*Measures the change in frequency of a capacitor using soil medium as a dielectric. Up to 70 % volumetric water content (0.70 kg water kg^{-1}soil) can be measured. It is relatively inexpensive, can be automated with irrigation, and is stable in different soil types and over a large range of moisture contents. It requires soil-specific calibration for accuracy, samples only a small volume of soil (about 4 in, 10 cm, radius around probe), and is sensitive to soil air gaps, saline soils and temperature.*

Other quantitatively methods of determining soil moisture include neutron (or scatter) probe (expensive) and electrical conductivity probes (limited accuracy).

Qualitatively (or tensiometric):

Tensiometers—*Measure how tightly (the "tension") water is held by soil from a range of 0 to −0.08 MPa (0 to −80 kPa). It provides a direct readout of soil water potential (or tension), is inexpensive, can be automated with irrigation, relatively reliable, good accuracy, and unaffected by soil salinity. Limitations include soil moisture retention curve needed to relate to soil water content, samples a small area near cup thus multiple samples are needed in larger areas, doesn't measure soil salinity content, and involves exposed gauges, sensitive to disturbance and soil air gaps.*

2. The following water content values were generated for two sands being used for a 30 *cm* (12 *in*) rootzone. If the $ET_p = 0.5\ cm\ day^{-1}$ ($0.2\ in\ day^{-1}$), approximately how many days' supply of water would be expected to be stored in each?

	Moisture content at		
Sand sample	Field capacity	Wilting point	Rooting depth (*cm*)
1 (medium)	0.400 $cm^3\ cm^{-3}$	0.050 $cm^3\ cm^{-3}$	20
2 (course)	0.400 $cm^3\ cm^{-3}$	0.150 $cm^3\ cm^{-3}$	12

Available water = (rooting depth, cm) × (FC–WP, $cm^3 cm^{-3}$)

***Sand 1**: 20 cm × (0.400–0.050 $cm^3\ cm^{-3}$) = 7.0 cm (or 2.8 in).*
***Sand 2**: 12 cm × (0.400–0.150 $cm^3\ cm^{-3}$) = 3.0 cm (or 1.2 in).*

Days of Water Stored:
For Sand 1: 7.0 cm available water ÷ 0.5 cm $day^{-1} ET_p \approx$ 14 days.

For Sand 2: 3.0 cm available water ÷ 0.5 cm $day^{-1}ET_p \approx 6$ *days.*

Even though both sands held similar amounts of moisture at field capacity, the medium-sized sand has a deeper rooting depth and should not wilt until soil reached a lower moisture content (0.050 $cm^3\ cm^{-3}$*) compared to the shallower rooting depth in the coarser sand sample which has a higher moisture content at its wilting point (0.150* $cm^3\ cm^{-3}$*).*

3. A golf green 5495 ft^2 (510 m^2) in area, 1 *ft* deep (30.5 *cm*) with a bulk density of 1.40 $g\ cm^{-3}$ starts to wilt when the TDR probe averages 12 % volumetric water content. If the superintendent wishes to increase the soil moisture content to 16 % of the whole soil depth, how many gallons of water are necessary?

$$\text{equivalent depth of water, } D_e \text{ at } 12\% = \text{soil depth} \times \text{volumetric water content}$$
$$= 30.5\,cm \times 0.12\,cm^3 \text{ water } cm^{-3} \text{ soil}$$
$$= 3.66\,cm$$
$$\text{equivalent depth of water, } D_e \text{ at } 16\% = 30.5 \times 0.16\,cm^3 \text{ water } cm^{-3} \text{ soil}$$
$$= 4.88\,cm$$
$$\text{additional depth } (cm) \text{ of water needed} = 4.88\,cm - 3.66\,cm$$
$$= 1.22\,cm (0.48\,in)$$
$$\text{additional water } (gal) \text{ needed} = 0.48\,in \times \frac{ac}{43{,}560\,ft^2} \times \frac{5495\,ft^2}{\text{green}} \times \frac{27{,}154\,gal}{ac\text{-}in}$$
$$= 1645\,gal\ (6227\,L)$$

4. The can test was performed with 20 cans spaced 5 *ft* (1.5 *m*) apart in a grid system. After a set time period (15 *min*), the depths in all cans were recorded. Calculate the irrigation system's distribution uniformity from the following values (in *inches*) caught:

1.	0.40	6.	0.25	11.	0.34	16.	0.39
2.	0.22	7.	0.28	12.	0.19	17.	0.37
3.	0.15	8.	0.30	13.	0.23	18.	0.35
4.	0.41	9.	0.31	14.	0.25	19.	0.34
5.	0.33	10.	0.21	15.	0.35	20.	0.33

After a 15-min run cycle, the average depth in the 5 ***least*** *filled cans was 0.20 in (0.5 cm). The average depth measured in all cans was 0.33 in (0.84 cm). The DU value is determined by the formula:*

$$DU = \frac{\textit{average } \textbf{least} \textit{ amount of water depth collected in } 25\% \textit{ of all cans}}{\textit{average amount of water collected in all cans}}$$
$$= \frac{0.2\,in}{0.3\,in}$$
$$= 0.67\ (or\ 67\%)$$

The irrigation application rate may then be calculated as:

$$\frac{0.3\,in}{15\,\mathrm{min}} \times \frac{60\,\mathrm{min}}{1} = 1.25\,in\,h^{-1}\ \left(3.2\,in\,hr^{-1}\right)$$

5. If pan evaporation is measured at 1.60 *in week*$^{-1}$, using a K_c value for bermudagrass of 75 % and a DU value for the irrigation system of 60 %, determine the actual irrigation amount needed to uniformly apply the weekly water requirement.

   a) Calculate weekly potential evapotranspiration (ET_p) for this turf.

$$ET_p = pan \times K_c = 1.60\,in \times 75\ \% = 1.20\,in$$

   b) Calculate total irrigation depth needed to apply minimum ET_p over entire area.

$$\begin{aligned} actual\ irrigation\ needed &= \frac{ET_p}{Distribution\ Uniformity} \\ &= \frac{1.20\ in}{60\%} \\ &= 2.0\ in \end{aligned}$$

Therefore, 2.0 in (5.0 cm) of total 'applied' water is required to uniformly apply a minimum of 1.2 in (3.0 cm) over the whole turf area.

6. A sand soil has a volumetric water content of 11 % at field capacity, a rooting depth of 5 *in* (12 *cm*), and a summer daily ET rate of 0.22 *in day*$^{-1}$ (5.6 *mm day*$^{-1}$). Determine the appropriate time (days) between irrigation cycles.

$$\begin{aligned} irrigation\ interval\ (days) &= \frac{soil\ water\ content\ at\ field\ capacity \times rooting\ depth\ (in)}{ET\ rate\ (in\ day^{-1})} \\ &= \frac{0.11\ \times 5\ in}{0.22\ in\ day^{-1}} \\ &= 2.5\ days\ between\ irrigation\ cycles,\ which\ brings \\ &\quad the\ soil\ back\ to\ field\ capacity \end{aligned}$$

7. List and briefly discuss the steps in formulating an irrigation strategy.

   a. *Calibrate an irrigation system's output and distribution uniformity (or DU).*
   b. *Determine daily ET rates or soil moisture status by one of the methods discussed. A reasonable estimate of daily summer mean ET rates for various grasses are provided in* Table 4.3
   c. *Accurately track daily rainfall and ET rates so a water budget can be set-up and followed.*

d. *When irrigation is needed, use the appropriate crop coefficient (0.75 to 0.85) to find daily ET rate and incorporate distribution uniformity (DU) of the irrigation system as shown earlier and below.*
e. *Make adjustments for rainfall, varying microclimates, and forecasted weather.*

8. An irrigation zone applies 0.66 *in* h^{-1} and the projected ET rate for the next 24 *h* is 0.22 *in*. After doing an irrigation audit, you determined its distribution uniformity is 73 %. Calculate how long the irrigation system should run to uniformly apply the 0.22 *in*.

$\frac{0.22in}{73\%} = 0.30\ in$ needed to apply at least 0.22 *in* over irrigation zone with a DU of 73%.

$$0.30\ in \times \frac{h}{0.66\ in} = 0.46\ h$$

$$0.46\ h \times \frac{60\ min}{h} = 27\ min$$

Therefore, the system would need to operate 27 min to apply 0.30 *in* of water.

9. Water use engineers employed at a municipality require a golf course to justify their water use permit in terms of total amount of water requested and how they determined this value.

A. Determine average yearly ET rate from one of the methods listed previously. In this example, 56.37 *in* (4.7 *ft*, 1.4 *m*) is used.
B. Determine normal yearly precipitation rate. In this example, 10.67 *in*. (27 *cm*) is used.
C. Area of irrigated turfgrass. In this example, 110 *ac* is used (3.1 *ac* for greens, 3.7 for tees, 43.7 for fairways, and 59.5 for roughs).
D. Determine the irrigation efficiency (DU). In this example, 70 % is used.

	Greens	Tees	Fairways	Roughs
E. turf area (*ac*)	3.1	3.7	43.7	59.5
F. turfgrass	Bentgrass	Bermuda overseeded Oct–May	Bermuda overseeded Oct–May	Bermuda
G. K_c (crop coefficient)	0.8	0.75	0.75	0.65
H. Turf Water Use*: [A × G] (which is ET × K_c)*	*45.1*	*42.3*	*42.3*	*36.6*
I. 25 % precipitation (*in*): *[B × 0.25]*[a]	*2.7*	*2.7*	*2.7*	*2.7*
J. Water use adjusted for 25 % precipitation (*in*): *[H – I]*	*42.4*	*39.6*	*39.6*	*33.9*
K. Irrigation water use (*in*): *[J/D]*	*60.6*	*56.6*	*56.6*	*48.4*
L. K converted to feet: *[K/12] (12 in = 1 ft)*	*5.1*	*4.7*	*4.7*	*4.0*

(continued)

	Greens	Tees	Fairways	Roughs
E. turf area (*ac*)	3.1	3.7	43.7	59.5
M. Annual irrigation use (*ac-ft*): *[E × L]*	*15.8*	*17.4*	*205.4*	*238.0*
N. Annual irrigation water use: *sum M for all turf areas*	*477 ac-ft (or 155,430,927 gal)*			

[a]Water use regulators often use a precipitation efficiency adjustment value to reflect the amount (percentage) of usable precipitation by plants. Rainfall is often at inefficient amounts (too high or low) or at the wrong agronomic time.

In the above example, to compare calculated annual irrigation use to the overall formula, ET × area, the following was determined:

O. ET × 110 *ac*: *A* (*ft*) × *C* (total turfgrass area) or 4.7 *ft* × 110 *ac* = 517 *ac-ft* predicted by the simple formula,

P. Calculation efficiency for water budget: *N*/*O* × *100* or 477 *ac-ft* ÷ 517 *ac-ft* × 100 = 92 %. This value indicates the simple formula of ET × area overestimated water needs by 8 % compared to the Water Budgeting process above.

10. List and briefly discuss the necessary laboratory tests for soil and water quality:

- *Water soluble salts (or Salinity drought hazard)—Total salt content as measured by the electrical conductivity (EC_w) or total dissolved salts (TDS) of water. Excessive salts produces plant physiological drought.*
- *Sodium status—Soil sodium level proportionally to Ca^{+2} and Mg^{+2} ions as measured by sodium adsorption ratio (SAR), exchangeable sodium percentage (ESP), or adjusted SAR (adj. SAR). SAR also is used to assess the sodium levels of water. Excessive sodium causes soil structure deterioration.*
- *Specific ions toxicity—Toxic ion levels, especially boron, chloride, fluoride, sulfate and nitrate-nitrogen.*
- *Alkalinity—Bicarbonates and carbonates as measured by residual sodium carbonate (RSC).*
- *pH and lime requirement.*
- *Suspended solids, as measured by total suspended solids (TSS).*
- *Soil nutrient imbalance based on:*
  - *Sufficiency levels of available nutrients and cation ratio.*
  - *Soil cation exchange capacity (CEC).*
  - *Percent base saturation.*
  - *Percent organic matter.*

11. How much salt is applied per *ac* if 1 *in* (2.5 *cm*) of water with salinity levels of 1.0 $dS\ m^{-1}$ (~640 *ppm*) is used? 1 *in* = 0.083 *ft*; 1 *gal* pure water weighs ~8.34 *lb*; 1 *ac-ft* water = 325,851 *gal*.

$$640 \times \frac{2.72}{ac-ft} \times 1 \quad ac-in \times \frac{1}{12\ in} = 145$$

12. An irrigation source has an EC_w of 0.53 *mmhos cm*$^{-1}$. What would the EC_w be in *dS m*$^{-1}$, *μmhos cm*$^{-1}$, and *ppm* TDS?

    a. *Convert mmhos to dS m*$^{-1}$*: (1 dS m*$^{-1}$ *= 1 mmhos cm*$^{-1}$*), so 0.53 mmhos cm*$^{-1}$ *= 0.53 dS m*$^{-1}$

    b. *Convert mmhos cm*$^{-1}$ *to μmhos cm*$^{-1}$.

$$0.53\ mmhos\ cm^{-1} \times \frac{1000\ \mu mhos\ cm^{-1}}{1\ mmhos\ cm^{-1}} = 530\ \mu mhos\ cm^{-1}$$

    c. *Convert mmhos cm*$^{-1}$ *to ppm:*

$$0.53 \quad cm^{-1} \times 640 = 339 \quad TDS$$

13. The salt content of a water sample is 1,121 *mg L*$^{-1}$ TDS. What is the salt content in *dS m*$^{-1}$ and *μmhos cm*$^{-1}$?

    a) To convert TDS (mg L^{-1} *or ppm) to dS m*$^{-1}$*, divide by 640 or multiply by 0.0016:*

$$1,121\ mg\ L^{-1} \div 640 = 1.75\ dS\ m^{-1}$$

$$or$$

$$1,121\ mg\ L^{-1} \times 0.0016 = 1.75\ dS\ m^{-1}$$

    b) To convert dS m^{-1}*(or mmhos cm*$^{-1}$*) to μmhos cm*$^{-1}$*, multiply by 1000:*

$$1.75\ dS\ m^{-1}\left(or \quad cm^{-1}\right) \times 1,000 = 1,750\ \mu mhos\ cm^{-1}$$

14. What is the TDS and EC of water containing 250 *μmhos cm*$^{-1}$ Ca^{+2}, 325 *μmhos cm*$^{-1}$ Mg^{+2}, and 480 *μmhos cm*$^{-1}$ Na^{+}?

    Convert each value to dS m^{-1} *by dividing by 1000:*

$$ECw = 0.25\ dS\ m^{-1}\ Ca^{+2} + 0.325\ dS\ m^{-1}\ Mg^{+2} + 0.48\ dS\ m^{-1}\ Na^{+} = 1.1\ meq \quad L^{-1}$$

$$TDS + 1.1\ dS\ m^{-1} \times 640 = 704\ mg\ L^{-1}$$

15. A water sample test reports 1000 *mg L*$^{-1}$ Na^{+}, 200 *mg L*$^{-1}$ Ca^{+2}, and 100 *mg L*$^{-1}$ Mg^{+2}. What is the SAR value for this water?

    step 1: *calculate the number of meq L*$^{-1}$ *of each ion:*

$$meq\ L^{-1} = \frac{concentration\ \left(mg\ L^{-1} or\ ppm\right)}{equivalent\ weight\ \left(mg\ meq^{-1}\right)}$$

$$Na^{+} : 1000\ mg\ L^{-1} \div \frac{23\ mg}{meq} = 43.5\ meq\ L^{-1}$$
$$Ca^{+2} : 200\ mg\ L^{-1} \div \frac{20\ mg}{meq} = 10\ meq\ L^{-1}$$
$$Mg^{+2} : 100\ mg\ L^{-1} \div \frac{12.2\ mg}{meq} = 8.2\ meq\ L^{-1}$$

step 2: *inset these values into the SAR equation as:*

$$\mathrm{SAR} = \frac{\mathrm{Na}^{+}}{\frac{\sqrt{Ca^{+2}+\mathrm{Mg}^{+2}}}{2}} = \frac{43.5}{\frac{\sqrt{10+8.2}}{2}} = 14.4 \quad meq \quad L^{-1}$$

16. A water analysis indicates a sodium concentration of 85 $meq\ L^{-1}$, a Ca^{+2} concentration of 33.3 $meq\ L^{-1}$ and a Mg^{+2} concentration of 7.1 $meq\ L^{-1}$. What is the SAR value for this water?

$$\mathrm{SAR} = \frac{Na^{+}}{\frac{\sqrt{Ca^{+2}+\mathrm{Mg}^{+2}}}{2}} = \frac{85}{\frac{\sqrt{33.3+7.1}}{2}} = 18.9 \quad meq \quad L^{-1}$$

17. An irrigation source containing 75 $mg\ L^{-1}$ Ca^{+2} and 30 $mg\ L^{-1}$ Mg^{+2}. How much each would be supplied in each *ac-ft* of irrigation applied?

    Pounds of salt applied per acre = irrigation water salinity level (ppm or mg L^{-1}) × 2.72 million lb (weight of water per ac-ft) × ac-ft water applied, therefore for each ac-ft:

    Ca^{+2}: 75 mg L^{-1} × 2.72 = 204 lb Ca supplied per ac-ft irrigation water applied.

    Mg^{+2}: 30 mg L^{-1} × 2.72 = 82 lb Mg per ac-ft irrigation water applied.

18. What is the exchangeable sodium percentage (ESP) of a soil with 15 meq 100 g^{-1} Na^{+} and a CEC of 150 meq 100 g^{-1}?

$$ESP = \frac{exchangeable\ sodium\ \left(meq\ L^{-1}\right)}{cation\ exchange\ capacity\ \left(meq\ L^{-1}\right)} \times 100$$
$$= \frac{15\ meq\ 100\ g^{-1}}{150\ meq\ 100\ g^{-1}}$$
$$= 10\%$$

19. The EC of an irrigation water is 0.9 $dS\ m^{-1}$ while the salinity tolerance of tall fescue is approximately 6 $dS\ m^{-1}$. What would be the leaching requirement for this irrigation water to maintain the soil salinity level near its current level?

$$leaching\ requirement = \frac{EC_w}{EC_{dw}} \times 100$$
$$= \frac{0.9\ dS\ m^{-1}}{6\ dS\ m^{-1}} \times 100$$
$$= 15\%$$

This means 15 % additional water is needed above normal turfgrass water needs to prevent salts from accumulating.

20. From the previous example, 15 % extra water was determined necessary for the tall fescue turf to leach salts from the rootzone. If the "average" ET loss [or ET (target)] for tall fescue in summer is 0.325 *in day*$^{-1}$, determine how much water is needed to meet the needs of the fescue and accomplish the leaching required.

$$Total\ amount\ of\ water\ to\ apply = \frac{ET(target)}{1 - (leaching\ requirement)}$$
$$= \frac{0.325\ in\ day^{-1}}{1 - 0.15}$$
$$= 0.38\ in\ day^{-1}(9.7\ mm\ day^{-1})$$

This indicates 0.38 in day^{-1}*(9.7 mm day*$^{-1}$*) is needed to meet the summer turfgrass water needs and to prevent salts from accumulating.*

21. Water source 1 has an EC_w of 2.8 $dS\ m^{-1}$ while water 2 has an EC_w of 0.6 $dS\ m^{-1}$.

   (a) If water 2 is mixed in a 3 to 1 ratio to water 1, what would the EC_w of the blended water be?
   (b) If water 2 is mixed equally with water 1, what would the EC_w of the blended water be?

a) $$EC_w\ (blend) = \frac{volume\ (water\ 1) \times EC_w\ (water\ 1) + volume\ (water\ 2) \times EC_w\ (water\ 2)}{volume\ (water\ 1) + volume\ (water\ 2)}$$
$$= \frac{[1 \times 2.8\ dS\ m^{-1} + 3 \times 0.6\ dS\ m^{-1}]}{1+3}$$
$$= \frac{4.6\ dS\ m^{-1}}{4}$$
$$= 1.15\ dS\ m^{-1}$$

b) $$EC_w\ (blend) = \frac{volume\ (water\ 1) \times EC_w\ (water\ 1) + volume\ (water\ 2) \times EC_w\ (water\ 2)}{volume\ (water\ 1) + volume\ (water\ 2)}$$
$$= \frac{[1 \times 2.8\ dS\ m^{-1} + 3 \times 0.6\ dS\ m^{-1}]}{1+3}$$
$$= \frac{4.6\ dS\ m^{-1}}{4}$$
$$= 1.15\ dS\ m^{-1}$$

22. The following water quality analysis report was generated for a potential irrigation source.

   a. Find the missing SAR value.

      step 1: *The units must be converted to meq* L^{-1}*: Na (78), Ca (6.6), Mg (15.6), P (1.55), K (73),* CO_3*(3.9),* HCO_3*(10)*
      step 2: *Insert the values into the SAR equation:*

$$\mathrm{SAR} = \frac{\mathrm{Na}^{+}}{\sqrt{\frac{Ca^{+2}+\mathrm{Mg}^{+2}}{2}}} = \frac{78}{\sqrt{\frac{6.6+15.6}{2}}} = 23.4\,meq\,L^{-1}$$

   b. If 7.5 *ac-in* of water is applied per month, how many pounds of sodium are being applied yearly?

$$\frac{1.8\ g\ Na}{L} \times \frac{1\ lb\ Na}{454\ g\ Na} \times \frac{7.5\ ac{-}in\ water}{month} \times \frac{12\ months}{1\ yr} \times \frac{27,154\ gal}{1\ ac-in} \times \frac{3.785\ L}{1\ gal}$$
$$= 36,674\ Na\ yr^{-1}$$

   c. If your 7.8 *ac* are irrigated on average of 4.5 *ac-in* every month, how much nitrogen is being applied?

      step 1: The amount of nitrogen in nitrate (NO_3) *must be determined: (molecular weights N = 14 g, O = 16)*

$$NO_3 = 14 + (3 \times 16) = 62\quad g\quad \%N = (14 \div 62) \times 100 = 22.6\%\ N$$

      From the analysis, 6 ppm NO_3*is in the water, thus, this is multiplied by 22.6 % to obtain ppm N.*
      6 ppm × 22.6 % = 1.36 ppm N is being applied in the 6 ppm NO_3

      step 2: Now determine how much nitrogen is being applied each month:

$$\frac{1.36\,lb N}{1,000000\,lb H_2O} \times \frac{8.33\,lb H_2O}{gal} \times \frac{27,154\,gal}{ac-in} \times \frac{4.5\,ac\text{-}in}{month} \quad \text{(or)}$$
$$\times\, 7.8\,ac = \ \sim 11\,lb N\ applied\ monthly\ over\ 7.8\ ac$$

$$\frac{1.36\,mg N}{L} \times \frac{1g}{1,000\,mg} \times \frac{lb}{454\,g} \times \frac{3.755\,L}{gal} \times \frac{27,154\,gal}{ac\text{-}in} \times \frac{4.5\,ac\text{-}in}{month}$$
$$\times\, 7.8\ ac = \ \sim 11\ lb\ N$$

23. Water conservation involves numerous activities and practices. List the three activities when developing BMPs for Turfgrass water conservation.

   (a) *Site assessment and initial planning* (i.e., *documenting grasses, soils, microclimate, and existing management practices*).

Sample no.	Na^+	Ca^{+2}	Mg^{+2}	P	K^+	Cl^-	Carbonate (CO_3)	Bicarbonate (HCO_3)	Nitrate (NO_3)	Sulfate (SO_4)	TDS	pH	Conductivity (*mmhos* cm^{-1})	SAR
	ppm													
1	1800	131	190	1.55	73	6800	117	610	6	86	5120	8.4	8.0	–

(b) *Evaluating and implementing water conservation strategies.*
(c) *Analyzing benefits and costs of water conservation measures.*

24. Define hydrophobic soils and potential cause(s) of their development.

Hydrophobic soils are "water-hating" or -repelling from the coating of sand particles of acids (humic and fulvic). Acids are produced from:

Natural breakdown of soil organic matter.
Certain fairy ring fungi producing a mat of below-ground hyphae which often becomes hydrophobic.
Exudates from turfgrass roots.
Lipids from decomposing organic matter.
Surface waxes from cuticles of turfgrass plant leaves.

25. Within the nonionic surfactant chemistry, two main groups of wetting agents are currently available. List and discuss these.

1. *Soil Penetrants. These are often characterized as "water-moving" chemistry, characterized by having ethylene oxide terminal functional groups. Ethylene oxide groups are hydrophilic, being able to attract or disperse water molecules. Soil penetrating wetting agents generally increase water infiltration and percolation through the rootzone, providing more uniform soil moisture distribution within the profile leading to "fast and firm" playing conditions.*
2. *Soil water retainers. These are often characterized as "water-holding" chemistry, containing propylene oxide terminal functional groups. Propylene oxide groups are hydrophobic, thus repel water molecules. These are used where moisture retention is needed, especially sand-based rootzones with little organic matter and high infiltration and percolation rates. These are especially useful to help retain moisture during drought periods.*

Appendix A: Unit Analysis

Calculations and formulas for various shapes	
Rectangle, square or parallelogram	area = length (L) × width (W)
Trapezoid	area = [a + (b × h)] ÷ 2
Circle	area = radius $(r)^2 \times 3.1416$ (or π) = diameter $(d)^2 \times 0.7854$
	radius = d ÷ 2
	diameter = r × 2
	circumference = $\pi \times$ d
Sphere	volume = $r^3 \times 4.1888 = d^3 \times 0.5236$
Triangle	area = 1/2(b × h)
Cylinder	volume = $r^2\pi L$
Cone	area = 1/3 ($\pi r^2 h$)
Cube	volume = length × L × L

Approximate Weight of Dry Soil

	Bulk density	Weight		
Type	$g\ cm^{-3}$	$lb\ ft^{-3}$	$kg\ m^{-2}$	$lb\ ac^{-1}$ (6 *in* deep)
sand	1.6	100 (or 2700 $lb\ yd^{-1}$)	1623	2,143,000
loam	1.3–1.55	80-95	1299–1542	1,714,000
clay or silt	1.0–1.30	65-80	1055–1299	1,286,000
muck	0.65	40	649	860,000
peat (compact)	0.325	20	325	430,000

Sand weights (*tons*): $= yd^3 \times 1.3$
Gravel weights (*tons*): $= ft^3 \times 110$

−0.5 to 1 *in* diametergravel = 2,700 $lb\ yd^{-1}$
−0.25 to 0.375 *in* diametergravel = 3,000 $lb\ yd^{-1}$

L.B. McCarty et al., *Applied Soil Physical Properties, Drainage, and Irrigation Strategies*, DOI 10.1007/978-3-319-24226-2

Approximate Organic Materials for 6-in depth per 1000 ft^2 (weight variance in materials may occur)

Organic material volume in mix, %	Approximate thickness applied to soil surfaces		Organic material needed	
	in	*cm*	yd^3 1000 ft^{-2}	m^3 100 m^{-2}
5	0.33	0.84	1.0	0.83
10	0.67	1.70	2.0	1.70
15	1.00	2.54	3.0	2.48
20	1.33	3.38	4.0	3.30
25	1.67	4.24	5.0	4.16
30	2.00	5.08	6.0	4.95

Example: If 10 % organic materials is incorporated into the top 6 *in* of a 1000 ft^2 area, the organic material is applied to a depth of 0.67 *in* and 2.0 yd^3 will be needed (1.7 *cm* and 1.7 m^3 100 m^{-2}).

Peat Moss Coverage

Depth (*in*)	Coverage (ft^2) 5.6 ft^3 Bale (compressed) covers	4.0 ft^3 Bale (compressed) covers
0.25	480	346
0.50	240	173
1.00	120	86
2.00	60	43
3.00	40	29
4.00	30	22
6.00	20	14

Conversions for determining irrigation needs
$1\,ac-in = 27,154\,gal = 43,560\,in^3 = 3630\,ft^3$
$1\,in\,1000\,ft^{-1} = 620\,gal = 83\,ft^3$
$1\,gal = 0.134\,ft^3 = 8.34\,lb$
$1\,\text{million}\,gal = 3.07\,ac\text{-}ft$
$7\frac{1}{2}\,gal = 1\,ft^3 = 231\,in^3$
$1\,ac\text{-}ft = 325,851\,gal = 43,560\,ft^3$
$1\,lb\,\text{water} = 0.1199\,gal$
$\text{Precipitation rate}\,(in\,h^{-1}) = \frac{gpm \times 96.3}{\text{area}\,(ft^2)}$

Water and soil calculations
$1\,mmhos\,cm^{-1} = 1{,}000\,\mu mhos\,cm^{-1} = 1\,dS\,m^{-1} = 0.1\,S\,m^{-1} = 1\,mS\,cm^{-1} = 10\,me\,q\,L^{-1}$
$1\,meq\,L^{-1} = 1\,mmol\,L^{-1} = 1\,mol\,m^{-3}$
$1\,meq\,100\,g^{-1} = 1\,mmol\,100\,g^{-1} = cmol\,kg^{-1}$
Electrical conductivity $(mmhos\,cm^{-1}$ or $dS\,m^{-1}) \times 640 =$ Total dissolved salts $(mg\,L^{-1}$ or $ppm)$
Total dissolved salts $(mg\,L^{-1}$ or $ppm) \times 0.0016 =$ Electrical conductivity $(mmhos\,cm^{-1}$ or $dS\,m^{-1})$

Energy
1 calorie $(cal) = 4.184$ Joule (J)
Joule $(J) = 1\,kg\,m^2\,s^{-2}$
$1\,kcal = 4.184\,kJ$

Slopes

10 % = 6E = 10:1	33 % = 18E = 3:1
18 % = 10E = 6:1	50 % = 26E = 2:1
25 % = 14E = 4:1	100 % = 45E = 1:1

Decimal and millimeter length equivalents

Fraction (*in*)	Decimals (*in*)	Millimeters (*mm*)
1	*1.00*	*25.4*
15/16	0.9375	23.812
7/8	0.875	22.225
13/16	0.8125	20.638
¾	*0.75*	*19.05*
11/16	0.6875	17.462
5/8	0.625	15.875
9/16	0.5625	14.288
½	*0.5*	*12.70*
7/16	0.4375	11.112
3/8	0.3750	9.525
11/32	0.34375	8.731
5/16	0.3125	7.938
9/32	0.28125	7.144
¼	*0.25*	*6.350*
15/64	0.234375	5.953
7/32	0.21875	5.556

(continued)

Fraction (*in*)	Decimals (*in*)	Millimeters (*mm*)
13/64	0.203125	5.159
1/5	0.200	5.08
3/16	0.1875	4.762
23/128	0.1797	4.564
11/64	0.171875	4.366
1/6	0.167	4.242
21/128	0.1641	4.168
5/32	0.15625	3.969
1/7	0.143	3.633
19/128	0.1484	3.769
9/64	0.140625	3.572
1/8	*0.1250*	*3.175*
7/64	0.109375	2.778
1/10	0.100	2.540
3/32	0.09375	2.381
5/64	0.078125	1.984
1/16	0.0625	1.588
3/64	0.046875	1.191
1/32	0.03125	0.794
1/64	0.015625	0.397

Surface Area Impacted and Topdressing Sand Needed to Fill Aerification Holes

					Dry sand to fill holes 3 *in* depth	
Spacing, *in*	Tine diameter, *in*	Tine diameter, *mm*	Holes, ft^{-2}, no.	Surface area impacted, %	~ft^3 1000ft^{-2}	~*lb* 1000ft^{-2}
1.0 × 1.0	0.250	6.350	144	4.91	12.3	1227
	0.375	9.525	144	11.04	27.6	2761
	0.500	12.700	144	19.63	49.1	4909
	0.625	15.875	144	30.68	76.7	7670
	0.750	19.050	144	44.16	110.4	11,040
	1.000	25.400	144	78.50	196.4	19,640
1.0 × 2.0	0.250	6.350	72	2.45	6.1	614
	0.375	9.525	72	5.52	13.8	1381
	0.500	12.700	72	9.82	24.5	2454
	0.625	15.875	72	15.34	38.4	3855
	0.750	19.050	72	22.09	55.2	5520
	1.000	25.400	72	39.27	98.2	9820

(continued)

Spacing, *in*	Tine diameter, *in*	Tine diameter, *mm*	Holes, ft^{-2}, no.	Surface area impacted, %	Dry sand to fill holes 3 *in* depth	
					~ft^3 $1000 ft^{-2}$	~*lb* $1000 ft^{-2}$
1.5 × 1.5	0.250	6.350	64	2.18	5.5	550
	0.375	9.525	64	4.91	12.3	1230
	0.500	12.700	64	8.72	21.8	2180
	0.625	15.875	64	13.63	34.1	3410
	0.750	19.050	64	19.63	49.1	4910
	1.000	25.400	64	34.89	87.3	8730
2.0 × 2.0	0.250	6.350	36	1.23	3.1	307
	0.375	9.525	36	2.76	6.9	690
	0.500	12.700	36	4.91	12.3	1227
	0.625	15.875	36	7.67	19.2	1917
	0.750	19.050	36	11.04	27.6	2760
	1.000	25.400	36	19.63	49.1	4910
7.0 × 7.0 (drill and fill)	0.750	19.050	2.94	0.90	2.3	230
	1.000	25.400	2.94	1.60	4.0	400

Appendix B: Unit Conversions

Metric prefix definitions (basic metric unit = 1)					
tera	=	10^{12}	deci	=	10^{-1}
giga	=	10^{9}	centi	=	10^{-2}
mega	=	10^{6}	milli	=	10^{-3}
kilo	=	10^{3}	micro	=	10^{-6}
hecto	=	10^{2}	nano	=	10^{-9}
deca	=	10^{1}	pico	=	10^{-12}

Metric prefix example (weight)									Metric prefix example (volume)				
$1\,kg$	=	$10^{3}\,g$	=	$10^{6}\,mg$	=	$10^{9}\,\mu g$	=	$10^{12}\,ng$	$1\,L$	=	$10^{3}\,ml$	=	$10^{6}\,\mu l$
$1\,g$	=	$10^{-3}\,kg$	=	$10^{3}\,mg$	=	$10^{6}\,\mu g$	=	$10^{9}\,ng$	$1\,ml$	=	$10^{-3}\,L$	=	$10^{-6}\,\mu l$
$1\,mg$	=	$10^{-6}\,kg$	=	$10^{-3}\,g$	=	$10^{3}\,\mu g$	=	$10^{6}\,ng$	$1\,\mu L$	=	$10^{-6}\,L$	=	$10^{-3}\,ml$
$1\,\mu g$	=	$10^{-9}\,kg$	=	$10^{-6}\,g$	=	$10^{-3}\,mg$	=	$10^{3}\,ng$					
$1\,ng$	=	$10^{-12}\,kg$	=	$10^{-9}\,g$	=	$10^{-6}\,mg$	=	$10^{-3}\,\mu g$					

L.B. McCarty et al., *Applied Soil Physical Properties, Drainage, and Irrigation Strategies*, DOI 10.1007/978-3-319-24226-2

Area equivalents

1 *acre* (*ac*)	=	43,560 ft^2	=	4840 yd^2	=	0.4047 hectares (*ha*)	=	160 rod^2	=	4047 m^2	=	0.0016 mi^2
1 *ha*	=	10,000 m^2	=	100 *are*	=	2.471 *ac*	=	107,639 ft^2				
1 yd^2	=	9 ft^2	=	0.836 m^2				1 yd^3	=	27 ft^3	=	0.765 m^3
1 ft^2	=	144 in^2	=	929.03 cm^2	=	0.09290 m^2		1 m^2	=	10,000 cm^2		
1 ft^3	=	1728 in^3	=	0.037 yd^3	=	0.02832 m^3	=	28,320 cm^3				
1 *ac-in*	=	102.8 m^3	=	27,154 *gal*	=	3630 ft^3						

Liquid equivalents

1 *gal*	=	4 *qt*	=	8 *pt*	=	16 *cup*	=	128 *fl oz*	=	8.337 *lb*				1 *barrel*	=	42 *gal*
	=	231 in^3	=	256 *tbsp*	=	0.134 ft^3	=	3.785 *L*	=	3785 *ml*						
1 *qt*	=	0.9463 *L*	=	2 *pt*	=	4 *cup*	=	32 *fl oz*	=	64 *tbsp*	=	57.75 in^3	=	946.4 *ml*		
1 *L*	=	2.113 *pt*	=	1000 *ml*	=	1.057 *qt*	=	33.8 *fl oz*	=	0.26 *gal*	=	0.0001 m^2	=	1000 cm^3		
1 *pt*	=	16 *fl oz*	=	2 *cup*	=	473.2 *ml*	=	32 *tbsp*	=	0.125 *gal*	=	0.5 *qt*				
1 *cup*	=	8 *fl oz*	=	0.5 *pt*	=	16 *tbsp*	=	236.6 *ml*		1 *tbsp*	=	14.8 *ml*	=	3 *tsp*	=	0.5 *fl oz*
1 *fl oz*	=	29.57 *ml*	=	2 *tbsp*	=	6 *tsp*	=	0.0313 *qt*		1 *tsp*	=	4.93 *ml*	=	0.1667 *fl oz*	=	80 *drops*
1 ft^3 H_2O	=	7.5 *gal*	=	62.4 *lb*	=	28.3 *L*				1 *ml*	=	1 cm^3	=	0.034 *fl oz*	=	0.002 *pt*

Pressure equivalents								Temperature equivalents				
1 *mm Hg*	=	133.32 *Pa*	=	0.133 *kPa*	=	133,333 *MPa*		°C	=	(°F – 32)	×	5/9
1 *Pa*	=	10^{-3} *kPa*	=	10^{-6} *MPa*				°F	=	(°C × 9/5)	+	32
1 *psi*	=	6.9 *kPa*	=	2.31 *ft* head								
1 *MPa*	=	10^{3} *kPa*	=	10^{6} *Pa*	=	10 *bar*	=	10.2 $kg\ cm^{-2}$	=	100 $N\ cm^{-2}$		
1 *atm*	=	760 *mmHg*	=	29.92 *in Hg*	=	1.013 × 10^{5} *Pa*	=	1.013 *bar*	=	14.69 *psi*	=	33.89 *ft* H_2O
1 *kPa* =	0.001 *mPa*	= 10 *cm* H_2O	=	10 *mbar*	=	0.01 *bar*	=	1 $J\ kg^{-1}$	=	0.0099 *atm*	=	0.145 *psi*

Length equivalents

km	=	0.621 statute *mile*	=	1000 *m*	=	100,000 *cm*	=	3281 *ft*	=	39,370 *in*
m	=	3.28 *ft*	=	39.4 *in*	=	100 *cm*	=	1.094 *yd*	=	1000 *mm*
cm	=	0.3937 *in*	=	0.01 *m*	=	0.03281 *ft*				
in	=	2.54 *cm*	=	25.4 *mm*	=	0.0254 *m*	=	0.08333 *ft*		
ft	=	0.3048 *m*	=	30.48 *cm*	=	12 *in*				
yd	=	0.9144 *m*	=	3 *ft*	=	91.44 *cm*				
statute *mile*	=	1760 *yd*	=	5280 *ft*	=	1.61 *km*	=	1609 *m*		

Mixture ratios					Flow						
$1\,mg\,g^{-1}$	=	$1000\,ppm$	=	$1\,fl\,oz\,gal^{-1}$	$7490\,ppm$	=	$1\,gpm$	=	$0.134\,ft^3\,min^{-1}$	=	$0.06308\,L\,sec^{-1}$
$1\,fl\,oz\,100\,gal^{-1}$	=	$75\,ppm$	=	$1\,qt\,100\,gal^{-1}$	$2\,tbsp\,gal^{-1}$	=	$1\,ft^3\,min^{-1}$	=	$448.83\,gal\,h^{-1}$	=	$7.481\,gal\,min^{-1}$
$1\,pt\,100\,gal^{-1}$	=	$1\,tsp\,gal^{-1}$			$1\,ft^3 s^{-1}$	=	$448.83\,gal\,min^{-1}$				

Weight equivalents

1 ton (US)	=	2000 *lb*	=	0.907 *metric ton*	=	907.2 *kg*								
1 *metric ton*	=	10^6 *g*	=	1000 *kg*	=	2205 *lb*								
1 *lb*	=	16 *oz*	=	453.6 *g*	=	0.4536 *kg*			1 *oz* (wt)	=		28.35 *g*	=	0.0625 *lb*
1 *g*	=	1000 *mg*	=	0.0353 oz	=	0.001 *kg*	=	0.002205 *lb*	1 *mg*	=		0.001 *g*		
1 *kg*	=	1000 *g*	=	35.3 *oz*	=	2.205 *lb*			1 *µg*	=		10^{-6}*g*	=	0.001 *mg*
ng	=	10^{-9}*g*	=	0.001 micro-grams (*µg*)					*picogram*	=		10^{-12}*g*		
1 % (v/v)	=	1.28 *fl oz gal*$^{-1}$	=	1 *gal* 100 *gal*$^{-1}$	=	10,000 *ppm*	=	10 *g L*$^{-1}$	=	1 g 100 *ml*$^{-1}$	=	1.33 *oz* (wt) *gal*$^{-1}$	=	8.34 *lb* 100 *gal*$^{-1}$
1 *ppm*	=	0.0001 %	=	1 *mg kg*$^{-1}$	=	1 *mg L*$^{-1}$	=	1 *µg g*$^{-1}$	=	1 *µl L*$^{-1}$	=	1 *µg ml*$^{-1}$		
	=	0.379 *g* 100 *gal*$^{-1}$	=	8.34 × 10^{-6}*lb gal*$^{-1}$	=	0.013 *fl oz* 100 *gal*$^{-1}$								
10 *ppm*	=	0.001 %	=	10 *mg L*$^{-1}$										
100 *ppm*	=	0.01 %	=	100 *mg L*$^{-1}$										
1000 *ppm*	=	1 *mg g*$^{-1}$	=	0.1 %	=	1000 *mg L*$^{-1}$								
1 *ppb*	=	1 *µg kg*$^{-1}$	=	1 *µg L*$^{-1}$	=	1 *ng ml*$^{-1}$	=	1 *ng* 1,000,000,000^{-1}				1 *ppt*	=	1 *pico-gram g*$^{-1}$

Water potential and mathematical units	
When performing mathematical calculations concerning water flow, potential or pressure terms are mostly used. Tensions, stress, and suction are some of the terms used to express potential. The more common mathematical units associated with these terms include: bars, centimeters of water (cm H_2O), centimeters of mercury (cm Hg), inches of water, atmospheres (atm), centibars (cb), millibars (mb), Joules per kilogram (J kg^{-1}), kilopascals (kPa), pounds per square inch (psi), ergs per gram (ergs g^{-1}), and dynes per square centimeter (dynes cm^{-2}). Bars and kilopascals (kPa) are commonly used units. Relationships between units include:	
1 bar $= 1020\,cm\,H_2O$ (or $\sim 1000\,cm\,H_2O$) $= 75.01\,cm\,Hg$ $= 401.4\,in\,H_2O$ at 4 °C $= 0.9869\,atm$(or $\sim 1\,atm$) $= 100\,cb = 1000\,mb$ $= 100\,J\,kg^{-1}$ $= 14.50\,psi$ $= 10^6 ergs\,g^{-1} = 10^6\,dynes\,cm^{-2}$ $\times 14.5 = psi$ $\times 1019.7 = g\,cm^{-2}$ $\times 29.53 = in\,Hg$ at 0 °C $\times 75 = cm\,Hg$ at 0 °C $\times 0.10 = MPa$ $\times 100 = kPa$ $\times 100{,}000 = Pa$	$1\,kPa = 1\,cb$ $= 0.001\,MPa = 1000\,Pa$ $= 10\,cm\,H_2O$ $= 0.75\,cm\,Hg$ at 0 °C $= 10\,mbar = 0.01\,bar$ $= 1\,J\,kg^{-1}$ $= 0.0099\,atm$ (or $\sim 0.01\,atm$) $= 0.145\,psi$ $= 10{,}000\,dynes\,cm^{-2}$ $\times 1 = J\,kg^{-1}$ $\times 1 = 0.01\,bar$ $\times 0.01 = bar$ $\times 0.145 = psi$ $\times 4.01 = in\,H_2O$ at 4 °C $\times 10.2 = cm\,H_2O$ at 4 °C
Numerous web sites exist dealing with unit conversions. One is www.unitconversion.org.	

Metric conversion

To convert	Multiply by	To obtain
Acres (ac)	0.4047	hectare (ha)
ac	43,560	sq. feet (ft^2)
ac	0.00405	sq. kilometer (km^2)
ac	4047	sq. meter (m^2)
ac	4840	sq. yards (yd^2)
Acre-feet (ac-ft)	325,851	sq. feet (ft^2)
ac-ft	43,560	cu. feet (ft^3)
ac-ft	1233.5	cu. meter (m^3)
ac-in	102.8	m^3
Bar	14.5	lb/in^2
Bar	1019.7	g/cm^3
Bar	29.53	inches Hg at 0 °C
Bar	75	cm Hg at 0 °C
Bar	0.001	J/kg
Bar	100	kPa
Bushels (dry)	0.03524	m^2
Bushels	1.245	ft^3
Calorie (cal)	4.184	Joules (J)
Centimeters (cm)	0.03281	feet (ft)

(continued)

Metric conversion		
To convert	Multiply by	To obtain
cm	0.3937	inches (in)
cm	0.1094	yards (yd)
cm	0.01	meters (m)
cm	10	millimeters (mm)
cm/s = cm sec^{-1} = cm per sec	1.9685	ft/min
cm/s	0.0223694	miles per hour (MPH)
cm^2 (square centimeters)	0.001076	ft^2
cm^2	0.1550	in^2
cm^2	0.01	sq. decimeter
cm^3 (cubic centimeters)	0.0610237	in^3
cm^3	0.0338	fl oz
cm^3	0.001057	qt^3
cm^3	0.000264172	gal
cm^3	0.001	cu. decimeter
Cup	8	fl oz
Cup	236.6	cm^3
Feet (ft)	30.48	cm
ft	0.3048	m
ft	305	mm
ft^2 (square feet)	929	cm^2
ft^2	0.0929	m^2
ft^2	9.294×10^{-6}	hectares (ha)
ft^2	144	in^2
ft^3 (cubic feet)	0.0283	m^3
ft^3	7.4805	gallons
ft^3	1728	cubic inches (in^3)
ft^3	0.037	cubic yards (yd^3)
ft^3	28.32	liters (L)
ft^3/1000 ft^2	0.030463	m^3/100 m^2
Feet per minute	0.01136	mph
Feet head of water	0.433	psi
Foot candle	10.764	lux
Gallons (gal)	3.785	liters
Gal	3785	ml
Gal	128	ounces (liquid)
Gal	0.13368	ft^3
Gal	231	in^3
Gal	3785	cm^3
Gal per acre (gpa)	9.354	L/ha
gpa	0.09354	L/100 m^2
gpa	2.938	oz/1000 ft^2 (liquid)
Gal/1000 ft^2	4.0746	L/100 m^2

(continued)

Metric conversion		
To convert	Multiply by	To obtain
Gal/min	2.228×10^{-3}	ft^3/s
Gal/min	0.06309	L/s
Gal/min	0.227125	m^3/h
Grams (g)	0.002205	lb
Gram	0.035274	oz
g/cm^3	0.036127	lb/in^3
g/cm^3	62.428	lb/ft^3
g/ft^2	96	lb/ac
g/ha	0.000893	lb/ac
g/ha	0.014275	oz/ac
g/kg	0.10	percent (%)
g/l	1000	PPM
g/l	10	%
g/l	0.00834595	lb/gal
g/l	0.13351	oz/gal
g/m^2	0.00020481	lb/ft^2
g/m^2	0.20481	$lb/1000\ ft^2$
Hectares (ha)	2.471	ac
Ha	107,639	ft^2
Ha	107.64	$1000\ ft^2$
horsepower (electrical or mechanical)	746	watts
hp	550	ft lb/s
hp	1.014	metric horsepower
hp	33,000	ft lb/min
Inches (in)	2.540	cm
in	0.0254	m
in	25.40	ml
Inches of mercury (in Hg)	3.4	kilopascals (kPa)
in/ft	0.083	mm/mm
in^2 (square inches)	6.4516	cm^2
in^3 (cubic inches)	16.3871	cm^3
in^3	0.55411	fl oz
in^3	0.01732	qt
in^3/h	0.00434	gal/h
Joules per kilograms (J/kg)	1	kPa
kilo Pascal (kPa)	1	J/kg
kPa	1	0.01 bar
kPa	0.01	bar
Kilograms (kg)	2.2046	lb
kg/ha	0.892	lb/ac
kg/ha	0.02048	$lb/1000\ ft^2$
$kg/100\ m^2$	2.048	$lbs/1000\ ft^2$

(continued)

Metric conversion		
To convert	Multiply by	To obtain
kg/L	8.3454	lb/gal
Kilometers (km)	100,000	cm
km	3281	ft
km	1000	m
km	0.6214	miles
km	1094	yd
km/h	0.62137	mph
km/h	54.6807	ft/min
Kilopascals (kPa)	0.145	lbs/in^2 (psi)
kPa	1	0.01 bar
kPa	1	J/kg
Liters (L)	0.2642	gallons
L	33.814	fl.oz.
L	2.113	pt
L	1.057	qt
L	0.035315	ft^3
L/m^2	3.2808	ft^3/1000 ft^2
L/100 m^2	0.2454	gal/1000 ft^2
L/100 m^2	1.9634	pt/1000 ft^2
Liters/ha	0.107	gal/ac
L/ha	0.0025	gal/1000 ft^2
L/ha	0.314	oz/1000 ft^2
L/ha	0.855	pt/ac
L/min	15.85	gal/h
Meters (m)	3.281	ft
Meters	39.37	in
Meters	1.094	yd
Meters	100	cm
Meters	0.001	km
Meters	1000	mm
Meters/s	2.2369	mph
M^2 (square meters)	10.764	ft^2
M^2	1550	in^2
M^2	1.196	yd^2
M^3 (cubic meters)	35.3147	ft^3
M^3	1.30795	yd^3
M^3	1000	L
M^3/ha	14.29	ft^3/ac
M^3/ha	0.0122	yd^3/1000 ft^2
M^3/ha	0.328	ft^3/1000 ft^2
mil	0.001	in
mil	0.0254	mm

(continued)

Metric conversion		
To convert	Multiply by	To obtain
Miles (nautical)	1.1508	miles (statute)
Miles (nautical)	6076	ft
Miles (statute)	160,900	cm
Miles	5280	ft
Miles	1.609	km
Miles	1760	yards
Miles per hour (mph)	1.467	ft/s
mph	88	ft/min
mph	1.61	km/h
mph	0.447	m/s
mg/kg	1	parts per million (ppm)
Milliequivalents per liter (meq/L)	1	millimoles per liter (mmol/L)
Milliequilvalents per 100 g (meq/100 g)	Eq. wt. × 10	parts per million (ppm)
Millimhos per centimeter (mmhos/cm)	1	decisiemens per meter (dS/m)
mmhos/cm	1000	micromhos per centimeter (μmhos/cm)
Milliliters (ml)	0.0338	oz (fluid)
ml	0.0002642	gal
ml/m^2	3.14	oz/1000 ft^2
ml/l	0.12793	oz/gal
ml/10,000 L	0.0128	fl oz/1000 gal
Millimeters (mm)	0.03937	in
1 mmHg at 0 °C	0.13332	kPa
1 mmHg	133333.3	mPa
Ounces (fluid) (oz)	0.02957	L
Ounces (fluid)	29.573	ml
Ounces (fluid)	0.03125	qt
Oz (fluid)/gal	7.81	ml/L
Ounces (fluid)/ac	0.0731	L/ha
Ounces (fluid)/ac	73.1	ml/ha
Ounces (fluid)/1000 ft^2	3.18	L/ha
oz (weight)	28.35	g
oz (weight)	0.0625	lb
oz (weight)/acre	0.07	kg/ha
oz (weight)/acre	70	g/ha
oz (weight)/1000 ft^2	3.05	kg/ha
oz (weight)/ft^2	305.15	g/m^2
oz (weight)/gal	7.5	g/L
oz (weight)/1000 ft^2	0.305	g/m^2
Percent (%)	10	g/kg
Pint (liquid) (pt)	0.473	liter
pt/ac	1.1692	L/ha

(continued)

Metric conversion		
To convert	Multiply by	To obtain
pt/ac	0.3673	oz/1000 ft^2
pt/1000 ft^2	0.50932	L/100 m^2
Parts per million (ppm)	2.719	lb ai/ac-ft of water
PPM	2.0	lbs/ac slice 7-*in* deep
PPM	2.25	kg/ha slice 7-*in* deep
PPM	0.001	g/L
PPM	8.34	lb/million gal
PPM	1	mg/kg
PPM	0.013	oz/100 gal of water
PPM	0.3295	gal/acre-foot of water
PPM	8.2897	lbs/million gal of water
Pounds (lb)	0.4536	kilograms (kg)
lb	453.6	g
lb/acre	1120	g/ha
lb /ac	1.12	kg/ha
lb /ac	1.0413	g/100 ft^2
lb /ac	0.02296	lb/1000 ft^2
lb /ac	0.112	g/m^2
lb /ac-ft	0.3682	g/m^3
lb /ac-ft	0.0003682	kg/m^3
lb /ft^2	4883	g/m^2
lb /ft^3	16.23	kg/m^3
lb /1000 ft^2	4.88	g/m^2
lb /1000 ft^2	48.83	kg/ha
lb /1000 ft^2	43.5597	lb/ac
lb /1000 ft^2	488	g/100 m^2
lb /1000 ft^2	0.4883	kg/100 m^2
lb /1000 ft^2	0.91	lbs/100 yd^2
lb /1000 ft^2	1.1	lbs/1000 ft^2
lb /yd^3	0.0005937	g/cm^3
lb /yd^3	594	g/m^3
lb /yd^3	0.5932	kg/m^3
lb /gal	0.12	kg/l
lb /1000 gal	0.12	g/1000 L
pounds per square inch (PSI)	6.89	kilopascals (kPa)
PSI	0.06895	bar
PSI	0.068046	atmosphere (atm)
PSI	2.31	feet head of water
Quarts (qt)	0.9463	L
Quarts	946	ml
Qt/A	2.3385	L/ha
Qt/A	0.7346	oz/1000 ft^2

(continued)

Metric conversion		
To convert	Multiply by	To obtain
Qt/100 gal	2.5	ml/L
Temperature, °C + 17.98	1.8	temperature, °F
Temperature, °F − 32	0.5555	temperature, °C
Ton (2000 lbs)	907	kg
Ton (2000 lbs)/ac	2240	kg/ha
Ton (2000 lbs)	0.907	ton (metric)
Ton (2000 lbs)/ac	2.241	ton (metric)/ha
Ton (metric)	2205	lb
Ton (metric)	1000	kg
Ton (metric)	1.102	ton (2000 lb)
Yards (yd)	91.44	cm
Yards	0.9144	m
Yards	914.4	mm
yd^2 (square yards)	0.836	m^2
yd^2	9	ft^2
yd^2	1296	in^2
yd^3 (cubic yards)	27	ft^3
yd^3	46,656	in^3
yd^3	0.7645	m^3
yd^3	765	L
$yd^3/1000\ ft^2$	0.825	$m^3/100\ m^2$
P_2O_5	0.437	P
K_2O	0.830	K
CaO	0.715	Ca
MgO	0.602	Mg
meq Ca^{+2}/100 g soil	400	lb Ca^{+2} per acre furrow slice
meq K^{+}/100 g soil	780	lb K^{+} per ac furrow slice
meq Na^{+}/100 g soil	460	lb Na^{+} per ac furrow slice
meq Mg^{+2}/100 g soil	109	lb Mg^{+2} per ac furrow slice
meq Fe^{+3}/100 g soil	372	lb Fe^{+3} per ac furrow slice
meq Zn^{+2}/100 g soil	654	lb Zn^{+2} per ac furrow slice
meq H^{+}/100 g soil	20	lb H^{+} per ac furrow slice
meq Al^{+3}/100 g soil	180	lb Al^{+3} per ac furrow slice
meq Ca^{+2}/100 g soil	9.2	lb Ca^{+2} per 1000 ft^2 furrow slice
meq K^{+}/100 g soil	18	lb K^{+} per 1000 ft^2 furrow slice
meq Na^{+}/100 g soil	10.6	lb Na^{+} per 1000 ft^2 furrow slice
meq Mg^{+2}/100 g soil	2.5	lb Mg^{+2} per 1000 ft^2 furrow slice
meq Fe^{+3}/100 g soil	8.5	lb Fe^{+3} per 1000 ft^2 furrow slice
meq Zn^{+2}/100 g soil	15	lb Zn^{+2} per 1000 ft^2 furrow slice
meq H^{+}/100 g soil	0.46	lb H^{+} per 1000 ft^2 furrow slice
meq Al^{+3}/100 g soil	4.1	lbs Al^{+3} per 1000 ft^2 furrow slice

American Society for Testing and Materials

ASTM reference	Analysis
C88-13	Standard Test Method for Soundness of Aggregates by Use of Sodium Sulfate or Magnesium Sulfate
C131/C131M-14	Standard Test Method for Resistance to Degradation of Small-Size Coarse Aggregate by Abrasion and Impact in the Los Angeles Machine
C136/C136M-14	Standard Test Method for Sieve Analysis of Fine and Coarse Aggregates
D75/D75 -14	Standard Practice for Sampling Aggregates
D422	Test Method for Particle-Size Analysis of Soils
D854-14	Standard Test Methods for Specific Gravity of Soil Solids by Water Pycnometer
D2974-14	Standard Test Methods for Moisture, Ash, and Organic Matter of Peat and Other Organic Soils
D4972 13	Standard Test Method for pH of Soils
D5550-14	Standard Test Method for Specific Gravity of Soil Solids by Gas Pycnometer
D5874-02	Standard Test Method for Determination of the Impact Value (IV) of a Soil
D6913-04 (2009)	Standard Test Methods for Particle-Size Distribution (Gradation) of Soils Using Sieve Analysis
F1551-09	Standard Test Methods for Comprehensive Characterization of Synthetic Turf Playing Surfaces and Materials
F1647-11	Standard Test Methods for Organic Matter Content of Athletic Field Rootzone Mixes
F1702-10	Standard Test Method for Measuring Impact-Attenuation Characteristics of Natural Playing Surface Systems Using a Lightweight Portable Apparatus
F1815-11	Standard Test Methods for Saturated Hydraulic Conductivity, Water Reten tion, Porosity, and Bulk Density of Athletic Field Rootzones
F1953-10	Standard Guide for Construction and Maintenance of Grass Tennis Courts
F2060-00 (2011)	Standard Guide for Maintaining Cool Season Turfgrasses on Athletic Fields
F2107-08	Standard Guide for Construction and Maintenance of Skinned Areas on Baseball and Softball Fields (See also WK39656 proposed revision)
F2269-11	Standard Guide for Maintaining Warm Season Turfgrasses on Athletic Fields

(continued)

L.B. McCarty et al., *Applied Soil Physical Properties, Drainage, and Irrigation Strategies*, DOI 10.1007/978-3-319-24226-2

ASTM reference	Analysis
F2270-12	Standard Guide for Construction and Maintenance of Warning Track Areas on Athletic Fields
F2396-11	Standard Guide for Construction of High Performance Sand-Based Rootzones for Athletic Fields
F2747-10	Standard Guide for Construction of Sand-based Rootzones for Golf Putting Greens and Tees
F3013-13	Standard Test Method for Density of Topsoil and Blended Soils In-place by the Core Displacement Method
WK32046	Revision of F1632-03 (2010)—Standard Test Method for Particle Size Analysis and Sand Shape Grading of Golf Course Putting Green and Sports Field Rootzone Mixes
WK37583	New Guide for Construction or Renovation of Native-soil Athletic Fields
WK35282	New Guide for Quality Control Protocols Related to Natural Turf Athletic Field Rootzone Constructions

These ASTM testing procedures are some of the more important ones for golf and sports fields. Refer to www.astm.org for the latest reference numbers and analytical techniques and methods.

Glossary

A

Acid Substance having a pH less than 7.0 or substance that releases hydrogen ions (H^+).

Acidity, active Hydrogen ion concentration in the aqueous phase of a soil. Expressed as a pH value.

Acidity, exchangeable Replaceable soil aluminum and hydrogen ions by an unbuffered salt solution such as KCl or NaCl.

Adhesion Molecular attraction and contact between the surfaces of two unlike substances or objects.

Adjuvant Substance in a formulation that enhances its effectiveness. Includes surfactants, crop oils, crop oil concentrates, anti-foaming agents, drift control agents, pH modifiers, and compatibility agents.

Absorption Uptake of ions or compounds by a substance.

Adsorption Bonding or adhering of ions or compounds to the surface of soil particles or plant parts.

Aerification In turf, a method of cultivation where hollow or solid tines or spoons are inserted into and removed from the turf to control soil compaction and increase water and fertilizer penetration; *hollow-tine* aerification removes soil plugs, *solid-tine* aerification does not remove plugs when holes are made, *deep-drill* aerification involves removing soil via long drill bits, and *hydro-, hydraulic-*, or *water-injection* aerification uses fine streams of high-pressure water to penetrate the soil surface.

Aggregate Collect together in tufts, groups, or bunches, such as soil clods.

Agronomy Science of crops and soils.

Air-entry point (or pressure) Point or pressure where gravity breaks the capillary tension of the largest soil pores and water begins to drain, and thus, permits air entry. This is also the depth of a perched water table and is often characterized with soil moisture retention curves. Below this point the soil is saturated while above it water aeration begins; also referred to as *critical tension*.

L.B. McCarty et al., *Applied Soil Physical Properties, Drainage, and Irrigation Strategies*, DOI 10.1007/978-3-319-24226-2

Air-filled porosity Fraction of soil bulk volume filled with air at any given time or under a given condition, such as a specified soil-water content.

Air porosity Portion of soil volume filled with air at a given situation such as a specified moisture potential; usually the larger pores.

Alkali Substance which has a pH > 7.0, also, when dissolved in water, forms a solution containing hydroxyl ions (OH^-). Also referred to as *base*.

Alkalinity Capacity of water to neutralize acids; a property imparted by carbonates, bicarbonates, hydroxides, and others.

Alluvium River floodplain composed of fragmental materials, broken down by weathering and erosion, transported by a stream or river and deposited.

Amendment Any material, such as sand, sawdust, gypsum, diatomaceous earth, peat, or calcined clay, added to soil to alter its chemical and/or physical characteristics.

Anaerobic Absence of molecular oxygen.

Anion Negatively charged ion attracted to a positively charged anode.

Atom Smallest unit of a chemical element that retains its characteristic properties.

Atomic number Number of protons in the nucleus of an atom.

Atomic weight Weight of an atom of an element relative to 1/12 of the weight of an atom of carbon ^{12}C, which has been assigned the value 12.

Available water Portion of soil water readily absorbed by plant roots; often considered to be the water held in soil between field capacity and permanent wilting point.

B

Bar In science, a unit of pressureused to express water potential ($1\,atm = 1.013\,bar = 0.1013\,mPa$).

Base *See* alkali.

Bedrock Solid, relatively un-weathered rock layer near the Earth's surface.

Bernoulli's principal Soil water potential decreases with increased rate of flow.

Bicarbonate (HCO_3^-) Ions in irrigation water that combines with calcium and magnesium to form insoluble lime (or calcite).

Bouyoucos Scale Gradation scale ($g$soil colloids L^{-1}) on a hydrometer used to measure the density of a suspension.

Brownian motion Random movement of microscopic particles suspended in a liquid or gas, caused by collisions with molecules of the surrounding medium; also called *Brownian movement*.

Buffer Substance that resists change in pH.

Buffer capacity Ability of soils to resist chemical change due to high cation exchange capacity and, in some cases, free calcium carbonate.

Buffersolution or pH buffer Substance used to determine buffering capacity of a soil for lime requirements.

Bulk density Mass (or weight) per unit volume of soil, often reported as grams soil per cubic centimeter ($g\,cm^{-3}$). A higher bulk density indicates aheavier or more compacted soil.

Bulk flow Overall movement of a liquid such as water; induced by gravity, pressure, or an interplay of both.

C

Calcareous soil Soil containing 10 to 1000 $g\ kg^{-1}$ of $CaCO_3$ (calcium carbonate, lime) equivalent.

Calcined clay Granular soil modification amendment consisting of highly fired or heated (termed *calcined*) clay minerals, such as montmorillonite and attapulgite, that are absorbent and stable.

Calcined diatomaceous earth Fired single-celled ocean organisms called *diatoms* that are absorbent and stable; often used as a granular soil modification amendment.

Calcite Crystalline form of calcium carbonate ($CaCO_3$) often formed when high bicarbonate-containing water is used for irrigation; also called *calcitic limestone*.

Calcium carbonate ($CaCO_3$) Aform of calcium combined with carbonate that remains insoluble until reacted with an acid. Occurs in nature as limestone, marble, chalk, marl, shells, and similar substances.

Calcium/magnesium carbonate ($CaCO_3 \cdot MgCO_3$) Calcium/magnesium combination precipitated from water high in calcium, magnesium, bicarbonates and carbonates that remains insoluble until reacted with an acid.

Calcium carbonate equivalent (CCE) Relative measurement of purity of a liming material compared to a value of 100 for pure calcium carbonate ($CaCO_3$).

Capillary fringe Zone immediately above the water table that is saturated, though under sub-atmospheric pressure (tension). Its height increases as the average soil pore size decreases. The top of the capillary fringe is referred to as the *air entry point*.

Capillary porosity Percentage of soil volume occupied by water due to capillary forces following gravity drainage.

Capillaryrise Height water will rise in a cylinder with its bottom in contact with a free water surface. This water rises against the pull of gravity due to the adhesive forces of water on the surface of the pore. Occurs in soil pores where they occur above each other to form an effective column with its bottom in contact with a free water surface.

Capillary water Water remaining in soil pores following drainage by gravity.

Cation Positively atomically charged ion.

Clay Soil particles $\leq$0.002 *mm* in diameter; also indicates a soil texture containing more than 40 % clay.

Coefficient of uniformity (design uniformity) In irrigation, efficiency of a sprinkler system based on precipitation rates at various points. In soil, quantifying the shape of the particle-size distribution curve, often using the ratio D_{60}/D_{10}.

Cohesion (or surface tension) Mutual attraction of molecules of the same substance.

Colloid Soil particles 0.1 to 0.001 μm in diameter, which may be a molecular aggregate.

Compaction Decreased volume of a given mass due to pressure. In soils, typically an unfavorable change due to an increase in soil bulk density ($g\ cm^{-3}$) and corresponding decrease in soil porosity.

Compressibility Soil property pertaining to a decrease in bulk volume when subjected to a load.

Conductivity Degree to which a soil-water extract or irrigation water facilitates the flow of electricity, indicating the level of ions in solution.

Contact angle Angle occurring when water is in contact with a solid surface.

Contours Outline of a figure, body, or mass; contour lines connect the points on a land surface with the same elevation.

Core aerification *See* aerification.

Core cultivation *See* aerification.

Coring *See* aerification.

Critical tension *See* air-entry point (or pressure). Also defined as the capillary rise in pores created by a particular particle diameter "dominating" the pore system.

D

D_{15} Particle diameter below which 15 % of soil particles are finer and 85 % are coarser. Used in turf to determine if a drainage stone (gravel) meets bridging requirements for use with a rootzone mix.

D_{85} Particle diameter below which 85 % of soil particles are finer and 15 % are coarser. Used in turf to determine if a rootzone mix meets bridging requirements for use with a coarser diameter gravel.

Darcy's Law Law describing proportional relationship of quantity of water flow through a saturated porous media to hydraulic gradient.

Deflocculation (or dispersion) Separation into individual components and spreading throughout a medium by chemical and/or physical means.

Density (of matter) Mass per unit volume.

Desalination Process of separating a saline solution into pure fresh water and brine.

Desiccation Plant moisture loss from hot, dry weather, fertilizers, or chemicals.

Diatomaceous earth Geologic deposit of siliceous skeleton material of diatoms (algae); used as an inorganic soil amendment.

Dissolved oxygen Atmospheric oxygen held in solution within water.

Drain line Underground pipe or other conduit that collects and removes excessive soil water.

Drainage Water movement out of the soil profile.

Drought Prolonged water stress that limits or prevents plant growth.

Drought avoidance Plant's ability to sustain internal water levels through morphological and physical growth features such as more efficient and deeper root systems, effective stomata closure, and thicker leaf cuticles.

Drought resistance Plant's ability to withstand drought conditions by combining drought avoidance and drought tolerance.

Drought tolerance Plant's ability to sustain internal water levels through biochemical and physiological processes.

E

Edaphic Pertaining to soil.

Effluent Water which has undergone one cycle of human or animal use and is partially or completely treated to make it suited for limited use. Also referred to as *recycled*, *reclaimed*, *wastewater*, and *treated sewage water*.

Electrical conductivity (EC) Measure of salinity using electrical conductance expressed as millimhos per centimeter ($mmhos\ cm^{-1}$) or decisiemens per meter ($dS\ m^{-1}$). EC_e is the electrical conductivity of soil from a saturated paste while EC_w is the electrical conductivity of water.

Electron Negatively charged subatomic particle that orbits the atom's positively charged nucleus, determining the atom's chemical properties.

Element Substance composed of only one kind of atom. These combine to compose all materials.

Eluviation Movement of humus, chemical substances, and mineral particles from upper soil layers to lower layers by the downward water movement; also referred to as *outwashing*.

Environment Conditions, influences, or forces affecting living forms.

Epsom salt Common name for magnesium sulfate ($MgSO_4$).

Equilibrium Physically or chemically balanced such that mass or energy transfer ceases.

Equivalent Amount of material that reacts with or provides one *gram* formula weight of hydrogen.

Equivalent weight (moles of ion charge) Amounts of substances that are equivalent to each other in chemical reactions; determined in an acid as the weight of substance furnishing one mole of hydrogen ions; determined in a base by the weight furnishing one mole of hydroxide (OH^-) ions. Also measured as the change in oxidation (valence) atoms undergo in a chemical reaction.

$$\text{equivalent weight} = \frac{\text{molecular weight}}{\text{number of H}^+\text{or OH}^-\text{per molecule}} \text{ or } \frac{\text{molecular weight}}{\text{valence}}$$

Erosion Movement of soil and rock by running water, wind, moving ice, or gravitational creep.

Evaporation Process of water returning to atmospheric air as vapor from land, water, and vegetation surfaces.

Evapotranspiration (ET) Combined loss of water from an area by evaporation from the soil surface and transpiration from plants; expressed as inches (*in*) or millimeters (*mm*) per day or week.

Exchangeable sodium percentage (ESP) Measure of excessive sodium hazard in the soil as the ratio of exchangeable sodium to the remaining exchangeable cations (Mg, Ca, and K).

F

Ferric iron (Fe_2O_3) Insoluble, oxidized (Fe^{+3}) iron source, often dense and hard, taking on an orange or rust color.

Ferrous iron (FeO) Soluble, reduced (Fe^{+2}) iron source, which forms under anaerobic conditions. Often has a black or dark grey color.

Fick's law Law describing ion or molecule movement by diffusion due to a concentration gradient.

Field capacity Water content of soil following drainage due to gravity; typically the water remaining in a soil 2 days following saturation.

Flocculation Aggregation or clumping together, especially clay, into larger clumps or aggregates.

Force Push or pull applied to an object causing motion or change in motion.

French drain Narrow trench backfilled with a porous medium, such as sand or gravel, used to intercept surface or lateral subsurface drainage water; also called *slit drain*.

G

Gibbs free energy (G) Energy available to do work ofsynthesis, transport, or movement.

Gley Mottled, patchwork of grey and rust colored soil from reduction of iron compounds by microorganisms in waterlogged conditions.

Gradient Change in magnitude of a property (i.e., temperature, pressure, or concentration) with distance or time.

Graduation index Expression of particle size uniformity of a sand in the middle 80 % of the range based on D_{90}/D_{10}. Lower values indicate more uniform particle size and less compaction potential.

Gravimetric water content (θ_m) Mass of water per mass of dry soil, often in units of *g* water g^{-1} soil.

$$\theta m = \frac{\text{mass of water}}{\text{mass of dry soil}} = \frac{\text{soil wet weight} - \text{soil dry weight}}{\text{soil dry weight}}.$$

Gravitational potential Portion of soil water potential due to differences in elevation of the reference pool of pure water.

Gravitational water Waterwhich moves through soil due to gravity.

Groundwater Subsurface water in the zone of saturation that moves freely; often horizontally.

H

Hardpan Soil layer that limits root penetration and water movement.

Hard water Water containing calcium, magnesium, or ferrous ions, which forms a precipitate with soap or crust.

Hooghoudt's equation Equation describing the relation between the depth and spacing of parallel subsurface drains, depth of the watertable, depth and hydraulic conducitivity of soils. Also called the *drainage equation*.

$$S = \frac{\sqrt{4Kh^2}}{v} \text{ or } S^2 = \frac{4Kh^2}{v}$$

where: S= Drain line spacing (in); the units used for h must be the same as those used for S.K_{sat}= Saturated hydraulic conductivity ($inhr^{-1}$) of the soil.h = Height of the (saturated) free-water zone midway between the two drains (in).v = Drain discharge rate, assumed to equal irrigation or rainfall rate ($inhr^{-1}$).Normally, the anticipated maximum rainfall or irrigation event rate is used here.

Humate (humin) Portion of soil organic matter insoluble in dilute alkali.

Humic acid Portion of humus that is water insoluble and extracted from soil with dilute alkali and precipitated upon acidification.

Humus Relatively stable, dark-colored colloidal soil organic matter containing no recognizable plant parts.

Hydrated lime Calcium hydroxide from reacting burnt lime (CaO) with water.

Hydraulic conductivity Ratio of flux of water flow in soil or other porous media to hydraulic gradient.

Hydrology Science of water distribution and movement.

Hydrolysis Splitting of one molecule by adding water.

Hydrometer Sealed cylinder with weighted bulb and graduated stem used to measure density of suspensions.

Hydrophilic Water-loving; attracting water.

Hydrophobic Water-hating; repelling water.

Hygroscopic water Unavailable soil water tightly held by bonding (absorption) to soil particles.

Hydroxyl group A OH^- group formed by the dissociation of a water molecule.

Hysteresis A nonunique relationship between two variables that changes depending on the sequences or starting point used to observe them. An example is different curves describing soil water content versus matric potential when a soil is gaining versus losing water.

I

Imbibition Water absorption into dry soil.

Infiltration Downward entry and movement of water into and through soil.

Infiltration rate Quantity of water entering soil per unit time.

Infiltrometer Device used for measuring the rate of water's entry into a soil.

Ions Electrically charged atoms resulting from the loss of electrons (cations) or gain of electrons (anions).

Irrigation Water application to a soil, usually for the purpose of crop production.

Iron oxides (FeO_x) Chemical compounds composed of iron and oxygen in soluble (ferrous or reduced iron) and insoluble (ferric or oxidized iron) forms. In sand-based soils, ferric iron is reduced to ferrous forms in anaerobic (lacking oxygen) conditions. The soluble ferrous forms then moves downward (leaches) until encountering free oxygen where it is oxidized or converted back to the insoluble ferric form. Ferrous forms often have a black or dark-gray color while the ferric forms turns orange or rust colored.

J

Jurin's equation Equation which describes the height a liquid rises in a capillary tube of radius to the liquid's surface tension. In general, the narrower the tube, the greater the rise of the liquid. This capillary rise is a result of *adhesion*, the attraction between the liquid and the tube, plus *cohesion*, the attraction of the liquid moles for one another. Also called *Capillary Rise equation.*

$$h = \frac{2T\cos\alpha}{gdr} = \frac{0.15}{r} = \frac{0.3}{D}$$

where: h = height of the water column (*cm*), T = surface tension of water (72 millinewtons or $mN\ m^{-1}$), g = acceleration due to gravity (980 $cm\ s^{-3}$), d = density of water (1 $g\ cm^{-3}$), r = radius of the water-containing pore (*cm*). D = diameter of the water-containing pore (*cm*).

K

Kinetic energy Energy due to motion. It is proportional to the mass and to the velocity squared.

L

Leaching Downward movement of soluble materials in a soil.

Leaching fraction In turf, fraction of water applied to soil that leaches below the rooting depth.

Leaching requirement Leaching fraction required to maintain rootzone salinity below a phytotoxic threshold value.

Lime Calcium oxide (CaO) and/or a variety of acid-neutralizing materials containing calcium or calcium and magnesium.

Limestone Sedimentary rock composed of more than half calcium carbonate ($CaCO_3$).

Loam Soil composed of 7–27 % clay, 28–50 % silt, and less than 52 % sand.

Localized dry spot (LDS) Soil that resists rewetting associated with thatch and/or organic acid coating of sand particles, buried debris, fairy ring fungi, or insufficient irrigation.

Lysimeter Device used in soil column to measure actual percolation (leaching) and evapotranspiration (ET) losses.

M

Matric (or capillary) potential A component of water potential due to adsorption of water to soil particles in unsaturated soil above the water table; carries a negative (minus) sign.

Milliequivalent One one-thousandth of an equivalent. One equivalent is one *gram* (*g*) hydrogen in one *liter* (*L*) of water, while one *milliequivalent* (*meq*) is 0.001 *g* (or 1 *mg*) hydrogen in one *L* of water. Milliequivalents per liter ($meq\ L^{-1}$) is the standard unit for dividing parts per million (*ppm*) by equivalent weight.

Mineral Naturally occurring component of rocks with a crystalline structure and a specific chemical composition.

Moisture retention curve *See* soil moisture retention curve.

Mole In chemistry, number of particles in one *mole* of any substance; always equal to Avogadro's number: 6.022×10^{23}.

Molecular weight Sum of relative weights of atoms in a molecule, with carbon atoms being the reference of 12.

Molecule Smallest possible unit of a compound, consisting of two or more atoms.

Muck *See* peat.

Muck soil Soil containing 20–50 % of well-decomposed organic matter.

N

Neutron probe Soil moisture measuring probe which measures reflection of scattered neutrons by hydrogen atoms in soil water.

Non-capillary (or aeration) porosity Percentage of a soil volume occupied by air at field capacity (or free drainage).

O

Organic matter Residual decomposition of plant or animal content in soil.

Organic soil Soil containing greater than 15 % organic matter.

Osmosis Diffusion of water from a region of greater water potential to one of lesser water potential across a selectively permeable membrane.

Osmotic (solute) potential Change in chemical potential (or free energy) of water produced by solutes being added to it; carries a negative (minus) sign; also called *solute potential.*

Osmotic pressure Pressure buildup from unequal salt concentrations across a cell wall or membrane. Water moves from lower salt concentration into areas with higher salt concentration, exerting additional pressure on the higher salt concentration side.

P

Particle density Mass (or weight) of dry soil per unit volume of soil solids (excluding pores).

Particle size Effective diameter of a particle measured by sedimentation or sieving methods.

Particle-size analysis Determination of the amounts of different soil separates in a soil sample, usually by sedimentation, sieving, or combinations of these methods.

Particle-size distribution Fractions of various soil separates in a soil sample.

Parts per million (ppm) Parts by weight of a compound in one million parts of the final mixture; $ppm = mg\ L^{-1}$.

Peat Partially decomposed organic matter accumulating under wet conditions. *Peat*refers to partially decomposed deposits while *muck* includes highly decomposed materials.

Peat moss Dried peat from various plants that is slow to decompose.

Peat soil An organic soil containing greater than 50 % organic matter.

Perched water table Saturated zone of fine-textured soil over an underlying coarser-textured soil. Water in the saturated zone will not move into the coarse-textured soil interface until sufficient water potential builds to overcome the attraction between water and fine-textured soil.

Percolation rate Downward movement of water through soil, especially saturated or near-saturated soil.

Perlite A light, expanded volcanic glass used as a soil amendment to increase porosity and water holding capacity. Expanded perlite has a typical bulk density range of 0.03–0.15 $g\ cm^{-3}$.

Permanent wilting point Soil water content at which plants wilt and do not recover; often considered to be the soil water content at –1.5 *MPa* (–15 *bar*) water potential.

Permeability Ease with which gas, liquids or plant roots penetrate or pass through a soil horizon.

Permeameter A device which confines a sample of soil and subjects it to fluid flow, in order to measure its hydraulic conductivity or intrinsic permeability.

pH Degree of acidity or alkalinity; defined as the negative logarithm of hydrogen ion activity. A scale of 0–14 is used where 7 is neutral, <7 is increasingly acidic, and >7 is increasingly basic (or alkaline).

Photosynthesis Process used by plants, algae, and cyanobacteria containing chlorophyll to convert light energy to chemical energy, using water and carbon dioxide to produce sugars, starches, and oxygen.

Piezometer Open-ended tube used to measure the liquid pressure head as the height to which the liquid rises against gravity.

Pipet method *See* Stokes' equation.

Pore Contiguous void betweenmineral and organic particles in soil, filled by air or water.

Pore-size distribution Relative abundance of various sized pores in a soil, expressed as percentages of the soil bulk volume.

Porosity Ratio of volume of pores (or voids) to total volume of soil.

Pressure potential Potential energy per unit volume of water due to the weight of water above a point. Also, pressure that develops in cells as water enters through cell walls; also called *turgor pressurel*.

R

Reduction (redox) Gain of an electron by an atom or molecule.

Relative humidity (RH) Ratio (expressed as a percentage) of water vapor in the atmosphere to the greatest possible quantity of water vapor in air at the same temperature.

Residual sodium carbonate (RSC) Measurement of sodium in irrigation water; used to evaluate whether sodium in it will cause soil structure problems from the potential precipitation of calcium and magnesium ions.

Rhizosphere Soil immediately adjacent to plant roots containing microorganisms that may differ from those in the general bulk soil.

Rootzone Soil portion from which plants absorb water and nutrients.

S

Saline-sodic soil Soil with enough soluble salts (>4 $dS\ m^{-1}$) and exchangeable sodium (>15 %) to impair its productivity.

Saline soil Soil with enough soluble salts (>4 $dS\ m^{-1}$) to impair its productivity.

Salt Compound containing positive ions from a base and negative ions from an acid, or that results from direct combination of metal and nonmetal (i.e., NaCl dissociates into Na^+ and Cl^- in water).

Sand Soil textural class consisting of particles between 0.05 and 2.0 *mm* in diameter.

Saturate Completely fill all voids.

Saturated flow Water movement in a soil filled to capacity with water. Often described by Darcy's Law.

Sediment Material derived from pre-existing rock deposited at, or near, the Earth's surface.

Sedimentation Process of sediment deposition.

Slit drain *See* French drain.

Sodic soil Nonsaline soil with sufficient sodium (exchangeable sodium percentage of ≥15 or sodium adsorption ratio of ≥13), a pH from 8.5 to 10, and dispersed soil colloids to reduce permeability.

Sodium adsorption ratio (SAR) Relative hazard of irrigation water from its sodium content relative to its amount of calcium (Ca^{+2}) and magnesium (Mg^{+2}), measured as millimoles of charge per liter ($mmol\ L^{-1}$).

Soil Upper layer of earth surface used as the natural medium for plant growth.

Soil compaction Increasing soil bulk density, and conversely decreasing soil porosity, by applying mechanical forces to the soil.

Soil conditioner Any material added to a soil to improve its physical properties, also called *soil amendment*.

Soil mix Prepared mixture of soil or sand plus amendments used as a growth medium.

Soil modification Artificial altering of soil by adding soil amendments to improve physical conditions.

Soil moisture retention curve Graph indicating soil water content at various tensions, suctions, or water potentials; also called *soil moisture release curve* or *soil water characteristic curve*.

Soil physical properties Soil characteristics, processes, or reactions caused by physical forces that can be described by, or expressed in, physical terms or equations. Examples include bulk density, hydraulic conductivity, porosity, particle-size distribution, etc.

Soil physics Science dealing with soil physical properties, especially the state and transport of water and energy in soil.

Soil pores Soil volume not occupied by soil particles; also referred to as *interstices* or *voids*.

Soil salinity Level of soluble salts in a soil. Often measured as the electrical conductivity of a saturation extract.

Soil test Test to determine chemical, physical, or microbiological property of a soil.

Soil texture Relative coarseness or fineness of a soil as determined by its proportions of sand, silt, and clay.

Soil water characteristic curve *See* soil moisture retention curve.

Soluble salts Total soluble ions in a soil measured as electrical conductivity (EC).

Solute A molecule dissolved in a solution.

Solute potential *See* osmotic potential.

Stokes' equation Equation used to determine soil texture from the settling rates of suspended sands, silts, and clay based on the diameter of the particle and viscosity of the suspension medium; also called the *Pipet method*.

Surfactant A material that improves the emulsifying, dispersing, spreading, or wetting properties of liquids. Acronym for **SURF**ace-**ACT**ive **A**ge**NTS**.

Surface tension *See* cohesion.

Suspension Soil particles in containment or support in a fluid media.

T

Tensiometer Tube-like device used for measuring matric potential or soil water tension.

Thatch Brown- to black-colored layer of dead leaves, stems, rhizomes, crowns, and stolons between the green vegetation and soil surface.

Thermocouple Device that responds to temperature differences between two junctionsof dissimilar metals.

Tile drain Concrete, ceramic, plastic, or related material pipe, placed at suitable depths and spacings in soil to enhance and/or accelerate soil water removal.

Time-domain reflectometry (TDR) Method of determining soil water content using the timing of wave reflections to determine the properties of various materials, such as the dielectric constant of soil.

Topsoil Uppermost layer of soil.

Turbidity Cloudiness or lack of transparency of water, often due to violent disturbance of sediment.

Turfgrass Grass species or cultivar maintained as a uniform, mowed vegetation.

Turgor pressure *See* pressure potential.

U

Uniformity coefficient Numerical expression of particle size uniformity defined as D_{60}/D_{10}, with an optimum range of 2–3. The higher the value, the less uniform the sand, therefore the greater potential for particle packing. Values less than 2 may pack insufficiently, resulting in unstable surfaces. For gravel, CU is defined as D_{90}/D_{15} and should be ≤ 2.5.

Unsaturated flow Water movement in a soil not filled to capacity with water. Frequently characterized by the capillary rise equation.

V

Vadose zone Aerated region of soil above the water table.

Van der Waals forces Relatively weak net attraction or repulsion of intermolecular forces between molecules, other than those due to covalent bonds.

Verticutting Slicing turf with a series of vertically mounted blades rotating on a shaft, usually to reduce thatch; also called *vertical mowing*.

Volumetric water content (θ_v) Water content of a soil by volume measured as the ratio of soil water volume to the soils total bulk volume.

W

Water balance Soil water storage based on water inputs and outputs.

Water drop penetration time (WDPT) Time required for water droplets to disperse on a surface, often used to measure soil water repellency.

Water hammer Shock waves in pipelines from water surges. A common source of pipe damage.

Water potential Potential (or gradient) energy per unit volume of water measured as the sum of gravimetric, osmotic, matric, and pressure potential; it is a negative value, and as the value is more negative, lower water potential occurs.

Water release characteristic curve *See* soil moisture retention curve.

Water retention curve *See* soil moisture retention curve.

Watershed Portion of a landscape that contributes water to a single discharge location; it is defined by its boundary or divide (ridges), across which no runoff occurs.

Water table Top level of groundwater saturation zone.

Wetting agent Substance that reduces surface tension and causes liquids to make better contact with treated surfaces. Commonly separated into four major groups, based on their ionization in water: (1) anionic; (2) cationic; (3) nonionic; and (4) amphoteric.

Wet wilt Plant wilting in the presence of water when evaporation exceeds root water uptake.

Wilt Plant collapse usually when evapotranspiration exceeds water uptake by roots.

Work Energy transfer required to move an object a certain distance.

X

Xylem Water-conducting tissue of a plant's vascular system.

References

Adams WA, Gibbs RF (1994) Natural turf for sport and amenity: science and practice. CAB Int, Univ Press, Cambridge

Baker SW, Richards CW (1993) Soil physical properties of soccer pitches: relationships between laboratory and field measurements. Int Turfgrass Res J 7:489–96

Baker SW, Richards CW (1997) Soil Physical properties of golf greens: relationships between laboratory and field measurements. Int Turfgrass Res J 8:47–58

Beard JB (1985) An assessment of water use by turfgrasses. In: Youngner VB, Cockerham ST (eds) Turfgrass water conversation Proceedings of the symposium ASPA, San Antonio. TX. Coop Ext Univ California, Oakland, CA, pp 47–60

Bigelow CA, Bowman DC, Cassel DK (2004) Physical properties of three sand size classes amended with inorganic materials or sphagnum peat moss for putting green rootzones. Crop Sci 44:900–7

Bingaman DE, Kohnke H (1970) Evaluating sands for athletic turf. Agron J 62:464–7

Blake GR (1980) Proposed standards and specifications for quality of sand for sand-soil-peat mixes. In: Beard JB (ed) Proceedings of the 3rd international turfgrass research conference, Munich, Germany. Turfgrass Society, and ASA, CSSA, and SSSA, Madison, WI, pp 195–203

Brady NC, Weil RR (2002) The nature and properties of soils, 13th edn. Prentice-Hall, Inc, Upper Saddle River, NJ

Brown KW, Duble RL (1975) Physical characteristics of soil mixtures used for golf green construction. Agron J 67:647–51

Carrow RN (1995) Drought resistance aspects of turfgrasses in the Southeast: evapotranspiration and crop coefficients. Crop Sci 35:1685–90

Davis WB, Paul JL, Bowman D (1990) The sand putting green—construction and management. University of California Division of Agriculture and Natural Resources Publication 21448

Elrick DE, Sheard RW, Baumgartner N (1980) A simple procedure for determining the hydraulic conductivity and water retention of putting green soil mixtures. Int Turfgrass Res J 4:189–200

Gee GW, Bauder JW (1986) Particle-size analysis. In: Klute A (ed) Method of soil analysis, part 1. ASA and SSSA, Madison, WI, pp 383–411

Green RL (2005) Golf course water use and regulation in California. Golf Course Manag 73(10):67–72

Habeck J, Christians N (2000) Time alters greens' key characteristics. Golf Course Manag 68(5):54–60

Handreck KA, Black ND (2007) Growing media for ornamental plants and turf, 3rd edn. New South Wales Univ Press, Kensington, NSW, pp 124–42

Hanks RJ, Ashcroft GL (1980) Applied soil physics. Springer, New York, NY, p 159

Harivandi MA (2007) Using recycled water on golf courses. Golf Course Manag 75(6):98–108

L.B. McCarty et al., *Applied Soil Physical Properties, Drainage, and Irrigation Strategies*, DOI 10.1007/978-3-319-24226-2

Hillel D (2004) Introduction to environmental soil physics. Elseiver Science, San Diego, CA, p 494

Hummel NW Jr (1998) Which root-zone recipe makes the best green? Golf Course Manag 66(12):49–51

Juncker PH, Madson JJ (1967) Soil moisture characteristics of sand-peat mixes. Soil Sci Soc Am 31:5–8

Karcher D (2013) Gadgets and gizmos: a short course to understanding measurement technology. Carolina GCSA conference and show manual. Carolina GCSA, Myrtle Beach, SC

Karcher D, Richardson M (2014) Wetting agents improve rootzone moisture distribution. Golf Course Manag 82(12):80–3

Karnok K, Tucker K (2002) Water-repellent soils part 1: where are we now? Golf Course Manag 70(6):59–62

Klute A (ed) (1986) Methods of soil analysis: part 1 – physical and mineralogical methods, 2nd edn. ASA, SSSA, SSSA, Madison, WI, Agronomy No. 9 (Part 1)

Luthin JN, Miller RD (1953) Pressure distribution in soil columns draining into the atmosphere. Soil Sci Soc Am Proc 17:329–33

McCarley EM, Quisenberry V, Smith BR (2001) Introductory soil science laboratory manual. Tavenner, Anderson, SC, p 108

McCarty B, Miller G, Waltz C, Hale T (2005) Designing, constructing and maintaining bermudagrass sports fields, 2nd edn. Clemson Univ Public Ser Public, Clemson, SC

McCarty LB (2011) Best golf course management practices, 3rd edn. Prentice-Hall, Inc, Upper Saddle River, NJ

McInnes KJ, Thomas JC (2012) Passive control of downward capillary wicking of water in sand-based rootzones. Hort Sci 47(2):275–9

McIntyre K, Jakobsen B (2000) Practical drainage for golf, sportsturf and horticulture. Ann Arbor Press, Chelsea, MI

Miller RW, Gardiner DT (2004) Soils in our environment, 10th edn. Prentice-Hall, Inc, Upper Saddle River, NJ

Oppold RS (1997) ISTRC's guidebook to your greens. Intern Sports Turf Research Center, Lenexa, KS, http://www.istrc.com/documents/ISTRCGuidebook.pdf

Pettyman GW, McCoy EL (2002) Effect of profile layering, rootzone texture, and slope on putting-green drainage rates. Agron J 94:358–64

Stewart VI (1994) Sports turf, science construction, and maintenance. Taylor & Francis, Abigndon

Taylor DH, Blake GR (1979) Sand content of sand-soil-peat mixtures for turfgrass. Agron J 43:394–7

Taylor DH, Blake GR, White DB (1987) Athletic field construction and maintenance. Ext Ser AG-BU-3125, 16, Minnesota, MN

USDA Soil Conservation Service (1971) Drainage of agricultural land. National Engineering Handbook Section 16, Washington, DC

Waltz FC Jr, Quisenberry VL, McCarty LB (2003) Physical and hydraulic properties of rootzone mixes amended with inorganics for golf putting greens. Agron J 95:395–404

Zontek SJ, Kostka SJ (2012) Understanding the different wetting agent chemistries. USGA Green Section Rec 50(15):1–6

Index

L.B. McCarty et al., *Applied Soil Physical Properties, Drainage, and Irrigation Strategies*, DOI 10.1007/978-3-319-24226-2

P

R

S

T

U

The manufacturer's authorised representative in the EU is Springer Nature Customer Service Centre GmbH, Europaplatz 3, 69115 Heidelberg, Germany. If you have any concerns regarding our products, please contact ProductSafety@springernature.com

Printed and bound by CPI Group (UK) Ltd, Croydon, CR0 4YY
22/09/2026
02221059-0011